Applied Mathematical Sciences
Volume 147

T0143022

Applied Mathematical Sciences

(continued after index)

Gilles Aubert   Pierre Kornprobst

Mathematical Problems in Image Processing

Partial Differential Equations and the
Calculus of Variations

Second Edition

 Springer

Gilles Aubert
Université de Nice Sophia-Antipolis
CNRS UMR 6621
Laboratoire J.A. Dieudonné
Parc Valrose
06108 NICE CX 2
France
gaubert@math.unice.fr

Pierre Kornprobst
INRIA, Projet Odyssée
2004 route des lucioles - BP 93
06902 SOPHIA ANTIPOLIS
France
Pierre.Kornprobst@sophia.inria.fr

Editors:

S.S. Antman
Department of Mathematics
and
Institute for Physical Science
and Technology
University of Maryland
College Park, MD 20742-4015
USA
ssa@math.umd.edu

J.E. Marsden
Control and Dynamical
Systems, 107-81
California Institute of
Technology
Pasadena, CA 91125
USA
marsden@cds.caltech.edu

L. Sirovic
Laboratory of Applied
Mathematics
Department of
Biomathematical
Sciences
Mount Sinai School
of Medicine
New York, NY 10029-6574
USA
chico@camelot.mssm.edu

Mathematics Subject Classification (2000): 35J, 35L, 35Q, 49J, 49N

ISBN   978-1-4419-2182-6   e-ISBN   978-0-387-44588-5

Printed on acid-free paper.

Printed in the United States of America.   (EB)

9 8 7 6 5 4 3 2 1

springer.com

To Jean-Michel Morel, whose ideas
have deeply influenced the mathematical
vision of image processing.

Foreword

Image processing, image analysis, computer vision, robot vision, and machine vision are terms that refer to some aspects of the process of computing with images. This process has been made possible by the advent of computers powerful enough to cope with the large dimensionality of image data and the complexity of the algorithms that operate on them.

In brief, these terms differ according to what kind of information is used and output by the process. In image processing the information is mostly the intensity values at the pixels, and the output is itself an image; in image analysis, the intensity values are enriched with some computed parameters, e.g., texture or optical flow, and by labels indicating such things as a region number or the presence of an edge; the output is usually some symbolic description of the content of the image, for example the objects present in the scene. Computer, robot, and machine vision very often use three-dimensional information such as depth and three-dimensional velocity and perform some sort of abstract reasoning (as opposed to purely numerical processing) followed by decision-making and action.

According to this rough classification this book deals with image processing and some image analysis.

These disciplines have a long history that can be traced back at least to the early 1960s. For more than two decades, the field was occupied mostly by computer scientists and electrical engineers and did not attract much interest from mathematicians. Its rather low level of mathematical sophistication reflected the kind of mathematical training that computer scientists and electrical engineers were exposed to and, unfortunately, still are: It is roughly limited to a subset of nineteenth-century mathematics.

This is one reason. Another reason stems from the fact that simple heuristic methods, e.g., histogram equalization, can produce apparently startling results; but these ad hoc approaches suffer from significant limitations, the main one being that there is no precise characterization of why and when they work or don't work. The idea of the proof of correctness of an algorithm under a well-defined set of hypotheses has long been almost unheard of in image processing and analysis despite the strong connection with computer science.

It is clear that things have been changing at a regular pace for some time now. These changes are in my view due to two facts: First, the level of mathematical sophistication of researchers in computer vision has been steadily growing in the last twenty-five years or so, and second, the number of professional mathematicians who develop an interest in this field of application has been regularly increasing, thanks maybe to the examples set by two Fields medallists, David Mumford and Pierre-Louis Lions. As a result of these facts the field of computer vision is going through a crucial mutation analogous to the one that turned alchemy into modern chemistry.

If we now wonder as to the mathematics relevant to image processing and analysis, we come up with a surprisingly long list: Differential and Riemannian geometry, geometric algebra, functional analysis (calculus of variations and partial differential equations), probability theory (probabilistic inference, Bayesian probability theory), statistics (performance bounds, sampling algorithms), and singularity theory (generic properties of solutions to partial differential equations) are all being successfully applied to image processing. It should be apparent that it is, in fact, the whole set of twentieth-century mathematics that is relevant to image processing and computer vision.

In what sense are those branches of mathematics relevant? As I said earlier, many of the original algorithms were heuristic in nature: No proof was in general given of their correctness, and no attempt was made at defining the hypotheses under which they would work or not. Mathematics can clearly contribute to change this state of affairs by posing the problems in somewhat more abstract terms with the benefit of a clarification of the underlying concepts, e.g., what are the relevant functional spaces, and what is the possibility of proving the existence and uniqueness of solutions to these problems under a set of well-defined hypotheses and the correctness of algorithms for computing these solutions? A further benefit of the increase of mathematical sophistication in machine vision may come out of the fact that the mathematical methods developed to analyze images with computers may be important for building a formal theory of biological vision: This was the hope of the late David Marr and should be considered as another challenge to mathematicians, computer-vision scientists, psychophysicists, and neurophysiologists.

Conversely, image processing and computer vision bring to mathematics a host of very challenging new problems and fascinating applications; they contribute to grounding them in the real world just as physics does.

This book is a brilliant "tour de force" that shows the interest of using some of the most recent techniques of functional analysis and the theory of partial differential equations to study several fundamental questions in image processing, such as how to restore a degraded image and how to segment it into meaningful regions. The reader will find early in the book a summary of the mathematical prerequisites as well as pointers to some specialized textbooks. These prerequisites are quite broad, ranging from direct methods in the calculus of variations (relaxation, Gamma convergence) to the theory of viscosity solutions for Hamilton–Jacobi equations and include the space of functions of bounded variations. Lebesgue theory of integration as well as Sobolev spaces are assumed to be part of the reader's culture, but pointers to some relevant textbooks are also provided.

The book can be read by professional mathematicians (who are, I think, its prime target) as an example of the application of different parts of modern functional analysis to some attractive problems in image processing. These readers will find in the book most of the proofs of the main theorems (or pointers to these in the literature) and get a clear idea of the mathematical difficulty of these apparently simple problems. The proofs are well detailed, very clearly written, and, as a result, easy to follow. Moreover, since most theorems can also be turned into algorithms and computer programs, their conclusions are illustrated with spectacular results of processing performed on real images. Furthermore, since the authors provide examples of several open mathematical questions, my hope is that this book will attract more mathematicians to their study.

It can also be read by the mathematically inclined computer-vision researcher. I do not want to convey the idea that I underestimate the amount of work necessary for such a person to grasp all the details of all the proofs, but I think that it is possible at a first reading to get a general idea of the methods and the main results. Hopefully, this person will then want to learn in more detail the relevant mathematics, and this can be done by alternating reading the textbooks that are cited and studying the proofs in the book. My hope is that this will convince more image-processing scientists that this mathematics must become part of the tools they use.

This book, written by two mathematicians with a strong interest in images, is a wonderful contribution to the mutation I was alluding to above, the transformation of image processing and analysis as well as computer, robot, and machine vision into formalized fields, based on sets of competing scientific theories within which predictions can be performed and methods (algorithms) can be compared and evaluated. This is hopefully a step in the direction of understanding what it means to see.

Sophia Antipolis, France   OLIVIER FAUGERAS

Preface to the Second Edition

During the four years since the publication of the first edition of this book there has been substantial progress in the range of image processing applications covered by the PDE framework. The main purposes of this second edition are to update the first edition by giving a coherent account of some of the recent challenging applications, to give the opportunity to the reader to make his own simulations easily, to update the existing material, and naturally to correct errors.

Review of Recent Challenging Applications_____

In Chapter 5, devoted to applications, we present four new topics.

Section 5.1 Reinventing some image parts by inpainting. Inpainting has broad applications in photo restoration, superresolution, primal-sketch-based perceptual image compression and coding, and the error concealment of image transmission.

Section 5.2 Decomposing an image into geometry and texture. The aim is to characterize strongly oscillating patterns (e.g., textures) so that an image may be decomposed as the sum of a geometric part and an oscillating one. Besides a better understanding of the nature of images and especially texture, this framework has direct applications in image compression and inpainting.

Section 5.3.4 Sequence restoration. In this section, we address the problem of video inpainting. We show that the simple extension of image inpainting methods (Section 5.1) to image sequences sequences is not sufficient: appropriate methods are needed. The film industry is the natural field of application: old movies need restoration and removal of scratches, rays, etc. In new movies one may require the removal of some visible objects out of the scene.

Section 5.5 Vector-valued images. In the first edition, we essentially considered gray-scale images, i.e., scalar images. In this new section we would like to give some elements to deal with vector-valued images. We present a framework that can be adapted and used in many applications such as color image restoration, inpainting, interpolation, and vector field visualization.

Make Your Own Simulations Easily

In the first edition, we proposed in the Appendix A an overview on finite difference approaches, to transform a continuous equation into a discretized one. As a further step, we would like to provide readers with the opportunity to make their own simulations with a minimal effort. To this end, we propose in this second edition a new Appendix B, where we present the programming tools that will allow the reader to implement and test easily some of the approaches presented in this book.

A web site is associated with this second edition. This web site contains some related links, complementary of information, and also source code that allow the reader to test easily some variational and PDE-based approaches.

http://www-sop.inria.fr/books/imath

Update the Existing Material

The core of the first edition, i.e., Chapters 2–4, was preserved. It is complemented with some recent contributions about algorithms, models, and theoretical studies.

Section 3.2.4 The projection algorithm by Chambolle is a convergent restoration algorithm to minimize the total variation with a quadratic fidelity attach term. This algorithm comes directly from the definition of the total variation and is based on duality arguments. It is an nice alternative to the half-quadratic approach.

Section 3.3.4 Neighborhood filters and nonlocal means filters. As a complement to Chapter 3, about image restoration, we describe how to extend the notion of Gaussian filtering. Interestingly, it is shown how these filters are indeed related to well-known PDEs.

Section 4.3.4 Theoretical justification of the reinitialization equation for the distance function. This equation is widely used when a curve evolution is implemented with a level set formulation, in order to avoid numerical problems. We prove in this section the existence and uniqueness of the solution in the framework of viscosity solutions, with a discontinuous Hamiltonian.

We welcome corrections and comments, which can be sent to our electronic mail address: gaubert@math.unice.fr. In due course, corrections will be placed on the book web site http://www-sop.inria.fr/books/imath.

For this second edition, we would like to express our deep gratitude to the following people for their various contributions:

- Antoni Buades, Marcelo Bertalmio, Vincent Caselles, Rachid Deriche, Gloria Haro Ortega, François Lauze, Patrick Perez, and David Tschumperlé for providing us with their experimental results.

- All the readers who sent us their comments and corrections.

- Gunnar Aronsson and Eli Shechtman for their valuable feedback on the first edition: They read the first edition thoroughly, used it in their classes, and provided many useful suggestions.

Sophia Antipolis, France   GILLES AUBERT
                                                                PIERRE KORNPROBST

Preface to the First Edition

It is surprising when we realize just how much we are surrounded by images. Images allow us not only to perform complex tasks on a daily basis, but also to communicate, transmit information, and represent and understand the world around us. Just think, for instance, about digital television, medical imagery, and video surveillance. The tremendous development in information technology accounts for most of this. We are now able to handle more and more data. Many day-to-day tasks are now fully or partially accomplished with the help of computers. Whenever images are involved we are entering the domains of computer vision and image processing. The requirements for this are reliability and speed. Efficient algorithms have to be proposed to process these digital data. It is also important to rely on a well-established theory to justify the well-founded nature of the methodology.

Among the numerous approaches that have been suggested, we focus on partial differential equations (PDEs), and variational approaches in this book. Traditionally applied in physics, these methods have been successfully and widely transferred to computer vision over the last decade. One of the main interests in using PDEs is that the theory behind the concept is well established. Of course, PDEs are written in a continuous setting referring to analogue images, and once the existence and the uniqueness have been proven, we need to discretize them in order to find a numerical solution. It is our conviction that reasoning within a continuous framework makes the understanding of physical realities easier and stimulates the intuition necessary to propose new models. We hope that this book will illustrate this idea effectively.

The message we wish to convey is that the intuition that leads to certain formulations and the underlying theoretical study are often complementary. Developing a theoretical justification of a problem is not simply "art for art's sake." In particular, a deep understanding of the theoretical difficulties may lead to the development of suitable numerical schemes or different models.

This book is concerned with the mathematical study of certain image-processing problems. Thus we target two audiences:

- The first is the mathematical community, and we show the contribution of mathematics to this domain by studying classical and challenging problems that come from computer vision. It is also the occasion to highlight some difficult and unsolved theoretical questions.

- The second is the computer vision community: we present a clear, self-contained, and global overview of the mathematics involved for the problems of image restoration, image segmentation, sequence analysis, and image classification.

We hope that this work will serve as a useful source of reference and inspiration for fellow researchers in applied mathematics and computer vision, as well as being a basis for advanced courses within these fields.

This book is divided into seven main parts. Chapter 1 introduces the subject and gives a *detailed plan of the book*. In Chapter 2, most of the mathematical notions used therein are recalled in an educative fashion and illustrated in detail. In Chapters 3 and 4 we examine how PDEs and variational methods can be successfully applied in the restoration and segmentation of one image. Chapter 5 is more applied, and some challenging computer vision problems are described, such as inpainting, sequence analysis, classification or vector-valued image processing. Since the final goal of any approach is to compute a numerical solution, we propose an introduction to the method of finite differences in the Appendix.

We would like to express our deep gratitude to the following people for their various contributions:

- The ARIANA group (joint project CNRS–INRIA–UNSA) from IN-RIA Sophia Antipolis and in particular Jean-Franis Aujol, Laure Blanc-Féraud and Christophe Samson for providing results regarding the classification problem.

- The ODYSSEE group (joint project INRIA–Ecole Normale Supieure–Ecole Nationale des Ponts et Chaussées) from INRIA Sophia Antipolis and especially Olivier Faugeras and Bertrand Thirion for their subsequent valuable comments.

- Agnés Desolneux, François Helt, Ron Kimmel, Etienne Mémin, Nikos Paragios, Luminita Vese, and Joachim Weickert for providing us with their experimental results.

Sophia Antipolis, France   GILLES AUBERT
                                                     PIERRE KORNPROBST

Contents

Guide to the Main Mathematical Concepts and Their Application

This book is principally organized by image processing problems and not by mathematical concepts. The aim of this guide is to highlight the different concepts used and especially to indicate where they are applied.

Direct Method in the Calculus of Variations (Section 2.1.2)

This terminology is used when the problem is to minimize an integral functional, for example

$$\inf\left\{F(u) = \int_\Omega f(x, u(x), \nabla u(x))\, dx, \ u \in V\right\}. \qquad (\mathcal{F})$$

The classical (or direct) method consists in defining a minimizing sequence $u_n \in V$, bounding u_n uniformly in V, and extracting a subsequence converging to an element $u \in V$ (compactness property, Section 2.1.1) and proving that u is a minimizer (lower semicontinuity property, Section 2.1.2). This technique has been applied in two cases:

- Image restoration (Section 3.2.3, Theorem 3.2.2)
- Sequence segmentation (Section 5.3.3)

Relaxation (Section 2.1.3)

When the direct method does not apply to a minimization problem $(\mathcal{F})$ (because the energy is not lower semicontinuous (l.s.c.) or the space is not reflexive, for example) it is then a classical approach to associate with $(\mathcal{F})$ another problem called $(R\mathcal{F})$ (relaxed problem), that is, another functional RF (relaxed functional). Then $(R\mathcal{F})$ is related to $(\mathcal{F})$ thanks to the following two properties: The first is that $(R\mathcal{F})$ is well posed that is $(R\mathcal{F})$ has solutions and $\min\{RF\} = \inf\{F\}$. The second is that we can extract from minimizing sequences of $(\mathcal{F})$ subsequences converging to a solution of $(R\mathcal{F})$. We have used this concept for the following:

- Image restoration for which the initial formulation was mathematically ill posed (Section 3.2.3, Theorem 3.2.1 and Section 3.3.1, Proposition 3.56).

Γ-convergence (Section 2.1.4)

The notion of Γ-convergence relates to convergence for functionals. It is particularly well adapted to deal with free discontinuity problems. Roughly speaking, if the sequence of functionals F_h Γ-converges to another functional F, and if u_h is a sequence of minimizers of F_h and u a minimizer of F, then (up to sequence) $\lim_{h \to 0} F_h(u_h) = F(u)$ and $\lim_{h \to 0} u_h = u$. This notion is illustrated in the following two cases:

- Approximation of the Mumford–Shah segmentation functional (Section 4.2.4, Theorem 4.2.8).

- Image classification (Section 5.4.1).

Viscosity Solutions (Section 2.3)

The theory of viscosity solutions aims at proving the existence and uniqueness of a solution for fully nonlinear PDEs of the form

$$\frac{\partial u}{\partial t} + F(x, u(x), \nabla u(x), \nabla^2 u(x)) = 0.$$

This is a very weak notion because solutions are expected to be only continuous. We have used this theory for the following:

- The Alvarez–Guichard–Lions–Morel scale space theory (Section 3.3.1, Theorem 3.3.2).

- Geodesic active contours and level set methods (Section 4.3.3, Theorem 4.3.2).

Notation and Symbols

About Functionals

For Ω an open subset of R^N we define the following real-valued function spaces:

$\mathcal{B}(\Omega)$	Borel subset of Ω.
S^{N-1}	Unit sphere in R^N.
dx	Lebesgue measure in R^N.
$\mathcal{H}^{N-1}$	Hausdorff measure of dimension $N-1$.
$BV(\Omega)$	Space of bounded variation.
$\mathrm{BV-w}^*$	The weak* topology of $BV(\Omega)$.
$C_0^p(\Omega)$	Space of real-valued functions, p continuously differentiable with compact support.
$C_0^\infty(\Omega)$	Space of real-valued functions, infinitely continuously differentiable with compact support.
$C^{0,\gamma}(\Omega)$	For $0 < \gamma \le 1$: space of continuous functions f on Ω such that $|f(x) - f(y)| \le C|x-y|^\gamma$, for some constant C, x, $y \in \Omega$. It is called the space of Hölder continuous functions with exponent γ.
$C^{k,\gamma}(\Omega)$	Space of k-times continuously differentiable functions whose kth partial derivatives belong to $C^{0,\gamma}(\Omega)$.

$(\mathcal{C}_0^\infty(\Omega))'$	Dual of $\mathcal{C}_0^\infty(\Omega)$, i.e. the space of distributions on Ω.
$L^p(\Omega)$	Space of Lebesgue measurable functions f such that $\int_\Omega |f|^p \, dx < \infty$.
$L^\infty(\Omega)$	Space of Lebesgue measurable functions f such that there exists a constant c with $|f(x)| \le c$, a.e. $x \in \Omega$.
$\mathcal{M}(\Omega)$	Space of Radon measures.
$SBV(\Omega)$	Space of special functions of bounded variation.
$LSC(\Omega)$	Space of lower semicontinuous functions on Ω.
$USC(\Omega)$	Space of upper semicontinuous functions on Ω.
$W^{1,p}(\Omega)$	With $1 \le p \le \infty$: Sobolev space of functions $f \in L^p(\Omega)$ such that all derivatives up to order 1 belong to $L^p(\Omega)$. $W^{1,\infty}(\Omega)$ is identified with the space of locally Lipschitz functions.
$W_0^{1,p}(\Omega)$	$\{u \in W^{1,p}(\Omega) : u|_{\partial\Omega} = 0\}$.

(Vector-valued spaces will be written in boldface: $\mathbf{BV}(\Omega)$, $\mathbf{C}_0^p(\Omega)$, $\mathbf{L}^p(\Omega)$, $\mathbf{M}(\Omega)$, $\mathbf{W}^{1,p}(\Omega)$, $\mathbf{SBV}(\Omega)$).

For X a Banach space with a norm $|.|_X$ and $v : (0,T) \to X$:

$C^m(0,T;X)$	With $m \ge 0$, $0 < T < \infty$: space of functions from $[0,T]$ to X m-times continuously differentiable. It is a Banach space with the norm $$|v|_{C^m(0,T;X)} = \max_{0 \le l \le m} \left(\sup_{0 \le t \le T} \left| \frac{\partial^l v}{\partial t^l}(t) \right|_X \right).$$
$L^p(0,T;X)$	With $1 \le p < \infty$: space of functions $v \to v(t)$ measurable on $(0,T)$ for the measure dt (i.e., the scalar functions $t \to |v|_X$ are dt-measurable). It is a Banach space with the norm $$|v|_{L^p(0,T;X)} = \left(\int_0^T |v|_X^p \, dt \right)^{1/p} < +\infty.$$
$L^\infty(0,T;X)$	Space of functions v such that $$|v|_{L^\infty(0,T;X)} = \inf_c \{|v|_X \le c, \text{ a.e. } t\}.$$

For a functional $F : X \to]-\infty, +\infty]$ where X is a Banach space:

$Argmin\ F$	$\{u \in X : F(u) = \inf_X F\}$.
$R_\tau(F), RF, \overline{F}$	Relaxed functional of F (for the τ-topology).

l.s.c.(sequentially)	Lower semicontinuous: F is called l.s.c. if for every sequence (u^n) converging to u we have $$\varliminf_{n \to +\infty} F(u^n) \geq F(u).$$
u.s.c.(sequentially)	Upper semicontinuous: F is called u.s.c. if for every sequence (u^n) converging to u we have $$\varlimsup_{n \to +\infty} F(u^n) \leq F(u).$$

About Measures

For μ and ν two Radon measures:

$|\mu|$	Total variation of the measure μ. If μ is vector-valued, we also denote $|\mu| = |\mu_1| + \cdots + |\mu_N|$.
$\nu \ll \mu$	ν is absolutely continuous with respect to μ if $\mu(A) = 0 \Rightarrow \nu(A) = 0$ for all Borel set $A \in R^N$.
$\nu \perp \mu$	ν is singular with respect to μ if there exists a Borel set $B \subset R^N$ such that $\mu(R^N - B) = \nu(B) = 0$.

About Functions

For a function $f : \Omega \subset R^N \to R$ and a sequence of functions $(f^n)_{n \in N}$ belonging to a Banach space X:

$f^n \xrightarrow[X]{} f$	The sequence (f^n) converges strongly to f in X.
$f^n \xrightharpoonup[X]{} f$	The sequence (f^n) converges weakly to f in X.
$f^n \xrightharpoonup[X]{*} f$	The sequence (f^n) converges to f for the weak* topology of X.
$\varlimsup_{n \to +\infty} f^n$	$\varlimsup_{n \to +\infty} f^n(x) = \inf_k \sup \{f^k(x), f^{k+1}(x), \ldots\}$.
$\varliminf_{n \to +\infty} f^n$	$\begin{aligned} \varliminf_{n \to +\infty} f^n(x) &= - \varlimsup_{n \to +\infty} f^n(x) \\ &= \sup_k \inf \{f^k(x), f^{k+1}(x), \ldots\}. \end{aligned}$
$|f|_X$	Norm of f in X.
$\mathrm{spt}(f)$	For a measurable function $f : \Omega \subset R^N \to R$, let $(w_i)_{i \in I}$ be the family of all open subsets such that $w_i \in \Omega$ and for each $i \in I$, $f = 0$ a.e. on w_i. Then spt (the support of f) is defined by $\mathrm{spt} f = \Omega - \bigcup_i w_i$.
Df	Distributional gradient of f.
$D^2 f$	Hessian matrix of f (in the distributional sense).
∇f	Gradient of f in the classical sense. It corresponds to the absolutely continuous part of Df with respect to the Lebesgue measure dx.

$\operatorname{div}(f)$ Divergence operator: $\operatorname{div}(f) = \sum_{i=1}^{N} \frac{\partial f}{\partial x_i}$.

$\nabla^2 f$ Hessian matrix of f in the classical sense: $(\nabla^2 f)_{i,j} = \frac{\partial^2 f}{\partial x_i \partial x_j}$.

$\triangle f$ Laplacian operator: $\triangle f = \sum_{i=1}^{N} \frac{\partial^2 f}{\partial x_i{}^2}$.

$\fint_\Omega f \, dx$ Mean value of f over Ω: $\fint_\Omega f \, dx = \frac{1}{|\Omega|} \int_\Omega f \, dx$.

$\overset{\bullet}{f}$ Precise representation of f.

$\mathcal{P}_\Omega^\pm$ "superjets."

For a function $\phi : R^N \to R$:

$\phi^*(.)$ The Fenchel–Legendre conjugate.

$\phi^\infty(.)$ The recession function defined by $\phi^\infty(z) = \lim_{s \to +\infty} \phi(sz)/s$.

Miscellaneous notation

$A \underset{\text{strong}}{\hookrightarrow} B$ A is relatively compact in B.

$A \underset{\text{weak}}{\hookrightarrow} B$ A is weakly relatively compact in B.

O^* The adjoint operator of O.

$|.|$ Euclidean norm in R^N.

G_σ The Gaussian kernel defined by $G_\sigma(x) = 1/(2\pi \sigma^2) \exp\left(-|x|^2/(2\sigma^2)\right)$.

$B(x,r) \subset R^N$ Ball of center x and radius r.

$\mathcal{S}(N)$ Subset of $N \times N$ symmetric matrices.

$SNR(I_1/I_2)$ *Signal-to-noise ratio*: used to estimate the quality of an image I_2 with respect to a reference image I_1. It is defined by $SNR(I_1/I_2) = 10 \log_{10}\left[\frac{\sigma^2(I_2)}{\sigma^2(I_1 - I_2)}\right]$, where σ is the variance.

$\alpha \vee s \wedge \beta$ Truncature function equal to α if $s \leq \alpha$, β if $s \geq \beta$, s otherwise.

$\operatorname{sign}(s)$ Sign function equal to 1 if $s > 0$, 0 if $s = 0$, and -1 if $s < 0$.

χ_R Characteristic function of R: $\chi_R(x) = \begin{cases} 1 \text{ if } x \in R, \\ 0 \text{ otherwise.} \end{cases}$

$\operatorname{Per}_\Omega(R)$ Perimeter of R in Ω defined as the total variation of χ_R.

Symbols for the Reader's Convenience

👁	Indicates general references, books, reviews, or other parts of the book where related information can be found.
☛	Summary of an *important idea.*
∎	Symbol marking the *end* of a proof, example or remark.
❋	Symbol indicating *unsolved or challenging unsolved problems* that need to be investigated further.
C++	Symbol indicating that some C++ code is available (see Appendix B for more information).

1
Introduction

1.1 The Image Society

Our society is often designated as being an "information society." It could also very well be defined as an "image society." This is not only because image is a powerful and widely used medium of communication, but also because it is an easy, compact, and widespread way to represent the physical world. If we think about it, it is indeed striking to realize just how much images are omnipresent in our lives, and how much we rely upon them: Just have a glance at the patchwork presented in Figure 1.1.

Advances made in acquisition devices are part of the origin of such a phenomenon. A huge amount of digital information is available. The second origin is naturally the increase in capacity of computers that enables us to process more and more data. This has brought about a new discipline known as computer vision.

For example, medical imaging has made substantial use of images from the earliest days. Many devices exist that are based on ultrasounds, X-rays, scanners, etc. Images produced by these can then be processed to improve their quality, enhance some features, or efficiently combine different pieces of information (fusion).

Another important field that concerns us directly is remote sensing. This designates applications where we need to analyze, measure, or interpret scenes at a distance. In addition to defense and video surveillance applications and road traffic analysis, the observation of the earth's resources is another important field. Image processing provides tools to track and

Figure 1.1. Illustration of some applications or systems that use image analysis. One may find in this patchwork examples from medicine, biology, satellite images, old movie restoration, forensic and video surveillance, 3-D reconstructions and virtual reality, robotics, and character recognition. Other applications include data compression, industrial quality control, fluids motion analysis, art (for instance for virtual databases or manuscript analysis), games and special effects.

quantify changes in forests, water supplies, urbanization, pollution, etc. It is also widely used for weather forecasting to analyze huge amounts of data.

Video processing is clearly becoming an important area of investigation. This is because in many applications we need to process not only still images but also sequences of images. Motion analysis and segmentation are two important cues necessary for analyzing a sequence. This is necessary, for instance, for forecasting the weather, by estimating and predicting the movement of clouds. It is also a determinant for more complex tasks such as compression. A clear understanding of the sequence in terms of background and foreground with motion information enables us to describe a sequence with less given information. New challenges and subsequent problems are arising as video and cinema become digital: storage, special effects, video processing like the restoration of old movies, etc.

Beyond these general themes, we could also mention many different applications where image processing is involved. These include "World Wide Web," character recognition, 3-D reconstruction of scenes or objects from images, quality control, robotics, fingerprint analysis, and virtual art databases.

Without necessarily knowing it, we are consumers of image processing on a daily basis.

1.2   What Is a Digital Image?

A digital image (also called a discrete image) comes from a continuous world. It is obtained from an analogue image by sampling and quantization. This process depends on the acquisition device and depends, for instance, on CCDs for digital cameras. Basically, the idea is to superimpose a regular grid on an analogue image and to assign a digital number to each square of the grid, for example the average brightness in that square. Each square is called a pixel, for picture element, and its value is the gray-level or brightness (see Figure 1.2).

Depending on the kind of image, the number of bits used to represent the pixel value may vary. Common pixel representations are unsigned bytes (0 to 255) and floating point. To describe a pixel, one may also need several channels (or bands): for example, a vector field has two components, a color image is described with three channels, red, green and blue.

The last important characteristic of an image is its size (or resolution). It is the number of rows and columns in the image. Just to give an idea, typical digital cameras now give images of size 320×240 and can reach 3060×2036 for professional ones. For a digital cinema we consider images of size 720×576 (standard video format), 1920×1440 (high definition) or higher; for medical imaging functional MRI images are about 128×128.

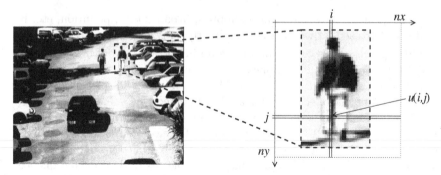

Figure 1.2. A digital image is nothing but a two-dimensional array of pixels with assigned brightness values.

In a way, the higher the resolution, the closer the digital image is to the physical world.

As in the real world, an image is composed of a wide variety of structures, and this is even more complex because of the digitalization and the limited number of gray levels to represent it. To give an idea, we show in Figure 1.3 an image and some close-ups on different parts. This shows the effects of low resolution (some areas would need more pixels to be represented) and low contrasts, different kind of "textures," progressive or sharp contours, and fine objects. This gives an idea of the complexity of finding an approach that allows to cope with different problems or structures at the same time.

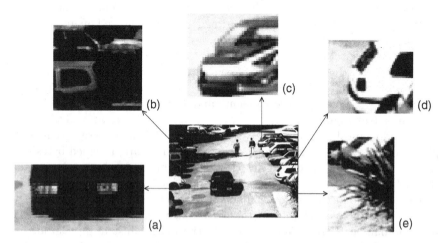

Figure 1.3. Digital image example: (a) low resolution, (b) low contrasts, (c) graduated shadings, (d) sharp transitions and (e) fine elements.

1.3   About Partial Differential Equations (PDEs)

Many approaches have been developed to process these digital images, and it is difficult to say which one is more natural than the other. Image processing has a long history. Maybe the oldest methods come from 1-D signal processing techniques. They rely on filter theory (linear or not), on spectral analysis, or on some basic concepts of probability and statistics. For an overview, we refer the interested reader to the book by Jain [188].

Today, more sophisticated tools have been developed. Three main directions emerge [104, 26]: stochastic modeling, wavelets, and partial differential equation (PDE) approaches. Stochastic modeling is widely based on Markov random field theory (see [215, 162, 160]). It deals directly with digital images. Wavelet theory is inherited from signal processing and relies on decomposition techniques (see the monograph by S. Mallat [224] and [93, 137, 138, 140, 141, 142]). We do not consider here these approaches and focus instead on PDE-based methods. They have been intensively developed in image analysis since the 1990s. Many textbooks are now available, see for example [266, 194, 293, 260, 300, 336].

PDEs, which belong to one of the most important parts of mathematical analysis, are closely related to the physical world. Every scientist comes across the wave equation or the heat equation, and the names of Euler, Poisson, Laplace, etc., are quite familiar. If PDEs originally come from physics and mechanics, one may encounter them more and more in other fields such as biology and finance and now in image analysis. One of the main interests in using PDEs is that the theory is well established. Of course, PDEs are written in a continuous setting, referring to analoguous images, and once the existence and the uniqueness of the solution have been proven, we need to discretize them in order to find a numerical solution. It is our conviction that reasoning in a continuous framework makes the understanding of physical realities easier and provides the intuition to propose new models. We hope that this book will illustrate this idea well.

1.4   Detailed Plan

This book is divided into five chapters and two appendices.

Chapter 2: Mathematical Preliminaries

We cover in this chapter most of the mathematics used in this book. We have tried to make it as complete and educative as possible. Examples are given to emphasize the importance of certain notions, or simply to illustrate them. This chapter should be read carefully. It is divided into six sections.

Section 2.1 The Direct Method in the Calculus of Variations.
Many problems in computer vision can be formulated as minimization problems. Among our first concerns, we are interested in the existence of a solution. The presentation of the methodology to prove existence reveals two major notions: coercivity and lower semicontinuity (abbreviated l.s.c.; see Definition 2.1.3). Counterexamples are proposed where these conditions are missing. Now, if the functional is not l.s.c., the idea is to look at the greatest l.s.c. functional less than or equal to the initial one. It is called the relaxed functional (Definition 2.1.6). Some interesting properties link the relaxed and the initial problem (Theorem 2.1.6). This notion is quite technical but will be very useful on many occasions. Finally, we recall the notion of Γ-convergence introduced by De Giorgi [163, 125] (Section 2.1.4, Definition 2.1.7). This permits us to define convergence for a sequence of functionals. For instance, this can be used to approximate a problem for which characterizing the solution is not easy (as a typical application, we refer to Section 3.2.4). The main properties are stated in Theorem 2.1.7. The link between Γ-convergence and pointwise convergence is clarified in Theorem 2.1.8.

Section 2.2 The Space of Functions of Bounded Variation. This is the functional space that is commonly used in image analysis. The main reason is that as opposed to classical Sobolev spaces [2], functions can be discontinuous across hypersurfaces. In terms of images, this means that images are discontinuous across edges. This is possible because derivatives are in fact measures. After we recall some basic definitions on measures in Section 2.2.1, the space $BV(\Omega)$ is defined in Section 2.2.2. This space has interesting properties that we recall in Section 2.2.3: semicontinuity, a notion of trace, compactness, and decomposability. The latter is perhaps the more specific: It says that the derivative of a function $u \in BV(\Omega)$ can be decomposed as

$$Du = \nabla u \, dx + (u^+ - u^-) \, n_u \, \mathcal{H}^{N-1}_{|S_u} + C_u,$$

that is, the sum of an absolutely continuous part with respect to the Lebesgue measure, a jump part, and a Cantor measure. We finally generalize this decomposition to give a sense to convex functions of measures (Section 2.2.4).

Section 2.3 Viscosity Solutions in PDEs. In many situations we will encounter equations that do not come from variational principles, such as in image restoration or segmentation by level sets. So we need to find a suitable framework to study these equations. To illustrate some of the problems, we consider in Section 2.3.1 the case of the 1-D eikonal equation:

$$\begin{cases} |u'(x)| = 1 & \text{in } (0,1), \\ u(0) = u(1) = 0. \end{cases}$$

Existence, uniqueness, and compatibility conditions are considered and reveal some difficulties. The framework of viscosity solutions is then presented (Section 2.3.2, Definition 2.3.1). Introduced in the 1980s for first-order PDEs by Crandall and P.L. Lions [119, 117, 217], the theory of viscosity solutions has enjoyed a very successful development and has been extended to second-order equations [118]. It is a notion of weak solutions that are continuous. To prove existence (Section 2.3.3), two approaches are classically used: the vanishing viscosity method (giving its name to the theory), and Perron's method. As for uniqueness (Section 2.3.4), it is usually based on the Crandall–Ishii lemma (Lemma 2.3.2). Since it is very technical, we try to illustrate its role in an easy example.

Section 2.4 Elements of Differential Geometry: the Curvature. Image analysis has obviously some connections with differential geometry. For instance, curvature is a very important cue, and its definition depends on the "object" that we are considering. We define here the notion of curvature depending on whether we are interested in parametrized curves, images, or 3-D surfaces. We refer to [218] for more details.

Section 2.5 Other Classical Results Used in This Book. We finally mention some important results used in this book in different proofs. The motivation is to help the reader to find most of the results in this monograph, which we have made as self-contained as possible.

Chapter 3: Image Restoration_____

Many applications are based on images and then rely on their quality. Unfortunately, those images are not always of good quality for various reasons (defects in the sensors, natural noise, interference, transmission problems, etc.). Some "noise" is introduced, and it is important to consider automatic and efficient approaches to removing it. It is historically one of the oldest concerns and is still a necessary preprocessing step for many applications. There exist many ways to tackle image restoration. Among others, we may quote the (linear) filtering approach [188], stochastic modeling [158, 68, 109, 131, 160], and the variational/PDE-based approaches. We focus here on the latter.

Section 3.1 Image Degradation. The notion of noise is quite vague, and it is often very hard to model its structure or its origins. Different models of degradation can be proposed, and we briefly review some of them. In all the sequel we will assume that the noise follows a Gaussian distribution.

Section 3.2 The Energy Method. Restoring an image can be seen as a minimization problem. We denote by u the original image and by u_0 the observed image, assuming the following model of degradation:

$$u_0 = R\,u + \eta,$$

where η stands for an additive white Gaussian noise and R is a linear operator representing the blur (usually a convolution). We show in Section 3.2.2 that the problem of recovering u from u_0 can be approximated by

$$\inf_{u\in\left\{u\in L^2(\Omega);\ \nabla u\in L^1(\Omega)^2\right\}} E\left(u\right) = \frac{1}{2}\int_{\Omega}|\,u_0 - R\,u|^2 dx + \lambda\int_{\Omega}\phi(\,|\nabla u|\,)dx,$$

where the function ϕ is convex and of linear growth at infinity, allowing the preservation of discontinuities. The choice of this function ϕ determines the smoothness of the resulting function u.

The mathematical study of this problem, considered in Section 3.2.3, reveals some major difficulties. The space $V = \left\{u \in L^2(\Omega);\ \nabla u \in L^1(\Omega)^2\right\}$ is not reflexive, and we cannot deduce any information from bounded minimizing sequences. So we relax the problem by studying it in the larger space $BV(\Omega)$, and the correct formulation of the problem is (Theorem 3.2.1)

$$\inf_{u\in BV(\Omega)}\ \overline{E}(u) \ = \ \frac{1}{2}\int_{\Omega}|u_0 - Ru|^2\ dx +$$

$$+ \lambda\int_{\Omega}\phi(|\nabla u|)\ dx + \lambda c\int_{S_u}(u^+ - u^-)d\mathcal{H}^1 + \lambda c\int_{\Omega - S_u}|C_u|.$$

We demonstrate the existence and the uniqueness of a minimizer for $\overline{E}(u)$ in Theorem 3.2.2.

Section 3.2.4 is about the numerical approximation of the solution. Although some characterization of the solution is possible in the distributional sense, it remains difficult to handle numerically. To circumvent the problem, two methods are presented here, namely a projection algorithm proposed by Chambolle [91], and the half-quadratic approach introduced Geman and Reynolds [159], and Charbonnier et al. [108]. The first relies on the idea that the minimization of the total variation can be viewed as a projection problem on a suitable convex set. In the second method, we consider sequences of close functionals, using Γ-convergence. This is quite standard and shows clearly the importance of the notion of Γ-convergence.

Section 3.2.5 investigates the role of the parameter λ with respect to the solution. This introduces the notion of scale, and we can wonder about the meaning of an "optimal" λ. This question is usually handled in a stochastic framework, and we mention some related work.

Until now we have always considered the convex case. Surprisingly, we can observe very good numerical results using nonconvex functions ϕ, for example with

$$\phi(s) = \frac{s^2}{1 + s^2}.$$

In the computer vision community, very little effort has been made on the interpretation of the result. It is a very hard problem indeed. For instance, we prove in Proposition 3.2.5 that the minimization problem

$$\inf_{u \in W^{1,2}(\Omega)} E(u) = \int_{\Omega} |u - u_0|^2 \, dx + \lambda \int_{\Omega} \frac{|\nabla u|^2}{1 + |\nabla u|^2} \, dx$$

has no minimizer in $W^{1,2}(\Omega)$ and that the infimum is zero (if u_0 is not a constant). It is still an open question to understand the numerical solution, and we conclude by discussing some recent related work.

Section 3.3 PDE-Based Methods. A restored image can be seen as a version of the initial image at a special scale. More precisely, we consider here models where an image u is embedded in an evolution process. We denote it by $u(t, \cdot)$. At time $t = 0$, $u(0, .) = u_0(.)$ is the original image. It is then transformed through a process that can be written

$$\frac{\partial u}{\partial t}(t, x) + \underbrace{F(x, u(t, x), \nabla u(t, x), \nabla^2 u(t, x))}_{\text{(A second-order differential operator)}} = 0 \quad \text{in} \quad \Omega,$$

which is a partial differential equation. In other words, the evolution of u may depend on its derivatives at different orders. This is a very generic form, and we show in this section some possibilities for F to restore an image. We distinguished three categories.

- Smoothing PDEs (Section 3.3.1). Perhaps the most famous PDE in image restoration is the heat equation:

$$\frac{\partial u}{\partial t}(t, x) - \Delta u(t, x) = 0. \tag{1.1}$$

  We recall some of its main properties: the equivalence with a Gaussian convolution, a low-pass filter, and some invariances of the operator T_t defined by $(T_t u_0)(x) = u(t, x)$, where $u(t, x)$ is the unique solution of (1.1). Because of its oversmoothing property (edges get smeared), it is necessary to introduce some nonlinearity. We then consider the model

$$\frac{\partial u}{\partial t} = \operatorname{div}\left(c(|\nabla u|^2) \nabla u\right), \tag{1.2}$$

  where here the function c is fixed so that the equation remains parabolic. If we compare it to the heat equation, this does not seem to introduce major changes, at least from a theoretical point of view. In order to preserve the discontinuities, we show that we would like to have $c(s) \approx 1/\sqrt{s}$ as $s \to +\infty$. Unfortunately, because of this degenerate behavior, it is no longer possible to apply general results from parabolic equations theory. A well-adapted framework to study this equation is nonlinear semigroup theory. The idea is to show that the divergence operator in (1.2) is maximal monotone. A convenient way to prove this is to identify the divergence operator with the sub-differential of a convex lower semicontinuous functional $\bar{J}(u)$ given in (3.58). We can then establish in Theorem 3.3.1 the existence and the uniqueness of a solution. The characterization of the solution,

$$u(t) \in \text{Dom}\,(\partial\bar{J}) \quad \text{and} \quad -\frac{du}{dt} \in \partial\bar{J}(u(t)),$$

is not very explicit, and we try in Proposition 3.3.4 to explain it better.

In 1992 Alvarez–Guichard–Lions–Morel introduced the notion of scale-space via PDEs [5, 6]. Given some axioms and invariance properties for an "image-oriented" operator T_t, the idea is to try to identify this operator. This is a major contribution and represents very original work. More precisely, under our previous assumptions, it can be established (Theorem 3.3.2) that $u(t,x) = (T_t u_0)(x)$ is the unique viscosity solution of

$$\frac{\partial u}{\partial t} = F(\nabla u, \nabla^2 u).$$

In other words, if T_t satisfy some natural assumptions, then it can be solved through a PDE depending only on the first and second derivatives of u. With more assumptions, F can be fully determined (Theorems 3.3.3 and 3.3.4).

We then present briefly Weickert's approach. We refer the interested reader to [336] for more details. Roughly speaking, it is a tensor-based version of equation (1.2) where the scalar coefficient c that controls the diffusion is replaced by a function of the diffusion *tensor:*

$$\nabla u \nabla u^t = \begin{pmatrix} u_x^2 & u_x u_y \\ u_x u_y & u_y^2 \end{pmatrix}.$$

Some Gaussian convolutions are introduced to take into account the scales. This approach permits us to better take into account the directions of diffusion and is theoretically justified [336].

Last but not least, we mention some contributions by El-Fallah et al. [145], Sochen et al. [306, 305], and Yezzi [345]. The specificity of these approaches is to consider an image as a surface. The differential operators are then based on the metric properties of the surface. Instead of recalling each model separately, we try to present them in a common framework.

- Smoothing-enhancing PDEs (Section 3.3.2). We consider equations that can behave locally as inverse heat equations. This is motivated by the original choices of c by Perona and Malik [275] in the equation

(1.2):

$$c(s) = \frac{1}{1 + s/k} \quad \text{or} \quad c(s) = e^{-s/k} \tag{1.3}$$

where k is a constant. If we denote by T and N the tangent and normal directions to the isophotes, the divergence term in (1.2) can be rewritten as

$$\text{div}(c(|\nabla u(t,x)|^2)\nabla u(t,x)) = c(|\nabla u(t,x)|^2)u_{TT} + b(|\nabla u(t,x)|^2)u_{NN},$$

where the function b depends on c. With choices of c as in (1.3), it can be verified that the function b may become negative. Classical arguments as in the previous section can no longer be applied. It is, in fact, the notion of "solution" itself that has to be defined differently. From Kichenassamy [191], a solution should consist of regions with "low" gradients separated by points of discontinuity where the gradient is infinite (Theorem 3.3.6). Therefore, the notion of solution must be understood in a measure sense, and it is still an open problem. Another possibility that we detail here is to introduce some regularization. Catté et al. [88] proposed to solve the following partial differential equation instead of (1.2):

$$\frac{\partial u}{\partial t}(t,x) = \text{div}(c(|(\nabla G_\sigma * u)(t,x)|^2)\nabla u(t,x)).$$

Interestingly, we show that this equation is now well posed, and we prove in Theorem 3.83 the existence and uniqueness of a solution in the distributional sense. The existence is shown using Schauder's fixed-point theorem.

Some challenging questions are still open, and these are mentioned at the end of this section.

• Enhancing PDEs (Section 3.3.3). In this section, we examine the case of deblurring (or enhancement). It is essentially devoted to the shock filter model proposed by Osher and Rudin [261] in the 1-D case:

$$u_t(t,x) = -|u_x(t,x)|\,\text{sign}\,(u_{xx}(t,x)). \tag{1.4}$$

To better understand the action of this equation, we consider the following simpler case:

$$u_t(t,x) = -|u_x(t,x)|\,\text{sign}(\,(u_0)_{xx}\,(t,x))$$

with $u_0(x) = \cos(x)$. Using the method of characteristics, we construct a solution explicitly. Although this cannot be done in the general case (1.4), it presents interesting calculations. The understanding of (1.4) from a theoretical point of view remains, to our knowledge, an open question.

- Neighborhood Filters, Non-Local Means Algorithm and PDE (Section 3.3.4). We close this chapter, devoted to image restoration, by describing some recent work that extends the notion of Gaussian filtering: neighborhood filters and nonlocal means filters, which are in fact related to PDEs [77, 78]. Here the notion of neighborhood must be understood broadly: neighboring pixels, neighboring or similar intensities, or "neighboring neighborhoods." Each of these meanings will correspond to a specific filter. We start with neighborhood filters (also called bilateral filtering), for which we consider that two pixels are close to each other not only if they occupy nearby spatial locations but also if they have some similarity in the photometric range:

$$F_{h,\rho}u(x) = \frac{1}{N(x)} \int_\Omega T_h(|u(y) - u(x)|) \, w_\rho(|x - y|)u(y)dy,$$

where $N(x)$ is a normalization factor and T_h, w_ρ are two decreasing functions such as Gaussian functions. If $T_h \equiv 1$, we recover the Gaussian filtering, and we study the asymptotic behavior of the solution obtained by iterating infinitely the filtering procedure: in the case of Gaussian filtering, we recover the solution of the heat equation (Lemma 3.3.2).

For general neighboring filters, Buades et al. [77, 78] have shown that the asymptotic behavior of the filters is related to well known differential operators such as the Perona–Malik model (see Theorem 3.3.9). Consequently, from a numerical point of view, this explains why neighborhood filters may also produce a staircase effect.

In order to suppress the staircase effect, Buades et al. [78] introduced an intermediate regression correction to better approximate the signal locally. Some results show the performance of this modified approach. Finally, we present the idea of nonlocal means filters, which take advantage of the high degree of redundancy of any natural image. Following [77] one defines the "neighborhood of a pixel x" as any set of pixels y in the image such that the intensity in a window around y looks like the intensity in a window around x. The similarity between two pixels x and y will be based on the similarity of the intensity gray level between their neighborhoods which can be estimated by a Gaussian distance. The nonlocal means filter can be written as

$$NL_h u(x) = \frac{1}{C(x)} \int_\Omega \exp(-(G_a \star (u(x + .) - u(y + .))^2)(0))u(y)dy.$$

Of course, due to the nature of this filter, most favorable cases for the NL-means are the periodic or the textured patterns, which will be illustrated by some examples.

Chapter 4: The Segmentation Problem_____

In this chapter we examine the segmentation problem, which with image restoration is one of the most important questions in image analysis. Though restoration and segmentation are not totally disconnected, image segmentation has its own objectives (see Section 4.1) and its own methodology. By segmentation we mean that we wish to know the constituent parts of an image rather than to improve its quality (which is a restoration task). The aim of Chapter 4 is to present two variational and PDE approaches to the segmentation problem. In the first approach, the idea is to consider that segmenting an image consists, from an initial image u_0, in constructing a new image u close to u_0 and made up of distinct homogeneous regions, the boundaries between these regions being sharp and piecewise regular. This is achieved by minimizing the Mumford–Shah functional (Section 4.2). The second approach (which can be seen as a point of view dual to the former) aims at detecting the contours of the objects lying in the image u_0 (or a smooth version of it). These contours are modeled by closed curves, which are to be identified. This edge detection method, called geodesic active contours, is presented in Section 4.3.

Section 4.1: Definition and Objectives. Based on some real examples, we suggest a definition of image segmentation. There are two different ways to consider it. The first is to have a simplified version of the original image, compounded of homogeneous regions separated by sharp edges. The second is to extract some "significant" contours. Still, in each case, the important features are edges, and so we briefly recall some of the earliest edge detectors that appeared in the literature.

Section 4.2: The Mumford–Shah Functional. In Section 4.2.1 we first present the formulation introduced by Mumford and Shah in 1985 [246]. It is an energy-based method. For a given $u_0(x)$ (the initial image), we search for a function $u : \Omega \to R$ and a set $K \subset \Omega$ (the discontinuity set) minimizing

$$E(u, K) = \int_\Omega (u - u_0)^2 \, dx + \alpha \int_{\Omega - K} |\nabla u|^2 \, dx + \beta \oint_K d\sigma.$$

The first term measures the fidelity to the data, the second imposes the condition that u be smooth in the region $\Omega - K$, and the third term establishes that the discontinuity set has minimal length and therefore is as smooth as possible. This type of functional forms part of a wider class of problems called "*free discontinuity problems.*" As we can imagine, there is an important literature related to the Mumford–Shah problem. We refer to the book of Morel and Solimini [245] for a complete review. Our aim is to present to the reader a clear overview of the main results.

Section 4.2.2 is concerned with the existence of a solution. As we can foresee, it is not an easy task, and the first question is: What is the good functional framework? Following Ambrosio [11, 12, 13], we show why the problem cannot be directly solved, and we demonstrate the necessity of defining an equivalent formulation involving only one unknown u. This is achieved by considering the functional

$$G(u) = \int_\Omega (u - u_0)^2 dx + \alpha \int_\Omega |\nabla u|^2 dx + \beta \, \mathcal{H}^{N-1}(S_u)$$

defined for $u \in SBV(\Omega)$ and where S_u stands for the discontinuity set of u, i.e., the complement, up to a set of $\mathcal{H}^{N-1}$ measure zero, of the set of Lebesgue points of u. It is proven that the problem inf $\{G(u), \quad u \in SBV(\Omega) \cap L^\infty(\Omega)\}$ admits at least a solution u and that the pair (u, K) with $K = \Omega \cap \overline{S_u}$ is a minimizer of the initial Mumford–Shah functional.

Once we obtain the existence of a minimizer, we would like to compute it. A natural way to do that is to search for optimality conditions. We establish these conditions (Theorem 4.2.3) assuming the following regularity hypotheses:

(C_1) K consists of a finite number of $C^{1,1}$-curves, meeting $\partial\Omega$ and meeting each other only at their endpoints.

(C_2) u is C^1 on each connected component of $\Omega - K$.

Section 4.2.3 investigates the regularity hypotheses (C_1), (C_2). In their seminal work [246], Mumford and Shah conjectured that the functional $F(u, K)$ admits a minimizer satisfying (C_1), (C_2), and then they proved that for such a minimizer, the only vertices of the curves γ_i forming K are the following:

(i) Points P on $\partial\Omega$ where one γ_i meets $\partial\Omega$ perpendicularly.

(ii) Triple points P where three γ_i meet with angles $2\pi/3$.

(iii) Crack-tips where a γ_i ends and meets nothing.

We first demonstrate properties (i), (ii), and (iii) qualitatively when K satisfies (C_1) and (C_2). Then, we state the important result from Bonnet [62, 64] where the hypotheses (C_1) and (C_2) are removed and replaced by a connectedness assumption.

The approximation of the Mumford–Shah functional is examined in Section 4.2.4. The lack of differentiability for a suitable norm does not allow us to use Euler equations directly. Therefore, it is natural to search for an approximation of $F(u, K)$ by a sequence F_ε of regular functionals defined on Sobolev spaces. There exist many ways to approximate $F(u, K)$, and we list some of them. We focus our attention on the Ambrosio–Tortorelli approximation [17], which is among the most used in image analysis. In their approach, the set K (or S_u) is replaced by an auxiliary variable v

(a function) that approximates the characteristic function $(1 - \chi_K)$. The sequence of functionals F_ε is defined by

$$F_\varepsilon(u, v) = \int_\Omega (u - u_0)^2 dx + \int_\Omega (v^2 + h(\varepsilon)) |\nabla u|^2 \, dx$$
$$+ \int_\Omega \left(\varepsilon |\nabla v|^2 + \frac{1}{4\varepsilon} (v - 1)^2 \right) dx,$$

where $h(\varepsilon)$ is a sequence of constants such that $\lim_{\varepsilon \to 0} h(\varepsilon) = 0$. The $F_\varepsilon(u, v)$ are elliptic functionals that Γ-converge to the Mumford and Shah functional.

We close Section 4.2 with some words about the numerical computation and some experimental results (Section 4.2.5). We refer the monograph of Braides [71] to anyone interested in the approximation of free discontinuity problems.

Section 4.3: Geodesic Active Contours and the Level-Set Method.
As suggested in Section 4.1, another approach for segmentation is to detect the "significant" contours, for example the contours of the different objects in a scene, rather than to construct a partition made of homogeneous regions. This is the active contour model. The idea of this approach is that contours are characterized by sharp variations of the image intensity, and so by infinite gradients. The principle consists in matching deformable curves to the contour objects by means of energy minimization.

In Section 4.3.1, we begin by describing the Kass–Witkin–Terzopoulos model [190], which is, to the best of our knowledge, one of the first in this direction. In this model the energy to be minimized is defined on a set C of closed, piecewise regular, parametric curves (called snakes) and is given by

$$J_1(c) = \int_a^b |c'(q)|^2 \, dq + \beta \int_a^b |c''(q)|^2 \, dq + \lambda \int_a^b g^2(|\nabla I(c(q))|) dq,$$

$c \in C$, where g is a decreasing monotonic function vanishing at infinity. Thanks to the third term, minimizing curves are attracted by the edges of the objects forming the image. We point out some drawbacks of this model (in particular, the nonintrinstic nature of J_1, which depends on the parametrization) that led Caselles, Kimmel, and Sapiro to propose their geodesic active contour model [87, 86] (Section 4.3.2). They observed that in the energy J_1 we can choose $\beta = 0$, and that, in a sense to be made precise (Definition 4.3.1), the minimization of $J_1(c)$ (with $\beta = 0$) is equivalent to the minimization of

$$J_2(c) = 2\sqrt{\lambda} \int_a^b g(\,|\nabla I(c(q))|\,)\,|c'(q)|\,dq.$$

Now, J_2 is invariant under a change of parametrization and can be seen as a weighted Euclidean length. This energy was also proposed in the same time by Kichenassamy et al. [193, 192]. Starting from an initial curve $c_0(q)$, we detect the image contours by evolving a family of curves $c(t, q)$ according to the decreasing gradient flow associated to J_2:

$$\frac{\partial c}{\partial t} = (\kappa\,g - \langle \nabla g, N \rangle)\,N, \tag{1.5}$$

where N is the unit outward normal to $c(t, q)$ and κ its curvature. Thanks to the definition of g, the evolution of $\{c(t, q)\}_{t \geq 0}$ is halted when edges are detected. However, due to its parametric formulation, this model still has some drawbacks. For example, a change of topology (corresponding to the detection of several objects) during the curve evolution is not allowed. To circumvent this type of problem, Osher and Sethian proposed in a pioneering work [262] the so-called level set method that we examine in Section 4.3.3. The idea is simple: A curve in R^2 can be seen as the zero-level line of a function u defined from R^2 to R. We reformulate the geodesic active contour model in terms of the function u, and we deduce that equation (1.5) can be rewritten as

$$\frac{\partial u}{\partial t} = \kappa\,g\,|\nabla u(t, x)| - \langle \nabla g, \nabla u \rangle. \tag{1.6}$$

Then we list the major advantages of this formulation, and we show that equations of the form (1.6) are well posed in the viscosity sense. We develop in detail the mathematical analysis of (1.6) (existence, uniqueness, maximum principles).

When implementing the equation (1.6), it is observed that the gradient of the level sets may become unbounded, which generates serious numerical problems. In order to avoid this difficulty, it is necessary to run the following reinitialization equation

$$\frac{\partial u}{\partial t} + \text{sign}(v(x, t_0))\,(|Du| - 1) = 0,$$

which is a discontinuous Hamilton–Jacobi equation. We show in Section 4.3.4 how this equation can be justified by extending the framework of viscosity solutions to the discontinuous case.

Section 4.3.5 is concerned with the experimental results obtained with (1.6). The main difficulty in the numerical approximation of (1.6) comes from the presence of hyperbolic terms. It is suggested why the discretization has to be carried out carefully, which leads to the introduction of entropic conservative schemes. The interested reader will find more details in Appendix A (Sections A.2 and A.3.4).

Finally, we point out in Section 4.3.6 how some limitations from the snakes or level set formulations may be overcome. Two recent contributions are presented. The first is concerned with the stopping criterion. Classically based on the intensity gradient (corresponding to sharp edges), this criterion may not be suitable for some kinds of images with soft contours or "perceptual" contours (imagine an image with dots not evenly distributed). The idea proposed by Chan and Vese [99] is to consider what is inside the regions instead of focusing only on the boundaries. The second contribution is related to the representation of curves using level sets. Although it is very convenient numerically, it can deal only with closed nonintersecting hypersurfaces. We present some recent developments [288, 169, 170], that describe a more flexible representation and should be investigated further in the future.

Chapter 5: Other Challenging Applications

The scope of this last chapter is more applied. Five kinds of recent applications are analyzed. Although some theoretical results are given and proved, the goal of this chapter is mainly to show how the previous material may be used for more "applied" and complex problems.

Section 5.1: Reinventing Some Image Parts by Inpainting. In this section we intend to illustrate a useful and relatively recent application of PDEs in image processing designated by inpainting or filling in. The goal of inpainting is to restore a damaged or corrupted image in which part of the information has been lost. Given a domain image D, a hole $\Omega \subset D$, and an intensity u_0 known over $D - \Omega$, we want to find an image u, an inpainting of u_0, that matches u_0 outside the hole and that has "meaningful" content inside the hole Ω. In this section, we will mainly focus on geometric images, i.e., without fine texture content, for which PDE-based approaches are particularly suitable. As we will see, most inpainting PDE methods are based on the simultaneous interpolation of isophotes and gray level intensities.

We review in Section 5.1.2 three variational approaches. The first one, the elastica model, is inspired by the pioneering work of Nitzberg, Mumford, and Shiota [255], who tried to identify occluding and occluded objects in a plane image, in order to compute the image depth map. When an object occludes another, the occluding and occluded boundaries form a particular configuration called a T-junction. The approach proposed by Masnou and Morel [230] is a variational model for finding the best curve fitting two given T-junctions T_1 and T_2 on $\partial\Omega$ with corresponding orientations θ_1 and θ_2 of

∇u. The model is to search for Γ among the minimizers of the criterion

$$\int_\Gamma (\alpha + \beta \kappa^p) d\mathcal{H}^1 + (\theta_1, N_1) + (\theta_2, N_2), \qquad (1.7)$$

where α, β are positive constants, $p \geq 1$, κ is the curvature of Γ, and the last two terms denote the angles between θ_i and N_i, the normal to Γ at T_i (for $i = 1, 2$). Γ is called a completion curve. The final model integrates (1.7) for all possible T-junctions.

The second variational model analyzed in Section 5.1.2 was proposed by Ballester et al. [34, 35, 36]. The authors consider a joint interpolation of vector fields and gray levels. Their approach is also inspired by the Elastica model and is closely related to the previous variational approach in [230]. In the model proposed by Ballester et al. there are two unknowns: the orthogonal direction of level lines θ and the gray levels u of the image. The authors propose to search for (u, θ) as a minimizer of the constrained problem

$$\min \int_\Omega |\operatorname{div}(\theta)|^p (a + b\nabla k \star u) \, dx,$$
$$|\theta| \leq 1, \ \theta \cdot \nabla u - |\nabla u| = 0 \text{ in } \Omega,$$
$$u = u_0 \text{ in } B, \ \theta \cdot N = \theta_0 \cdot N \text{ on } \partial\Omega,$$

where $p > 1, a > 0, b \geq 0, k$ denotes a regularizing kernel of class C^1 with $k(x) > 0$ a.e., and $N(x)$ denotes the outer unit normal at $x \in \partial\Omega$.

The third variational model is based on the total variation (TV). In [103, 95, 97], Chan and Shen proposed to minimize

$$F(u) = \int_D |Du| \, dx, \quad u = u_0 \text{ in } D - \Omega.$$

We consider a slightly modified version of the criterion and we show the existence of a solution that satisfies the maximum principle.

In Section 5.1.3 two PDE-based models are presented. The first one, by Bertalmio et al. [50], is inspired by basic techniques used by art conservators to inpaint when restoring real paintings. The resulting mathematical model is a third-order PDE that propagates the level lines arriving at the hole. The authors model the propagation of the Laplacian in the direction of the level lines by solving the PDE

$$\frac{\partial u}{\partial t}(x, y, t) = \nabla(\Delta u(x, y, t)) . \nabla u^\perp(x, y, t).$$

The second model, by Chan and Shen [98], comes from restoration problems and also involves third-order equations. Starting from previous work based on total variation [98], the authors proposed in [98] to introduce the curvature in the diffusion coefficient. They proposed the so-called curvature-driven diffusion model (CDD)

$$\frac{\partial u}{\partial t} = \text{div} \left(\frac{g(\kappa) \, \nabla u}{|\nabla u|} \right).$$

Finally, we point out in Section 5.1.4 the limitations of variational and PDE-based approaches. In particular, this kind of approach will perform correctly when the region to recover is small, and when the region to be filled in corresponds to a geometric part of the image: losses in texture parts cannot be recovered accurately. In this section we briefly describe some possible approaches to overcoming these limitations.

Section 5.2: Decomposing an Image into Geometry and Texture. In this section we present some recent contributions [235, 333] concerning decomposition models. These models aim at characterizing strongly oscillating patterns (e.g., textures) so that an image may be decomposed as the sum of a geometric part and an oscillating one. To do this, we introduce in Section 5.2.2 the functional space

$$G(\Omega) = \left\{ v \in L^2(\Omega) \, / \, v = \text{div}(\xi) \, , \, \xi \in L^\infty(\Omega, R^2) \, , \, \xi \cdot N = 0 \text{ on } \partial\Omega \right\},$$

endowed with the norm

$$|v|_{G(\Omega)} = \inf \left\{ |\xi|_{L^\infty(\Omega, R^2)} \, / v = \text{div}(\xi) \, , \, \xi \cdot N = 0 \text{ on } \partial\Omega \right\}.$$

We show that $G(\Omega)$ is well adapted to capture oscillating signals. The model proposed by Meyer in [235] is described in Section 5.2.3. Given an image f, the problem is to solve

$$\inf_{(u,v) \in BV(\Omega) \times G(\Omega)/f=u+v} \left(\int_\Omega |Du| + \alpha \, |v|_{G(\Omega)} \right),$$

where α is a weighting parameter.

The main difficulty in solving numerically Meyer's problem is that it involves a term coming from an L^∞-norm (in the G-norm), which is not easy to handle in practice. To overcome this difficulty, it is classical to introduce a new related problem that in some sense approximates the original one. So we present in Section 5.2.4 the approach by Aujol et al. [21], which essentially consists in replacing $|v|_{G(\Omega)}$ with a constraint $|v|_{G(\Omega)} \leq \mu$. More precisely, we consider the problem

$$\inf_{\substack{(u,v) \, \in \, BV(\Omega) \times G(\Omega) \\ |v|_{G(\Omega)} \leq \mu}} \left(\int_\Omega |Du| + \frac{1}{2\lambda} |f - u - v|^2_{L^2(\Omega)} \right), \qquad (1.8)$$

where λ, μ are two nonnegative parameters. We first show that if λ and μ are fixed, then the problem (1.8) admits a unique solution that is numerically computed using a projection-based algorithm. The convergence of the algorithm is rigorously proved. Then we examine the asymptotic case in which λ goes to zero. We find a limit variational problem whose minimizers are in some sense solutions of the original problem.

Some results on noisy images and images with texture are shown in Section 5.2.5.

Finally, we mention in Section 5.2.6 some recent extensions of this work. The first concern the model. In [30] the authors have proposed to look for a decomposition of the type $u + v + w$, where u is the geometric part, v is only the texture part, and w the noise. The second kind of extension concerns the applications of this formalism. They include new inpainting or image compression algorithms.

Section 5.3: Sequence Analysis. Since the term "sequence analysis" may appear quite wide and imprecise, we begin in Section 5.3.1 by giving some generalities about sequences. The aim is to make the reader aware that one uses all one's knowledge of the environment or scenario to analyze a sequence. Without this, it becomes a very difficult problem.

One of the first difficulties is motion estimation (Section 5.3.2). Beyond the fact that motions may be of different types and amplitudes, motion can be recovered only by way of intensity variations. This explains why it is possible to recover only an apparent motion (a priori different from the projection of the 3-D real motion field). This section is essentially a review of existing variational approaches for optical flow. We first present the so-called optical flow constraint, which is a scalar equation corresponding to the conservation of intensity along trajectories. It links the intensity of the sequence $u(t, x) : R^+ \times \Omega \to R$ (the data) to the instantaneous motion field at time $t = t_0$, $\sigma(x) : \Omega \to R^2$ (the unknown), by

$$\sigma(x) \cdot \nabla u(t, x_0) + u_t(t, x_0) = 0.$$

Since this is not sufficient to solve the problem (one equation for two unknowns: the two components of the flow field), many solutions have been proposed, and we recall some of them. We focus on regularization approaches and mention some of the large quantity of research based on this method. We then present in more detail a discontinuity-preserving approach by Aubert, Deriche, and Kornprobst [133, 24, 25] in which some theoretical results have been established. The problem is to minimize

$$E(\sigma) = \underbrace{\int_\Omega |\sigma \cdot Du + u_t|}_{A(\sigma)} + \alpha^s \underbrace{\sum_{j=1}^{2} \int_\Omega \phi(D\sigma_j)}_{S(\sigma)} + \alpha^h \underbrace{\int_\Omega c(Du)|\sigma|^2 \, dx}_{H(\sigma)},$$

where α^s, α^h are positive constants. From now on, unless specified otherwise, all derivatives are written in a formal setting (i.e., Du is the gradient of u in the distributional sense). Since we look for discontinuous optical flows, a suitable theoretical background for studying this problem will be the space of bounded variation $\mathbf{BV}(\mathbf{\Omega})$ (see Section 2.2). The energy is compounded of three terms:

- $A(\sigma)$ is the "L^1"-norm of the optical flow constraint. In fact, since it is formal here, it has to be interpreted as a measure.

- $S(\sigma)$ is the smoothing term, chosen as for image restoration (see Section 3.2) in order to preserve the discontinuities of the flow.

- $H(\sigma)$ is related to homogeneous regions. The idea is that if there is no texture, that is to say no gradient, there is no way to estimate the flow field correctly. Then one may force it to be zero.

We conclude this section by discussing the well-foundness of the optical flow constraint in some situations, and try to indicate some future paths for this still challenging problem.

The second problem that we consider in Section 5.3.3 is the problem of sequence segmentation. Here, segmentation means finding the different objects in a scene, and it is naturally carried out in relation to velocity estimation. Motion-based segmentation works well on some sequences (low noise, good sampling, ...) but will fail otherwise. Another idea is to use the background redundancy to segment the scene into layers (foreground/background). It is shown on a synthetic example how this idea may be considered and what the limits of the naive approach are. In particular, in the case of noisy sequences, obtaining a reference image is as difficult as segmenting the sequence. So it would certainly be more efficient to look for the reference image and the segmentation at the same time. This can be formalized in terms of PDEs, as done by Kornprobst et al. [206], where a coupled approach is proposed. Let $N(t, x)$ denote the given noisy sequence, for which the background is assumed to be static. We look simultaneously for:

- The restored background $B(x)$.

- The sequence $C(t, x)$ that indicates the moving regions. Typically, $C(t, x) = 0$ if the pixel x belongs to a moving object at time t, and 1 otherwise.

The minimization problem is

$$\inf_{B,C} \underbrace{\iint_V C^2(B-N)^2 + \alpha_c(C-1)^2 \; dx \, dt}_{A(\sigma)} + \underbrace{\alpha_b^r \int_\Omega \phi_1(DB) + \alpha_c^r \iint_V \phi_2(DC)}_{S(\sigma)},$$

where α_c, α_b^r, α_c^r are positive constants. The energy is compounded of two kinds of terms:

- $A(\sigma)$ realizes the coupling between the two variables B and C. The second term forces the function C to be equal to 1, which corresponds to the background. However, if the current image N is too different from the background (meaning that an object is present), then the first term will be too large, which forces C to be 0.

- $S(\sigma)$ is the smoothing term. As usual, the functions ϕ_i are chosen in order to preserve the discontinuities (see Section 3.2.2).

Although this functional is globally nonconvex and degenerate, existence and uniqueness can be proven on the space of bounded variation, with a condition on the coefficient α_c. The performance of the algorithm is illustrated on several real sequences.

We consider in Section 5.3.4 the problem of sequence restoration and more precisely movie restoration. Little research has been carried out on this topic, especially using PDEs. The aim of this section is essentially to make the reader aware of the difficulties of this task. It is mainly due to the wide variety of ways a film can be damaged (chemical, mechanical, handling, etc.). Faced with this variety of defects, we propose to classify them by taking into account the degree of human interaction that is necessary to detect and/or correct them. Interestingly, this degree of interaction/automation can also be found in the different systems that are proposed nowadays. Something important that comes out of this discussion is that no system is able to deal with every kind of defect. By trying to define a "perfect" restoration, it turns out that it has to be perceptually correct: It will be sufficient for a defect to be no longer perceived even if it has not been completely removed. More than any other domain in computer vision, movie restoration should benefit from advances in the study of human perception (see [189], for instance).

We end this section by presenting some PDE-based approaches for video inpainting. We show that the simple extension of image inpainting methods (Section 5.1) to image sequences sequences is not sufficient: appropriate methods are needed. This will be illustrated on real test sequences.

Section 5.4 Image Classification. The classification of images consists in assigning a label to each pixel of an image. This has rarely been introduced in a variational formulation (continuous models), mainly because the notion of classes has a discrete nature. Many classification models have been developed with structural notions such as region-growing methods, for example [272], or by a stochastic approach (discrete models) using Markov random field theory [53, 350]. It is considered that stochastic methods are robust but nonetheless time-consuming. The goal of this section is to show how PDEs and variational techniques can also solve some image classification problems. Two variational models are presented. For these two models it is assumed that each pixel is characterized only by its intensity level, that each class C_i has a Gaussian distribution of intensity $N(\mu_i, \sigma_i)$, and that the number of classes K and the parameters (μ_i, σ_i) are known (supervised classification) (see Section 5.4.1).

In Section 5.4.2 we present a level set formulation [290]. Here, classification is seen as a partitioning problem. If Ω is the image domain and u_0 is the data, we search for a partition of Ω defined by

$$\Omega = \bigcup_{i=1}^{K} (\Omega_i \cup \Gamma_i) \text{ and } \Omega_i \cap \Omega_j = \emptyset, \quad i \neq j,$$

where $\Gamma_i = \partial \Omega_i \cap \Omega$ is the intersection of the boundary of Ω_i with Ω and $\Gamma_{ij} = \Gamma_{ji} = \Gamma_i \cap \Gamma_j$, $i \neq j$, the interface between Ω_i and Ω_j. Moreover, to be admissible, the partitioning process has to satisfy some constraints:

- Taking into account the Gaussian distribution property of the classes.

- Ensuring the regularity of each interface.

This can be achieved by identifying each Ω_i with an upper level set of a signed function $\phi_i(.)$, i.e., $\phi_i(x) > 0$ if $x \in \Omega_i$, $\phi_i(x) = 0$ if $x \in \partial \Omega_i$, and $\phi_i(x) < 0$ otherwise.

These three requirements (partitioning, Gaussian distribution of classes, regularity of interfaces) can be satisfied through the minimization of the following global energy defined on Ω:

$$F(\phi_1, \ldots, \phi_K) = \int_{\Omega} \left(\sum_{i=1}^{K} H(\phi_i(x)) - 1 \right)^2 dx$$

$$+ \sum_{i=1}^{K} e_i \int_{\Omega} H(\phi_i(x)) \frac{(u_0(x) - \mu_i)^2}{\sigma_i^2} dx + \sum_{i=1}^{K} \gamma_i \int_{\phi_i = 0} ds,$$

where $H(s)$ is the Heaviside function: $H(s) = 1$ if $s > 0$ and $H(s) = 0$ if $s < 0$. Unfortunately, the functional F has some drawbacks (nondifferentiability, presence of a boundary term, etc.). Thus, instead of F, we propose to minimize an approximated F_α close to F (as $\alpha \to 0$), and we use several experimental results to demonstrate the soundness of the model.

Section 5.4.3 presents another model of classification coupled with a restoration process [291]. It is based on the theory of phase transitions developed in mechanics [4, 80]. The model still relies on the minimization of an energy whose generic form is

$$J_\varepsilon(u) = \int_{\Omega} (u(x) - u_0(x))^2 dx + \lambda^2 \varepsilon \int_{\Omega} \varphi(|\nabla u(x)|) dx + \frac{\eta^2}{\varepsilon} \int_{\Omega} W(u(x), \mu, \sigma) dx.$$

The first two integrals in J_ε are the usual terms in restoration (we can choose, for instance, $\phi(s) = s^2$ or $\phi(s) = \sqrt{1 + s^2}$). The third integral is a classification term. The role of the function $W(u)$ is to attract the values of u toward the labels of the classes. Since the chosen labels are the means μ_i of the classes, good candidates for W are those that satisfy $W(\mu_i) = 0$, $i = 1, \ldots, K$, $W(u) \geq 0$, for all u. Such functions are known as multiple wells potentials. The parameter $\varepsilon > 0$ is destined to tend to zero, and its contribution is a major one in the restoration-classification process. Roughly speaking, as ε decreases during the first steps of the algorithm, the weight of $\frac{1}{\varepsilon} \int_\Omega W(u(x), \mu, \sigma) \, dx$ is negligible with respect to the two others, and only the restoration process runs. As ε becomes smaller, the diffusion is progressively stopped while the classification procedure becomes active. This phenomenon is illustrated on experimental results.

Concerning the mathematical aspect of the approach, we show that the model is well posed when the regularization function is $\phi(s) = s^2$. We study the asymptotic behavior (as $\varepsilon \to 0$) of $J_\varepsilon(u)$ as well as the behavior of a sequence of minimizers u_ε of J_ε. The proofs rely on Γ-convergence theory and are borrowed from previous work related to the theory of phase transitions [239, 309, 310]. We conclude this section by giving some illustrative examples using synthetic and real images.

Section 5.5 Vector-Valued Images. In this book, we have mainly considered gray-scale images, i.e., scalar images. This section focus on vector-valued images $u : \Omega \mapsto R^p (p > 1)$. A typical example is color images $(p = 3)$. In fact, most algorithms and results given for scalar images can be extended to vector-valued images. However, extending an algorithm designed for scalar images is not generally as simple as applying the same algorithm to each component independently. This section presents an elegant and unified presentation by Tschumperlé et al. [324] that is well adapted for restoration tasks.

Section 5.5.2 introduces the structure tensor [322, 336, 347]

$$S(u) = \sum_{i=1}^{p} \nabla u_i \nabla u_i^T,$$

where each ∇u_i corresponds to the gradient of the ith canal. The structure tensor S is particularly interesting since its eigenvalues $\lambda_\pm$ define the local min/max vector-valued variations of u in the eigenvectors, directions θ_+ and θ_- respectively.

In Section 5.5.3 we formulate the restoration problem as a minimization problem, which is very similar to the scalar case (see Section 3.2). The problem is to minimize

$$E(u) = \int_\Omega \| u - u_0 \|^2 \, dx + \mu \int_\Omega \phi(\nabla u) \, dx,$$

where u_0 is the initial image to be reconstructed. The real function $\phi(\nabla u)$ is related to the local variations of the image through the eigenvalues $\lambda_\pm$ of the structure tensor S, and several choices are discussed.

Section 5.5.4 presents a generic vector-valued PDE approach [324]

$$\frac{\partial u_i}{\partial t} = \sum_{j=1}^{p} \operatorname{tr}\left(A^{ij} H^i\right) \quad \text{for } i = 1, \ldots, p,$$

where the (A^{ij}) are 2×2 matrices to be specified and the (H^i) denote the Hessian matrix of u_i. Again, coefficients A^{ij} can be specified through the structure tensor S and according to some qualitative properties. This formalism can be successfuly applied to several applications, and we illustrate some of them.

Appendix A: Introduction to Finite Difference Methods

The aim of this monograph is to explain how PDEs can help to model certain image processing problems and to justify the models. Until now, we have always considered a continuous setting ($x \in \Omega \subset R^2$) referring to analog images. However, working in this field also requires that one test the models in the digital (discrete) world. So we need to know how to discretize the equations. Although several approaches can be considered (such as finite elements or spectral methods), the success of finite differences in image analysis is due to the structure of digital images, with which we can associate a natural regular fixed grid. The aim of this chapter is to present the basis and the main notions of finite difference methods. It is also to provide the discretization of some complex operators presented in this book.

Section A.1: Definitions and Theoretical Considerations Illustrated by the 1-D Parabolic Heat Equation. This section gives the main notions and definitions for discrete schemes (convergence, consistency, and stability). Every notion is illustrated by explicit calculations for the case of the 1-D heat equation.

Section A.2: Hyperbolic Equations. This section is concerned with hyperbolic partial differential equations, which are the most difficult equations to discretize properly. Just to make the reader aware of it, we first consider the linear 1-D advection equation:

$$\frac{\partial u}{\partial t} + a\frac{\partial u}{\partial x} = 0,$$

where a is a constant. Interestingly, we show that if we do not choose a suitable discretization for the spatial derivative, then the scheme will be unconditionaly unstable. The right way is to choose an upwind scheme that takes into account the direction of the propagation.

We then consider the nonlinear Burgers equation

$$\frac{\partial u}{\partial t} + u\,\frac{\partial u}{\partial x} = 0$$

and show by the method of characteristics that a shock can appear even if we start from continuous initial data, and we show that this also brings some numerical difficulties. The notions of monotone and conservative schemes are then necessary for capturing the correct entropy solution.

Section A.3: Difference Schemes in Image Analysis. After introducing the principal notation we present the discretization of certain equations encountered in this book. Here the aim is to offer the reader the possibility of reimplementing these equations. The problems considered are the following:

- Restoration by energy minimization (from Section 3.2): We detail the discretization of a divergence term that can also be found for the Perona and Malik equation.

- Enhancement by Osher and Rudin's shock filters (from Section 3.3.3): The main interest is to use a flux limiter called minmod.

- Curve evolution with the level set method. Having in mind the geodesic active contours equation from Section 4.3.3, we decompose the problem by studying separately the different terms. We start with the classical mean curvature motion, and we show on a simple example that a reinitialization is in general necessary to preserve a distance function. The second example concerns motions with a constant speed. We mention the possibility for such evolutions (and more generally with monotone speed) of using a fast-marching approach. The third equation is the pure advection equation. Finally, we return to the segmentation problem with geodesic active contours. Some examples illustrate the equations.

We hope that this appendix will give the reader effective ideas for discretizing and implementing the PDEs studied throughout this monograph and those encountered in their own research.

Appendix B: Experiment Yourself!

In Appendix A we explained the main ideas of the finite difference methods, which allow one to discretize the continuous equations. A step further, we wish to provide our readers the programming tools to implement the approaches and test them on their own data.

> A web site is associated with this second edition. This web site contains some related links, complementary of information, and also source code that allow the reader to test easily some variational and PDE-based approaches.
>
> `http://www-sop.inria.fr/books/imath`

Section B.1: The CImg Library. In this section we justifiy the technical choices for the software development:

- The chosen programming language is the object-oriented language C++, which is freeware and a very efficient language.

- The chosen image processing library is the CImg library, which stands for "Cool Image," developed by David Tschumperlé in 2000. The CImg library is simple to use and efficient, and it offers a number of interesting functions.

Section B.2: What Is Available Online?. This section gives a nonexhaustive list of the C++ codes available online and shows an example of CImg code. The proposed codes correspond in general to approaches explained in the book. The symbol $\boxed{\texttt{C++}}$ written in the margin will indicate to the reader that some code is available. It is suggested that the reader regularly consult the book web site, which will contain updated information. Note that external contributors are encouraged to submit their own C++ source codes.

2
Mathematical Preliminaries

How to Read This Chapter

This chapter introduces the mathematics used in this book. It covers some of the main notions in the calculus of variations and the theory of viscosity solutions. It is introductory, and we have tried to made it as self-contained as possible. A careful reading is advised for a better understanding of the analysis to follow. The prerequisite for this chapter is a good course in advanced calculus.

- Section 2.1 introduces the basic tools concerning optimization problems in Banach spaces. We answer the following questions: What are good hypotheses for ensuring the existence and uniqueness of a minimizer? What can be said when some assumptions are missing? We also introduce the notion of relaxed problem when the original one is ill posed (Section 2.1.3). Section 2.1.4 concerns the notion of Γ-convergence, which is a notion of convergence for functionals. The theory of Γ-convergence is particularly useful to approximate free discontinuity or ill-posed problems. This notion will be used several times in all the book.

- Section 2.2 presents the space $BV(\Omega)$ of functions of bounded variation. This space appears to be a suitable functional space for image analysis, since it contains functions that can be discontinuous across curves (i.e., across edges). Note that many theoretical results given in this book will be established in a *BV* framework: See Sections 3.2, 4.2, and 5.3.

- Section 2.3 introduces the concept of viscosity solutions for nonlinear second-order PDEs. This notion is interesting, since it allows us to prove the existence of a solution in a very weak sense and also its uniqueness which is the most difficult part. An application example will be studied further in Section 4.3.

- Some basic elements of differential geometry used throughout this book are recalled in Section 2.4.

- We conclude in Section 2.5 by giving various results of interest (inequalities and theorems) to be as much as possible self-contained. This section may be consulted as needed.

For a systematic study of the classical mathematics presented in this book, the following references may be useful:

☞ For a general presentation on functional analysis: [74, 153, 286, 287, 346],

☞ For a general presentation on partial differential equations: [148, 161],

☞ To know more on Sobolev spaces: [2],

☞ To know more on integration: [285],

☞ For a general presentation on differential geometry: [218].

2.1 The Direct Method in the Calculus of Variations

2.1.1 *Topologies on Banach Spaces*

Let us introduce some definitions. Let $(X, |.|)$ denote a real Banach space.[1] We denote by X' the topological dual space of X:

$$ X' = \left\{ l : X \to R \text{ linear such that } |l|_{X'} = \sup_{x \neq 0} \frac{|l(x)|}{|x|_X} < \infty \right\}. $$

Classically, X can be endowed with two topologies (we work only with sequences):

Definition 2.1.1 (topologies on X)

---

[1]A Banach space is a complete, normed linear space. Complete means that any Cauchy sequence is convergent.

(i) *The strong topology, denoted by* $x_n \xrightarrow{X} x$, *is defined by*

$|x_n - x|_X \to 0 \ (n \to +\infty)$.

(ii) *The weak topology, denoted by* $x_n \xrightarrow{X} x$, *is defined by*

$l(x_n) \to l(x) \ (n \to +\infty)$ *for every* $l \in X'$.

Strong convergence implies weak convergence, but the converse is false in general.

Example Let us present a counterexample. We consider the sequence $f_n(x) = \sin(2\pi x n)$, $x \in (0,1)$, as $n \to \infty$. We are going to establish that $f_n(x) \xrightarrow{L^2(\Omega)} 0$ and that $f_n(x) \xrightarrow{L^2(\Omega)} 0$ is not true.

To prove weak convergence, we have for all $\varphi \in C^1(0,1)$, thanks to a classical integration by parts,

$$\int_0^1 \sin(2\pi x n)\varphi(x) \, dx = \frac{1}{2\pi n}\left[\varphi(0) - \varphi(1)\right] + \frac{1}{2\pi n}\int_0^1 \cos(2\pi x n)\varphi'(x) \, dx$$

So it is clear that $\langle f_n, \varphi \rangle \to 0$ as $n \to \infty$ for all $\varphi \in C^1(0,1)$, where $\langle ., . \rangle$ is the usual scalar product in $L^2(0,1)$. By density, this result can be generalized for all $\varphi \in L^2(0,1)$.

Now, to prove that there is no strong convergence, we observe that

$$\int_0^1 \sin^2(2\pi x n) \, dx = \frac{1}{2}\int_0^1 (1 - \cos(4\pi x n)) \, dx = \frac{1}{2}.$$

∎

The dual X' can also be endowed with the strong and the weak topologies:

Definition 2.1.2 (topologies on X')

(i) *The strong topology, denoted by* $l_n \xrightarrow{X'} l$, *is defined by*

$|l_n - l|_{X'} \to 0$, *or equivalently,* $\sup\limits_{x \neq 0} \dfrac{|l_n(x) - l(x)|}{|x|_X} \to 0 \ (n \to +\infty)$.

(ii) *The weak topology, denoted by* $l_n \xrightarrow{X'} l$, *is defined by*

$z(l_n) \to z(l) \ (n \to +\infty)$ *for every* $z \in (X')'$, *the bidual space of X.*

In some cases it is more convenient to equip X' with a third topology:

(iii) *The weak* topology, denoted by* $l_n \xrightarrow[X']{*} l$, *is defined by*

$l_n(x) \to l(x) \ (n \to +\infty)$ *for every* $x \in X$.

The interest of the weak* topology will be clear later (see Theorem 2.1.1). We recall that the space X is called *reflexive* if $(X')' = X$ and that X is *separable* if it contains a countable dense subset.

Examples Let Ω be an open subset of R^N.

- $X = L^p(\Omega)$ is reflexive for $1 < p < \infty$ and separable for $1 \le p < \infty$. The dual space of $L^p(\Omega)$ is $L^{p'}(\Omega)$ for $1 \le p < \infty$ with $1/p+1/p' = 1$.

- $X = L^1(\Omega)$ is nonreflexive and $X' = L^\infty(\Omega)$.   ∎

The main properties associated with these different topologies are summarized in the following theorem.

Theorem 2.1.1 (weak sequential compactness)

(i) *Let X be a reflexive Banach space, $K > 0$, and $x_n \in X$ a sequence such that $|x_n|_X \le K$. Then there exist $x \in X$ and a subsequence x_{n_j} of x_n such that $x_{n_j} \xrightarrow[X]{} x$ $(n \to +\infty)$.*

(ii) *Let X be a separable Banach space, $K > 0$, and $l_n \in X'$ such that $|l_n|_{X'} \le K$. Then there exist $l \in X'$ and a subsequence l_{n_j} of l_n such that $l_{n_j} \xrightarrow[X']{*} l$ $(n \to +\infty)$.*

The interest of the weak* topology is that it allows one to obtain compactness results even if X is not reflexive. Notice that nothing can be said about the strong convergence of the sequences.

2.1.2 Convexity and Lower Semicontinuity

Let X be a Banach space, $F : X \to R$, and consider the minimization problem

$$\inf_{x \in X} F(x).$$

Let us first consider the problem of the existence of a solution. Proving it is usually achieved by the following steps, which constitute the direct method of the calculus of variations:

(A) One constructs a *minimizing sequence* $x_n \in X$, i.e., a sequence satisfying $\lim_{n \to +\infty} F(x_n) = \inf_{x \in X} F(x)$.

(B) If F is *coercive* $\left(\lim_{|x| \to +\infty} F(x) = +\infty \right)$, one can obtain a uniform bound $|x_n|_X \le C$. If X is reflexive, then by Theorem 2.1.1 one deduces the existence of $x_0 \in X$ and of a subsequence x_{n_j} such that $x_{n_j} \xrightarrow[X]{} x_0$.

(C) To prove that x_0 is a minimum point of F it suffices to have the inequality $\varliminf_{x_{n_j} \to x_0} F(x_{n_j}) \ge F(x_0)$, which obviously implies that
$$F(x_0) = \min_{x \in X} F(x).$$

This latter property, which appears here naturally, is called weak *lower semicontinuity*. More precisely, we have the following definition:

Definition 2.1.3 (lower semicontinuity)
F is called lower semicontinuous (l.s.c.) for the weak topology if for all sequence $x_n \rightharpoonup x_0$ we have

$$\lim_{x_n \longrightarrow x_0} F(x_n) \geq F(x_0). \tag{2.1}$$

The same definition can be given with a strong topology.

☛ *In the direct method, the notion of weak l.s.c. emerges very naturally.*

Unfortunately, it is difficult in general to prove weak l.s.c.. A sufficient condition that implies weak l.s.c. is convexity:

Definition 2.1.4 (convexity) *F is convex on X if*

$$F(\lambda x + (1 - \lambda)y) \leq \lambda F(x) + (1 - \lambda)F(y)$$

for all x, $y \in X$ and $\lambda \in [0,1]$.

Theorem 2.1.2 (l.s.c. strong and weak) *Let $F : X \to R$ be convex. Then F is weakly l.s.c. if and only if F is strongly l.s.c.*

This theorem is useful, since in most cases strong l.s.c. is not very hard to prove.

If F is an integral functional, we can even say more about the link between convexity and l.s.c. Let $\Omega \subset R^N$ be a bounded open set, and let $f : \Omega \times R \times R^N \to R$ be a continuous function satisfying

$$0 \leq f(x,u,\xi) \leq a(x,|u|,|\xi|), \tag{2.2}$$

where a is increasing with respect to $|u|$ and $|\xi|$, and integrable in x. Let $W^{1,p}(\Omega)$ be the Sobolev space

$$W^{1,p}(\Omega) = \{u \in L^p(\Omega),\ Du \in L^p(\Omega)\},$$

where Du is the distributional gradient of u. (Notice that in this case Du is a function and we can also denote it ∇u.) For $u \in W^{1,p}(\Omega)$ we consider the functional

$$F(u) = \int_\Omega f(x,u(x),Du(x))\ dx.$$

Theorem 2.1.3 (l.s.c. and convexity) *$F(u)$ is (sequentially) weakly l.s.c. on $W^{1,p}(\Omega)$, $1 \leq p < \infty$ (weakly* l.s.c. if $p = \infty$), if and only if f is convex in ξ.*

Remarks

- We emphasize that convexity is a sufficient condition for existence. There exist nonconvex problems admitting a solution [27].

- Theorem 2.1.3 has been established assuming that $u : \Omega \subset R^N \to R^M$ with $M = 1$. In dealing with gray-scale images we are in this situation, since $N = 2$, $M = 1$. This result is also true in the case $N = 1$ and $M > 1$. Both cases can be referred to as scalar (either N or M is equal to 1). However, this theorem is no longer true in the vectorial case, when $N > 1$ and $M > 1$. A weaker definition of convexity has to be introduced (quasiconvexity), and similar results can be obtained. We refer the interested reader to [121] for more details (see also Section 5.5).

■

Therefore, in the scalar case, that is, $u(x) \in R$, the natural condition to impose on the integrand $f(x, u, \xi)$ to obtain the existence of a minimizer for F is convexity in ξ. More precisely, we have the following theorem:

Theorem 2.1.4 *Let $\Omega \subset R^N$ be bounded and $f : \Omega \times R \times R^N \to R$ continuous satisfying*

(i) $f(x, u, \xi) \geq a(x) + b |u|^p + c |\xi|^p$ *for every (x, u, ξ) and for some $a \in L^1(\Omega)$, $b > 0$, $c > 0$, and $p > 1$.*

(ii) $\xi \to f(x, u, \xi)$ *is convex for every (x, u).*

(iii) *There exists $u_0 \in W^{1,p}(\Omega)$ such that $F(u_0) < \infty$. Then the problem*
$$\inf \left\{ F(u) = \int_\Omega f(x, u(x), \nabla u(x)) \, dx, \ u \in W^{1,p}(\Omega) \right\}$$ *admits a solution. Moreover, if $(u, \xi) \to f(x, u, \xi)$ is strictly convex for every x, then the solution is unique.*

In Theorem 2.1.4 the coercivity condition (i) implies the boundness of the minimizing sequences. Condition (ii) permits us to pass to the limit on these sequences. Condition (iii) ensures that the problem has a meaning. This can be summarized as follows:

☞ *Convexity is used to obtain l.s.c., while coercivity is related to compactness.*

Before going further, let us illustrate on three examples the importance of coercivity, reflexivity, and convexity.

Examples Let $\Omega = \,]0, 1]$. We propose below some classical examples where either coercivity, reflexivity, or convexity is no longer true:

(A) Weierstrass ($N = M = 1$).
Let f be defined by $f(x, u, \xi) = x\xi^2$ and let us set:

$$m = \inf \left\{ \int_0^1 x \left(u'(x)\right)^2 dx \text{ with } u(0) = 1 \text{ and } u(1) = 0 \right\}.$$

Then, we can show that this problem does not have any solution. The function f is convex, but the $W^{1,2}(\Omega)$-coercivity with respect to u is not satisfied because the integrand $f(x, \xi) = x\xi^2$ vanishes at $x = 0$. Let us first prove that $m = 0$. The idea is to propose the following minimizing sequence:

$$u_n(x) = \begin{cases} 1 & \text{if } x \in \left(0, \dfrac{1}{n}\right), \\ -\dfrac{\log(x)}{\log(n)} & \text{if } x \in \left(\dfrac{1}{n}, 1\right). \end{cases}$$

It is then easy to verify that $u_n \in W^{1,\infty}(0,1)$, and that

$$F(u_n) \equiv \int_0^1 x \left(u_n'(x)\right)^2 dx = \frac{1}{\log(n)} \to 0.$$

So we have $m = 0$. If there exists a minimum $\bar{u}$, then we should have $F(\bar{u}) = 0$, that is, $\bar{u}' = 0$ almost everywhere (a.e.) in $(0,1)$, which is clearly incompatible with the boundary conditions.

(B) Minimal surfaces ($p = 1$).
Let f be defined by $f(x, u, \xi) = \sqrt{u^2 + \xi^2}$. Then the associated functional F is convex and coercive on the *nonreflexive Banach space* $W^{1,1}(\Omega)$ ($F(u) \geq \frac{1}{2}|u|_{W^{1,1}(\Omega)}$). This example shows the importance of reflexivity.
Let us set:

$$m = \inf \left\{ \int_0^1 \sqrt{u^2 + u'^2} dx \text{ with } u(0) = 0 \text{ and } u(1) = 1 \right\}.$$

Let us prove that $m = 1$. First, we can observe that

$$F(u) \equiv \int_0^1 \sqrt{u^2 + u'^2} dx \geq \int_0^1 |u'| \, dx \geq \int_0^1 u' dx = 1.$$

So we have $m \geq 1$. Then, if we consider the sequence

$$u_n(x) = \begin{cases} 0 & \text{if } x \in \left(0, 1 - \dfrac{1}{n}\right), \\ 1 + n(x - 1) & \text{if } x \in \left(1 - \dfrac{1}{n}, 1\right), \end{cases}$$

we can verify that $F(u_n) \to 1$ ($n \to +\infty$). So $m = 1$. If we assume the existence of a solution $\bar{u}$, then we should have

$$1 = F(\bar{u}) = \int_0^1 \sqrt{\bar{u}^2 + \bar{u}'^2} dx \geq \int_0^1 |\bar{u}'| \, dx \geq \int_0^1 \bar{u}' dx = 1 \Rightarrow \bar{u} = 0,$$

which obviously does not satisfy the boundary conditions. As a conclusion, there is no solution to this problem.

(C) Bolza.

Let f be defined by $f(x, u, \xi) = u^2 + (\xi^2 - 1)^2$. The Bolza problem is

$$\inf \left\{ F(u) = \int_0^1 ((1 - u'^2)^2 + u^2) \, dx \; ; \; u \in W^{1,4}(0, 1) \right. \tag{2.3}$$
$$\left. u(0) = u(1) = 0 \right\}.$$

The functional is clearly *nonconvex*. It is easy to see that inf $F = 0$. Indeed, for n an integer and $0 \le k \le n - 1$, if we choose

$$u_n(x) = \begin{cases} x - \dfrac{k}{n} & \text{if } x \in \left(\dfrac{2k}{2n}, \dfrac{2k+1}{2n} \right), \\ -x + \dfrac{k+1}{n} & \text{if } x \in \left(\dfrac{2k+1}{2n}, \dfrac{2k+2}{2n} \right), \end{cases}$$

then $u_n \in W^{1,\infty}(0, 1)$ and

$$0 \le u_n(x) \le \frac{1}{2n} \text{ for every } x \in (0, 1),$$
$$|u'_n(x)| = 1 \text{ a.e. in } (0, 1),$$
$$u_n(0) = u_n(1) = 0.$$

Therefore,

$$0 \le \inf_u F(u) \le F(u_n) \le \frac{1}{4n^2}.$$

Letting $n \to +\infty$, we obtain $\inf_u F(u) = 0$. However, there exists no function $u \in W^{1,4}(0, 1)$ for which $u(0) = u(1) = 0$ and $F(u) = 0$. So the problem (2.3) does not have a solution in $W_0^{1,4}(0, 1)$. ∎

Once we have the existence of a minimum, the natural second step is to write the optimality conditions. For that we need the definition of the Gâteaux derivative.

Definition 2.1.5 (Gâteaux derivative) *Let X be a Banach space and $F : X \to R$. We call*

$$F'(u; v) = \lim_{\lambda \to 0^+} \frac{F(u + \lambda v) - F(u)}{\lambda}$$

the directional derivative of F at u in the direction v the limit if it exists. Moreover, if there exists $\tilde{u} \in X'$ such that $F'(u; v) = \tilde{u}(v)$, $\forall v \in X$, we say that F is Gâteaux differentiable at u and we write $F'(u) = \tilde{u}$.

If F is Gâteaux differentiable and if the problem $\inf_{v \in X} F(v)$ has a solution u, then we have

$$F'(u) = 0.$$

Conversely, if F is convex, then a solution u of $F'(u) = 0$ is a solution of the minimization problem. The equation $F'(u) = 0$ is called an *Euler–Lagrange equation* (also referred to as an Euler equation). Let us write it explicitly for a functional F defined by

$$F(u) = \int_\Omega f(x, u(x), \nabla u(x))\, dx,$$

where f is of class C^1 with respect to (u, ξ) and satisfies conditions (i) and (iii) of Theorem 2.1.4 and the following growth condition for the derivatives:

$$\begin{aligned}\left| \frac{\partial f}{\partial u}(x, u, \xi) \right| &\leq a'(1 + |u|^{p-1} + |\xi|^p), \\ \left| \nabla_{|\xi} f(x, u, \xi) \right| &\leq a''(1 + |u|^p + |\xi|^{p-1}),\end{aligned} \qquad \text{a.e. } x,\; \forall(u, \xi), \qquad (2.4)$$

for some constants $a', a'' > 0$. Then we can prove that for $u \in W^{1,p}(\Omega)$

$$F'(u) = \frac{\partial f}{\partial u}(x, u, \nabla u) - \sum_{i=1}^{i=N} \frac{\partial}{\partial x_i}\left(\frac{\partial f}{\partial \xi_i}(x, u, \nabla u) \right). \qquad (2.5)$$

We leave the proof to the reader as an exercise.

Remark Notice that the growth conditions (2.4) allow us to apply the Lebesgue dominated convergence theorem to prove (2.5). ∎

2.1.3 Relaxation

In this section we examine the case where the functional F is not weakly l.s.c. As we will see with a counterexample, there is no hope of obtaining, in general, the existence of a minimum for F. We could, however, associate with F another functional RF whose minima should be weak cluster points of minimizing sequences of F. This idea is important and will be used several times in this book.

As an illustration, let us consider again the Bolza problem (2.3). In this example, the integrand $f(u, \xi) = (1 - \xi^2)^2 + u^2$ is nonconvex in ξ, and the functional F is not weakly l.s.c. on $W^{1,4}(0, 1)$. In such a situation, it is classical to define the lower semicontinuous envelope (or relaxed function) $R_\tau F$ of F.

☞ To know more about this concept of relaxation: [67, 42, 66, 79, 122, 144, 121].

Let X be a Banach space and $F : X \to \overline{R}$. In the sequel, we equally denote by τ the strong or the weak topology of X.

Definition 2.1.6 (relaxed functional) *The τ-lower semicontinuous envelope (also called relaxed functional) $R_\tau F$ of F is defined for every $x \in X$ by $R_\tau F(x) = \sup \{G(x), \; G \in \Gamma\}$ where Γ is the set of all τ-lower semicontinuous functions on X such that $G(y) \le F(y)$ for every $y \in X$.*

To compute $R_\tau F$, the following characterization is useful:

Theorem 2.1.5 (characterization of the relaxed functional)
$R_\tau F$ is characterized by the following properties:

(i) *For every sequence x_n τ-converging to x in X,*
$$R_\tau F(x) \le \lim_{n \to +\infty} F(x_n).$$

(ii) *For every x in X there exists a sequence x_n τ-converging to x in X such that $R_\tau F(x) \ge \varlimsup_{n \to +\infty} F(x_n).$*

We consider now the relation between the original problem $\inf \{F(x), \; x \in X\}$ and the relaxed problem $\inf \{RF_\tau(x), \; x \in X\}$.

Theorem 2.1.6 (main properties) *Let X be a reflexive Banach space and let τ be the weak topology. Assume that $F : X \to \bar{R}$ is coercive. Then the following properties hold:*

(i) *$R_\tau F$ is coercive and τ-lower semicontinuous.*

(ii) *$R_\tau F$ has a minimum point in X.*

(iii) *$\min_{x \in X} R_\tau F(x) = \inf_{x \in X} F(x).$*

(iv) *Every clusterpoint of a minimizing sequence for F is a minimum point for $R_\tau F$.*

(v) *Every minimum point for $R_\tau F$ is the limit of a minimizing sequence for F.*

In summary, starting with a minimization problem that has no solution, we can define a relaxed problem whose connections with the original one are clearly stated in Theorem 2.1.6. For integral functionals, one possibility to compute the relaxed functional is to use the polar and bipolar functions. Let $f : R^N \to R$. We define the polar of f (also called Legendre–Fenchel transform), the function $f^* : R^N \to R$, as

$$f^*(\eta) = \sup_{\xi \in R^N} \{\eta \cdot \xi - f(\xi)\},$$

and the bipolar of f as

$$f^{**}(\xi) = \sup_{\eta \in R^N} \{\eta \cdot \xi - f^*(\eta)\}.$$

Since f^* and f^{**} are the suprema of affine functions, it turns out that f^* and f^{**} are convex. In fact, from convex analysis results f^{**} is the convex envelope of f, i.e., the greatest convex function less than f.

Notice that the function f^* has an interesting geometric interpretation. The definition of f^* implies

$$f(\xi) \geq \eta \cdot \xi - f^*(\eta) \quad \text{for all } \xi \in R^N;$$

that is, the affine function $h(\xi) = \eta \cdot \xi - f^*(\eta)$ is everywhere below the graph of f. If this supremum is reached, for instance at $\xi = \xi_1$, then $-f^*(\eta)$ is the intersection of this function with the vertical axis.

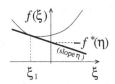

Examples

- If $f(\xi) = \dfrac{1}{p}|\xi|^p$, $1 < p < \infty$, then $f^*(\eta) = \dfrac{1}{q}|\eta|^q$, with $\dfrac{1}{p} + \dfrac{1}{q} = 1$.

- If $f(\xi) = |\xi|$, then $f^*(\eta) = \begin{cases} 0 & \text{if } |\eta| \leq 1, \\ +\infty & \text{otherwise.} \end{cases}$

- If $f(\xi) = e^\xi$, $\xi \in R$, then $f^*(\xi) = \begin{cases} \eta \log(\eta) - \eta & \text{if } \eta > 0, \\ 0 & \text{if } \eta = 0, \\ +\infty & \text{if } \eta < 0. \end{cases}$ ■

Let us mention that the notion of polarity also exists in infinite-dimensional spaces. So, let us consider the functional

$$F(u) = \int_\Omega f(x, \nabla u(x)) \, dx,$$

where Ω is a bounded open subset of R^N and $f : \Omega \times R^N \to R$ is a continuous function such that for every $\xi \in R^N$ and a.e. x,

$$a|\xi|^p \leq f(x, \xi) \leq b(|\xi|^p + 1), \quad 1 < p < \infty,$$

for some constants $a, b > 0$. If τ is the weak topology of $W^{1,p}(\Omega)$, then

$$R_\tau F(u) = \int_\Omega f^{**}(x, \nabla u(x)) \, dx,$$

where the polar functions are always computed with respect to the gradient variable. To illustrate this, we compute in the example below the relaxed functional for the Bolza problem.

Example For the previous Bolza problem,

$$\inf \left\{ F(u) = \int_0^1 ((1 - u'^2)^2 + u^2) \, dx \; ; \; u \in W^{1,4}(0,1), u(0) = u(1) = 0 \right\}.$$

The lack of weak l.s.c., as mentioned above, is due to the presence of the nonconvex function $\xi \rightarrow (1 - \xi^2)^2$. The second term $\int_0^1 u^2 \, dx$ is weakly continuous according to the compact embedding $W^{1,4}(0,1) \subset L^2(0,1)$. Then, the relaxed functional of F is

$$RF(u) = \int_0^1 (((u'^2 - 1)_+)^2 + u^2) \, dx,$$

where $t_+ = t$ if $t \geq 0$ and $t_+ = 0$ otherwise.

The problem $\inf\{RF(u), u \in W^{1,4}(0,1), u(0) = u(1) = 0\}$ admits a unique solution u_0, and the set $E = \{x \in (0,1); |u_0'(x)| < 1\}$ has a positive Lebesgue measure (otherwise, if $|E| = 0$, then $|u_0'(x)| \geq 1$ a.e. x and F would have a minimizer, which is false). ∎

2.1.4 About Γ-Convergence

We recall in this section the main results concerning Γ-convergence. Introduced by De Giorgi [163, 125], its aim is to give a meaning to the convergence of a sequence of functionals.

👁 To know more on Γ-convergence: [163, 20, 122].

This notion will be particularly useful for approximating nonconvex problems in numerous applications throughout this book.

Let X be a separable Banach space endowed with a topology τ and let $F_h : X \rightarrow \overline{R}$ be a sequence of functionals.

Definition 2.1.7 (Γ-convergence) *We say that F_h Γ-converges to F ($F = \Gamma - \lim F_h$) for the topology τ if:*

(i) *For every x in X and for every sequence x_n τ-converging to x in X,*

$$F(x) \leq \varliminf_{h \rightarrow +\infty} F_h(x_h).$$

(ii) *For every x in X there exists a sequence x_n τ-converging to x in X such that*

$$F(x) \geq \varlimsup_{h \rightarrow +\infty} F_h(x_h).$$

Definition 2.1.8 (equicoercivity) *We say that the sequence of functionals F_h is equicoercive if for every $t \geq 0$ there exists K_t a compact subset of X such that $\{x \in X; F_h(x) \leq t\} \subset K_t$ for all h.*

The main properties of Γ-convergence are summarized in the following theorem.

Theorem 2.1.7 (main properties of the Γ-limit) *Let X be a separable Banach space and let F_h be a sequence of equicoercive functionals from X into $\overline{R}$. Then:*

(i) *The Γ-limit of F_h, if it exists, is unique and l.s.c.*

(ii) *There exists a subsequence F_{h_k} and F such that $F = \Gamma-\lim F_{h_k}$.*

(iii) *If $F = \Gamma-\lim F_h$, then $F + G = \Gamma-\lim_{h \to +\infty} (F_h + G)$ for all continuous $G : X \to R$.*

(iv) *Set $F = \Gamma-\lim F_h$ and let us suppose that $F(x)$ admits a unique minimum point x_0 on X and let $x_h \in X$ be such that $\left| F_h(x_h) - \inf_{x \in X} F_h(x) \right| \le \varepsilon^h$ where $\varepsilon^h \to 0$ $(h \to +\infty)$ then x_h converges to x_0 in X and $\lim_{h \to +\infty} F_h(x_h) = F(x_0)$.*

In general, there is no relation between Γ-convergence and pointwise convergence. However, there exist some connections.

Theorem 2.1.8 (Γ-convergence and pointwise convergence)

(i) *If F_h converges to F uniformly, then F_h Γ-converges to F.*

(ii) *If F_h is a decreasing sequence converging to F pointwise, then F_h Γ-converges to RF, the lower semicontinuous envelope of F.*

Now we illustrate these ideas by giving an example [122].

Example Let Ω be an open of R^N and let a_h be a sequence of functions satisfying, for all h, $0 < c_1 \le a_h(x) \le c_2$ a.e. $x \in \Omega$ for some constants c_1 and c_2. Up to subsequences. from Theorem 2.1.1 there exist $a, b \in L^\infty(\Omega)$ such that $a_h \xrightarrow[L^\infty(\Omega)]{*} a$ and $1/a_h \xrightarrow[L^\infty(\Omega)]{*} b$. Then let us consider the sequence of functionals $F_h : L^2(\Omega) \to R$ defined by

$$F_h(u) = \int_\Omega a_h \, u^2 \, dx.$$

We claim that F_h Γ-converges to $F(u) = \int_\Omega u^2/b \, dx$ for the weak topology of $L^2(\Omega)$. According to the definition, we have to prove the following:

(i) For every $u \in L^2(\Omega)$ and for every $u_h \xrightarrow[L^2(\Omega)]{} u$, $\lim_{h \to +\infty} F_h(u_h) \ge F(u)$.

(ii) For every $u \in L^2(\Omega)$ there exists a sequence $u_h \in L^2(\Omega)$, $u_h \xrightarrow[L^2(\Omega)]{} u$, such that $\overline{\lim}_{h \to +\infty} F_h(u_h) \le F(u)$.

We begin by examining (ii).

Let $u \in L^2(\Omega)$ and let us define $u_h = u/b\,a_h$. Since $1/a_h \xrightarrow[L^\infty(\Omega)]{*} b$, it is clear that $u_h \xrightarrow[L^2(\Omega)]{} u$ and

$$\overline{\lim_{h \to +\infty}} \; F_h(u_h) = \overline{\lim_{h \to +\infty}} \int_\Omega \frac{u^2}{b^2\,a_h}\,dx = \int_\Omega \frac{u^2}{b}\,dx,$$

so $\overline{\lim_{h \to +\infty}} \; F_h(u_h) \leq F(u)$ (in fact, equality holds).

For proving (i), let $u \in L^2(\Omega)$ and let v_h be any sequence such that $v_h \xrightarrow[L^2(\Omega)]{} u$. Let $u_h = u/b\,a_h$. Then from the inequality $a_h\,(v_h - u_h)^2 \geq 0$, we deduce

$$a_h\,v_h^2 \geq a_h\,u_h^2 + 2\,a_h\,u_h(v_h - u_h) = -\,a_h\,u_h^2 + 2\,\frac{u}{b}v_h,$$

which yields

$$F_h(v_h) \geq -F_h(u_h) + 2\int_\Omega \frac{u}{b}\,v_h\,dx.$$

Therefore

$$\lim_{h \to +\infty} F_h(v_h) \geq -\overline{\lim_{h \to +\infty}} \; F_h(u_h) + 2\int_\Omega \frac{u^2}{b}\,dx \geq F(u),$$

which concludes the proof. ∎

Remark As an exercise, the reader can see for himself that the Γ-limit of F_h for the strong topology of $L^2(\Omega)$ is $G(u) = \int_\Omega a\,u^2\,dx$! ∎

2.2   The Space of Functions of Bounded Variation

In most computer-vision problems the discontinuities in the images are significant and important features. So we need to be able to represent discontinuous functions. Unfortunately, classical Sobolev spaces do not allow one to take into account such phenomena since the gradient of a Sobolev function is a function. When u is discontinuous, the gradient of u has to be understood as a measure, and the space $BV(\Omega)$ of functions of *bounded variation* [15, 149, 164, 130, 167] is well adapted for this purpose. In this section we recall some basic definitions about measures. Then $BV(\Omega)$ is defined, and its main properties are examined. We end this section by introducing the notion of convex functions of measures.

2.2.1 Basic Definitions on Measures

Definition 2.2.1 (algebra) *Let X be a nonempty set and let $\Im$ be a collection of subsets of X.*

(i) *We say that $\Im$ is an algebra if $\emptyset \in \Im$ and $E_1 \cup E_2 \in \Im$, $X - E_1 \in \Im$, whenever $E_1, E_2 \in \Im$.*

(ii) *We say that an algebra $\Im$ is a σ-algebra if for any sequences $(E_h) \subset \Im$ their union $\bigcup_h E_h$ belongs to $\Im$. σ-algebras are closed under countable intersections.*

(iii) *If (X, τ) is a topological space, we denote by $B(X)$ the σ-algebra generated by the open subsets of X (the smallest σ-algebra containing the open subsets).*

We can now give the definition of a positive measure.

Definition 2.2.2 (positive measure) *Let $\mu : \Im \to [0, +\infty]$. We say that μ is a positive measure if $\mu(\emptyset) = 0$ and μ is σ-additive on $\Im$, i.e., for any sequences (E_h) of pairwise disjoint elements of $\Im$:*

$$\mu\left(\bigcup_{h=0}^{\infty} E_h\right) = \sum_{h=0}^{\infty} \mu(E_h).$$

We say that μ is bounded if $\mu(X) < \infty$.

We will also use the notion of vector-valued measures.

Definition 2.2.3 (vector-valued measure) *Let X be a nonempty set and $\Im$ be a σ-algebra on X, $m \in N$, $m \geq 1$. We say that $\mu : \Im \to R^m$ is a measure if $\mu(\emptyset) = 0$ and for any sequences (E_h) of pairwise disjoint elements of $\Im$,*

$$\mu\left(\bigcup_{h=0}^{\infty} E_h\right) = \sum_{h=0}^{\infty} \mu(E_h).$$

If $m = 1$, we say that μ is a real or a signed measure, and if $m > 1$, we say that μ is a vector-valued measure. When $X = R^N$ we will denote by M the space of vector-valued Radon measures.[2] If μ is a measure, we define its total variation $|\mu|$ for every $E \in \Im$ as follows:

$$|\mu|(E) = \sup\left\{\sum_{h=0}^{\infty} |\mu(E_h)| \,;\, E_h \in \Im \text{ pairwise disjoint, } E = \bigcup_{h=0}^{\infty} E_h\right\}.$$

Then $|\mu|$ is bounded measure.

---

[2] A Radon measure on R^N is a measure that is finite in each compact set $K \subset R^N$.

Definition 2.2.4 (μ-negligible) *Let μ be a positive measure. We say that $A \subset X$ is μ-negligible if there exists $E \subset \Im$ such that $A \subset E$ and $\mu(E) = 0$. A property $P(x)$ depending on the point $x \in X$ holds μ-a.e. in X if the set where P fails is a μ-negligible set.*

We end this subsection by recalling what is perhaps the most important tool in integration theory: Lebesgue decomposition. Let us first introduce some definitions. We suppose now that $X = R^N$ and $\Im = B(R^N)$.

Definition 2.2.5 (Radon–Nikodym derivative) *Let μ be a bounded positive measure, and let ν be a measure. Let $B(x, r)$ be the ball of center x and radius r. We set*

$$\Delta(x, r) = \begin{cases} \dfrac{\nu\,(B(x, r))}{\mu(B(x, r))} & \text{if } \mu\,(B(x, r)) > 0, \\ +\infty & \text{if } \mu(B(x, r)) = 0, \end{cases}$$

and

$$\overline{D_\mu \nu}\,(x) = \overline{\lim_{r \to 0}}\,\Delta(x, r),\ \underline{D_\mu \nu}\,(x) = \varliminf_{r \to 0}\,\Delta(x, r).$$

If $\overline{D_\mu \nu}\,(x) = \underline{D_\mu \nu}\,(x) < \infty$, we say that ν is differentiable with respect to μ in x and we call

$$\frac{d\nu}{d\mu}(x) = \overline{D_\mu \nu}\,(x) = \underline{D_\mu \nu}\,(x)$$

the Radon–Nikodym derivative of ν with respect to μ.

Definition 2.2.6 (absolutely continuous, mutually singular) *Let μ be a positive measure, and let ν be a measure. We say that ν is absolutely continuous with respect to μ and we write $\nu \ll \mu$ if $\mu\,(E) = 0 \Rightarrow \nu\,(E) = 0$. We say that μ and ν are mutually singular and we write $\mu \perp \nu$ if there exists a set E such that $\mu\,(R^N - E) = \nu\,(E) = 0$.*

We have the following theorem:

Theorem 2.2.1 (Lebesgue decomposition) *Let μ be a positive bounded measure on $(R^N, B(R^N))$ and ν a vector-valued measure on $(R^N, B(R^N))$. Then there exists a unique pair of measures ν_{ac} and ν_s such that*

$$\nu = \nu_{ac} + \nu_s, \qquad \nu_{ac} \ll \mu, \qquad \nu_s \perp \mu.$$

Moreover,

$$\frac{d\nu}{d\mu} = \frac{d\nu_{ac}}{d\mu}, \quad \frac{d\nu_s}{d\mu} = 0,\, \mu\text{-a.e.} \quad \text{and} \quad \nu(A) = \int_A \frac{d\nu}{d\mu}\, d\mu + \nu_s(A)$$

for all $A \in B(R^N)$, where ν_{ac} and ν_s are the absolutely continuous part and the singular part of ν.

2.2.2 Definition of $BV(\Omega)$

Let Ω be a bounded open subset of R^N and let u be a function in $L^1(\Omega)$. We set

$$\int_\Omega |Du| = \sup\left\{\int_\Omega u\,\text{div}\varphi\,dx;\ \varphi = (\varphi_1, \varphi_2, \ldots, \varphi_N) \in \mathcal{C}_0^1(\Omega)^N, |\varphi|_{L^\infty(\Omega)} \leq 1\right\},$$

where $\text{div}\varphi = \sum_{i=1}^N \dfrac{\partial \varphi_i}{\partial x_i}(x)$, dx is the Lebesgue measure, $C_0^1(\Omega)^N$ is the space of continuously differentiable functions with compact support in Ω, and

$$|\varphi|_{L^\infty(\Omega)} = \sup_x \sqrt{\sum_i \varphi_i^2(x)}$$

Examples

- If $u \in C^1(\Omega)$, then $\int_\Omega u\,\text{div}\varphi\,dx = -\int_\Omega \nabla u \cdot \varphi\,dx$ and $\int_\Omega |Du| = \int_\Omega |\nabla u(x)|\,dx$.

- Let u be defined in $(-1, +1)$ by $u(x) = -1$ if $-1 \leq x < 0$ and $u(x) = +1$ if $0 < x \leq 1$. Then $\int_{-1}^{+1} u\,\varphi'\,dx = -2\varphi(0)$ and $\int_{-1}^{+1} |Du| = 2$.
  We can remark that Du, the distributional derivative of u, is equal to $2\delta_0$, where δ_0 is the Dirac measure in 0. In fact, we have the decomposition $Du = 0dx + 2\delta_0$. ∎

Definition 2.2.7 ($BV(\Omega)$) *We define $BV(\Omega)$, the space of functions of bounded variation, as*

$$BV(\Omega) = \left\{u \in L^1(\Omega);\ \int_\Omega |Du| < \infty\right\}.$$

We are going to show that if $u \in BV(\Omega)$, then Du (the distributional gradient of u) can be identified to a Radon vector-valued measure. Let $u \in BV(\Omega)$ and let $L : C_0^1(\Omega)^N \to R$ be the functional defined by

$$L(\varphi) = \int_\Omega u\,\text{div}\varphi\,dx.$$

Then L is linear, and since $u \in BV(\Omega)$, we have

$$\sup\left\{L(\varphi);\ \varphi \in C_0^1(\Omega)^N, |\varphi|_{L^\infty(\Omega)} \leq 1\right\} = c < \infty,$$

where c is a constant depending only upon Ω and u. So for all $\varphi \in C_0^1(\Omega)^N$

$$|L(\varphi)| \leq c\,|\varphi|_{L^\infty(\Omega)}. \tag{2.6}$$

Now let $K \subset \Omega$ be a compact set and $\varphi \in C_0(\Omega)^N$, supp $\varphi \subset K$. We can always find a sequence $\varphi_k \in C_0^1(\Omega)^N$ such that

$$\varphi_k \to \varphi \text{ uniformly, } k \to +\infty,$$
$$|\varphi_k|_{L^\infty(\Omega)} \leq |\varphi|_{L^\infty(\Omega)}, \ \forall k.$$

Let $\overline{L}(\varphi) = \lim_{k \to +\infty} L(\varphi_k)$. From (2.6) this limit exists and is independent of the choice of the sequence φ_k. Therefore, L uniquely extends to a linear continuous functional:

$$\overline{L}: \ C_0(\Omega)^N \to R.$$

From the Riesz representation theorem [285] there exists a Radon measure μ (a positive measure finite on compact sets of R^N) and a μ-measurable function σ such that

$$|\sigma(x)| = 1 \quad \mu\text{-a.e.} \quad x,$$

$$\int_\Omega u \ \text{div}\varphi \ dx = -\int_\Omega \sigma \cdot \varphi \ d\mu \text{ for all } \varphi \in C_0^1(\Omega)^N,$$

which means that Du is a vector-valued Radon measure $(Du = \sigma \ d\mu)$.

An important example is the case $u = \chi_A$, the characteristic function of a subset A of R^N. Then

$$\int_\Omega |Du| = \sup \left\{ \int_A \text{div}\varphi \ dx; \ \varphi \in C_0^1(\Omega)^N, \ |\varphi|_{L^\infty(\Omega)} \leq 1 \right\}.$$

If this supremum is finite, A is called a set of finite perimeter in Ω, and we write

$$\int_\Omega |Du| = \text{Per}_\Omega(A).$$

If ∂A is smooth, then $\text{Per}_\Omega(A)$ coincides with the classical length $(N = 2)$ or surface area $(N = 3)$.

2.2.3 *Properties of* $BV(\Omega)$

We summarize below the main properties of $BV(\Omega)$ that we will use in the sequel. We assume that Ω is bounded and has a Lipschitz boundary.

(P_1) *Lower semicontinuity*

Let $u_j \in BV(\Omega)$ and $u_j \xrightarrow[L^1(\Omega)]{} u$. Then $\int_\Omega |Du| \leq \varliminf_{j \to +\infty} \int_\Omega |Du_j|$.

(P_2) *Trace*

The trace operator tr $: u \to u|_{\partial\Omega}$, from $BV(\Omega)$ to $L^1(\partial\Omega, \mathcal{H}^{N-1})$, is linear continuous for the strong topology of $BV(\Omega)$. Here $\mathcal{H}^{N-1}$ denotes the $(N-1)$-dimensional measure (see Definition 2.2.8).

(P_3) *A weak* topology*
   $BV(\Omega)$ is a Banach space endowed with the norm
   $|u|_{BV(\Omega)} = |u|_{L^1(\Omega)} + \int_\Omega |Du|$. We will not use this topology, which
   possesses no good compactness properties. Classically, in $BV(\Omega)$ one
   works with the BV$-$w* topology, defined as

$$u_j \xrightarrow[BV-w^*]{*} u \quad \Leftrightarrow \quad u_j \xrightarrow{L^1(\Omega)} u \text{ and } Du_j \xrightarrow[M]{*} Du, \qquad (2.7)$$

   where $Du_j \xrightarrow[M]{*} Du$ means $\int_\Omega \varphi Du_j \to \int_\Omega \varphi Du$ for all φ in
   $C_0(\Omega)^N$. Equipped with this topology, $BV(\Omega)$ has some interesting
   compactness properties.

(P_4) *Compactness*
   Every uniformly bounded sequence u_j in $BV(\Omega)$ is relatively compact
   in $L^p(\Omega)$ for $1 \le p < \frac{N}{N-1}$, $N \ge 1$. Moreover, there exist a subsequence
   u_{j_k} and u in $BV(\Omega)$ such that $u_{j_k} \xrightarrow[BV-w^*]{*} u$. We also recall that $BV(\Omega)$
   is continuously embedded in $L^p(\Omega)$ with $p = +\infty$ if $N = 1$, and
   $p = \frac{N}{N-1}$ otherwise.

(P_5) *Decomposability of $BV(\Omega)$*
   We are going to show that Du can be decomposed as the sum of a
   regular measure and a singular measure. Before doing so we need the
   definition of Hausdorff measure.

Definition 2.2.8 (Hausdorff measure) *Let $k \in [0, +\infty]$ and
$A \subset R^N$. The k-dimensional Hausdorff measure of A is given by*

$$\mathcal{H}^k(A) = \lim_{\delta \to 0} \mathcal{H}^k_\delta(A),$$

where for $0 < \delta \le \infty$, $\mathcal{H}^k_\delta(A)$ is defined by

$$\mathcal{H}^k_\delta(A) = \frac{w_k}{2^k} \inf \left\{ \sum_{i \in I} |\text{diam}(A_i)|^k, \ \text{diam}(A_i) \le \delta, \ A \subset \bigcup_{i \in I} A_i \right\}$$

*for finite or countable covers $(A_i)_{i \in I}$; $\text{diam}(A_i)$ denotes the diameter
of the set A_i, and w_k is a normalization factor equal to $\pi^{k/2} \Gamma(1+k/2)$
where $\Gamma(t) = \int_0^\infty s^{t-1} e^{-s} ds$ is the gamma function (w_k coincides with
the Lebesgue measure of the unit ball of R^k if $k \ge 1$ is an integer).
We define the Hausdorff dimension of A by*

$$\mathcal{H} - \dim(A) = \inf\{k \ge 0; \ \mathcal{H}^k(A) = 0\}.$$

Then $\mathcal{H}^k$ is a measure in R^N, $\mathcal{H}^N$ coincides with the Lebesgue measure dx, and for $1 \le k \le N$, k integer, $\mathcal{H}^k(A)$ is the classical
k-dimensional area of A if A is a C^1 k-dimensional manifold embedded
in R^N. Moreover, if $k > k' \ge 0$, then $\mathcal{H}^k(A) > 0 \Rightarrow \mathcal{H}^{k'}(A) = +\infty$.

Let us return to $BV(\Omega)$. If u belongs to $BV(\Omega)$ and if in Theorem 2.2.1 we choose $\mu = dx$, the N-dimensional Lebesgue measure, and $\nu = Du$, we get

$$Du = \nabla u\, dx + D_s u,$$

where $\nabla u(x) = \dfrac{d(Du)}{dx}(x) \in L^1(\Omega)$ and $D_s u \perp dx$. ∇u is also called the approximate derivative of u (see [11]). In fact, we can say more for $BV(\Omega)$ functions. In [11] Ambrosio showed that the singular part $D_s u$ of Du can be decomposed into a "jump" part J_u and a "Cantor" part C_u. Before specifying what J_u is exactly, we have to define the notion of approximate limit. Let $B(x,r)$ be the ball of center x and radius r and let $u \in BV(\Omega)$. We define the approximate upper limit $u^+(x)$ and the approximate lower limit $u^-(x)$ by (see Figure 2.1)

$$u^+(x) = \inf \left\{ t \in [-\infty, +\infty]\; ;\; \lim_{r\to 0} \frac{dx(\{u > t\} \cap B(x,r))}{r^N} = 0 \right\},$$

$$u^-(x) = \sup \left\{ t \in [-\infty, +\infty]\; ;\; \lim_{r\to 0} \frac{dx(\{u < t\} \cap B(x,r))}{r^N} = 0 \right\}.$$

If $u \in L^1(\Omega)$, then

$$\lim_{r\to 0} \frac{1}{|B(x,r)|} \int_{B(x,r)} |u(x) - u(y)|\,dy = 0 \quad \text{a.e. } x. \tag{2.8}$$

A point x for which (2.8) holds is called a Lebesgue point of u, and we have

$$u(x) = \lim_{r\to 0} \frac{1}{|B(x,r)|} \int_{B(x,r)} u(y)\, dy, \tag{2.9}$$

and $u(x) = u^+(x) = u^-(x)$. We denote by S_u the jump set, that is, the complement, up to a set of $\mathcal{H}^{N-1}$ measure zero, of the set of Lebesgue points

$$S_u = \left\{ x \in \Omega\; ;\; u^-(x) < u^+(x) \right\}.$$

Then S_u is countably rectifiable, and for $\mathcal{H}^{N-1}$- a.e. $x \in \Omega$, we can define a normal $n_u(x)$.

Therefore, the result proved in [11] is

$$Du = \nabla u\, dx + (u^+ - u^-)\, n_u\, \mathcal{H}^{N-1}_{|S_u} + C_u. \tag{2.10}$$

Here $J_u = (u^+ - u^-)\, n_u\, \mathcal{H}^{N-1}_{|S_u}$ is the jump part and C_u is the Cantor part of $D_s u$. We have $C_u \perp dx$, and C_u is diffuse, i.e., $C_u\{x\} = 0$. More generally, we have $C_u(B) = 0$ for all B such that $\mathcal{H}^{N-1}(B) < \infty$; that is, the Hausdorff dimension of the support of C_u is strictly greater than $N - 1$. From (2.10) we can deduce the total variation of

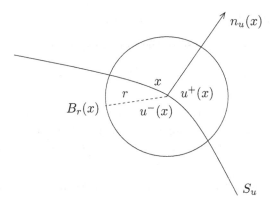

Figure 2.1. Definition of u^+, u^-, and the jump set S_u.

Du:

$$|Du|\,(\Omega) = \int_\Omega |Du|$$

$$= \int_\Omega |\nabla u|\,(x)\,dx + \int_{S_u} |u^+ - u^-|\,d\mathcal{H}^{N-1} + \int_{\Omega - S_u} |C_u|.$$

Remark An important fact about differentiation is that a summable function is "approximately continuous" at almost every point. This means that if u is simply in $L^1(\Omega)$, the right-hand side of (2.9) exists dx a.e. However, if u is also in $BV(\Omega)$, we can say more. Let us define

$$\overset{\bullet}{u}(x) = \frac{u^+(x) + u^-(x)}{2}, \qquad \mathcal{H}^{N-1} \text{ a.e. on } S_u. \qquad (2.11)$$

Then it can be shown [334, 149] that $\overset{\bullet}{u}$ is well-defined $\mathcal{H}^{N-1}$ a.e. on S_u. An interesting property of $\overset{\bullet}{u}$ is that we have the following approximation result:

$$\overset{\bullet}{u}(x) = \lim_{\varepsilon \to 0} \eta_\varepsilon \star u(x), \qquad \mathcal{H}^{N-1} \text{ a.e.}, \qquad (2.12)$$

where (η_ε) are the usual mollifiers (see Section 2.5.3). The function $\overset{\bullet}{u}$ is called the *precise representation* of u, since it permits us in some way to define u, $\mathcal{H}^{N-1}$ a.e. Note that $\overset{\bullet}{u}$ and u are in fact the same elements in $BV(\Omega)$ (they belong to the same equivalence class of dx a.e. equal functions), and therefore their distributional derivatives are the same. ∎

2.2.4 Convex Functions of Measures

We would like to give a sense to the formal expression:

$$\int_\Omega \Psi(Du)$$

when u is a $BV(\Omega)$ function, i.e., when Du is a measure. According to the above discussion we are for the moment able to define only the total variation of Du, i.e., when $\Psi(\xi) = |\xi|$. Let us extend this to more general Ψ.

☞ To know more on convex functions of measures: [167, 130].

Let $\phi : R \to R^+$ be convex, even, nondecreasing on R^+ with linear growth at infinity, and let ϕ^∞ be the recession function of ϕ defined by $\phi^\infty(z) = \lim_{s \to +\infty} \dfrac{\phi(s\,z)}{s}$. Then for $u \in BV(\Omega)$ and if $\Psi(\xi) = \phi(|\xi|)$, we set

$$\int_\Omega \Psi(Du) = \int_\Omega \phi(|\nabla u(x)|)\, dx + \phi^\infty(1) \int_{S_u} |u^+ - u^-|\, d\mathcal{H}^{N-1} + \phi^\infty(1) \int_{\Omega - S_u} |C_u|.$$

Of course, if $\Psi(\xi) = |\xi|$, this definition coincides with the total variation of Du. The main consequence of this definition is that

$$u \to \int_\Omega \Psi(Du) \quad \text{is l.s.c. for the BV$-$w* topology.}$$

We will use this notion in image restoration (Chapter 3).

2.3 Viscosity Solutions in PDEs

2.3.1 About the Eikonal Equation

Until now, we have recalled some mathematical tools necessary for tackling problems in computer vision from a variational point of view. But in many situations, e.g., nonlinear filtering in restoration, the equations we have to solve do not come from variational principles. They are PDEs that are not Euler–Lagrange equations of functionals, and so we need different tools.

To convince the reader of the difficulties in studying nonlinear PDEs, let us consider this very simple 1-D example:

$$\begin{cases} |u'(x)| = 1 & \text{in } (0,1), \\ u(0) = u(1) = 0, \end{cases} \tag{2.13}$$

which is the eikonal equation. Several questions arise:

- Existence. Clearly, (2.13) cannot admit a C^1 solution, since in this case from Rolle's theorem we would deduce the existence of $x_0 \in\,]0,1[$

such that $u'(x_0) = 0$, which is in contraction with $|u'(x_0)| = 1$. So the gradient of u has to "break" and a theory involving nonregular solutions has to be developed.

- Uniqueness. For example, the function $u^+(x) = \dfrac{1}{2} - \left|\dfrac{1}{2} - x\right|$ is a solution of (2.13) for almost every $x \in (0,1)$. But $u^-(x) = -u^+(x)$ is also a solution. In fact, there is an infinite number of solutions of the form

$$u_n(0) = u_n(1) = 0 \,,$$

$$u'_n(x) = 1 \quad \text{if} \ x \in \left]\dfrac{2k}{2^n}, \dfrac{2k+1}{2^n}\right], \qquad k = 0, \dots, 2^n - 1.$$

$$u'_n(x) = -1 \quad \text{if} \ x \in \left]\dfrac{2k+1}{2^n}, \dfrac{2k+2}{2^n}\right],$$

Some examples are given in Figure 2.2. Moreover $0 \le u_n(x) \le \dfrac{1}{2}^n$, so $u_n(x) \to 0$ uniformly $(n \to +\infty)$. However, 0 is not a solution of (2.13). Note that $u^+(x)$ is the greatest solution of (2.13).

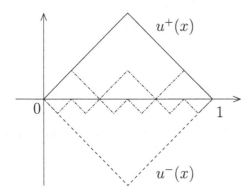

Figure 2.2. Examples of admissible solutions of the 1-D eikonal equation.

- Compatibility conditions. Let us consider the N-dimensional eikonal equation

$$\begin{cases} |\nabla u(x)| = f(x) & \text{in } \Omega, \text{ a bounded open set of } R^N, \\ u_{|\partial\Omega} = u_0(x). \end{cases} \qquad (2.14)$$

Let $x, y \in \bar{\Omega}$, and let $\xi(t) : [0, T] \to R^N$ be a Lipschitz path such that $\xi(0) = x$, $\xi(T) = y$, and $|\xi'(t)| \le 1$ a.e. $t \in (0, 1)$. We formally have

$$u(y) - u(x) = \int_0^T \nabla u(\xi(s)) \cdot \xi'(s) \, ds.$$

Thus

$$|u(y) - u(x)| \leq \int_0^T f(\xi(s))\, ds,$$

from which we deduce

$$|u(y) - u(x)| \leq L(x, y),$$

where

$$L(x, y) = \inf_{\xi, T} \left\{ \int_0^T f(\xi(s))\, ds;\ \xi(0) = x;\ \xi(T) = y; \right.$$

$$\left. |\xi'(t)| \leq 1;\ \xi(t) \in \overline{\Omega} \right\}.$$

Writing this necessary condition for x, y belonging to the boundary $\partial\Omega$ of Ω, we obtain

$$|\, u_0(y) - u_0(x)\,| \leq L(x, y), \qquad (2.15)$$

which is a compatibility condition on the data.

Remarks

- It can be shown [217] that in fact, (2.15) is also a sufficient existence condition.

- If $f \equiv 1$ and if Ω is convex, (2.15) can be written as $|\, u_0(y) - u_0(x)\,| \leq |y - x|$.

■

2.3.2 *Definition of Viscosity Solutions*

As is suggested in the previous example, there is a need to define a suitable and well-defined framework. Introduced in the 1980s for first-order PDEs by Crandall and P.L. Lions [119, 117, 217], the theory of viscosity solutions has demonstrated a very successful development and has been extended to second-order equations [118].

☞ For the general theory of viscosity solutions: [118, 39, 110].

Generally speaking, the theory of viscosity solutions deals with equations called *Hamilton–Jacobi equations* of the form

$$\frac{\partial u}{\partial t}(t, x) + H(t, x, u, \nabla u(x), \nabla^2 u(x)) = 0, \quad t \geq 0,\ x \in \Omega, \qquad (2.16)$$

with boundary and initial conditions. The map $H :]0,T] \times \Omega \times R \times R^N \times S^N \to \Im$ is called a *Hamiltonian*; S^N is the set of $N \times N$ symmetric matrices, and $\nabla^2 u$ stands for the Hessian matrix of u. We will always suppose that H is degenerate elliptic:

$$H(t,x,u,p,S) \geq H(t,x,u,p,S') \quad \text{if} \quad S \leq S' \tag{2.17}$$

(S^N is ordered with the natural order: $S \leq S' \Leftrightarrow \xi^t(S'-S)\xi \geq 0$ for all $\xi \in R^N$). The theory aims to define generalized solutions of (2.16), and particularly, solutions which are *a priori* only required to be continuous. The following theorem, stated for the stationary case, is useful for understanding the definition of viscosity solutions.

Theorem 2.3.1 *Let $H : \Omega \times R \times R^N \times S^N \to \Im$ be continuous and degenerate elliptic and let $u \in C^2(\Omega)$. Then u is a solution of $H(x,u(x),\nabla u(x),\nabla^2 u(x)) = 0$ in Ω if and only if*

(i) $\forall \phi \in C^2(\Omega)$, $\forall x_0 \in \Omega$ *a local maximum of* $(u - \phi)(x)$,

$$H(x_0,u(x_0),\nabla\phi(x_0),\nabla^2\phi(x_0)) \leq 0.$$

(ii) $\forall \phi \in C^2(\Omega)$, $\forall x_0 \in \Omega$ *a local minimum of* $(u - \phi)(x)$,

$$H(x_0,u(x_0),\nabla\phi(x_0),\nabla^2\phi(x_0)) \geq 0.$$

Proof Let $\phi \in C^2(\Omega)$, and let $x_0 \in \Omega$ be a local maximum of $(u-\phi)(x)$. Then from classical arguments,

$$\nabla u(x_0) = \nabla\phi(x_0),$$
$$\nabla^2 u(x_0) \leq \nabla^2\phi(x_0).$$

Hence from (2.17),

$$\begin{aligned} H(x_0,u(x_0),\nabla\phi(x_0),\nabla^2\phi(x_0)) &= H(x_0,u(x_0),\nabla u(x_0),\nabla^2\phi(x_0)) \\ &\leq H(x_0,u(x_0),\nabla u(x_0),\nabla^2 u(x_0)) = 0. \end{aligned}$$

The inequality (ii) is proven by similar arguments.

Reciprocally, if (i) and (ii) are true, by choosing $\phi = u$ and since each point $x \in \Omega$ is both a local maximum and a local minimum of $(u-\phi)(x) = 0$, we obtain

$$H(x_0,u(x_0),\nabla u(x_0),\nabla^2 u(x_0)) \leq 0,$$
$$H(x_0,u(x_0),\nabla u(x_0),\nabla^2 u(x_0)) \geq 0.$$

That is,

$$H(x_0,u(x_0),\nabla u(x_0),\nabla^2 u(x_0)) = 0.$$

■

☛ *In the former equivalence (Theorem 2.3.1), u need only be continuous. The derivatives are, in fact, evaluated on the test functions ϕ.*

This observation leads to the following definition:

Definition 2.3.1 (viscosity subsolution, supersolution, solution)
Let $H : \,]\,0,T\,] \times \Omega \times R \times R^N \times S^N \to \Im$ be continuous, satisfying (2.17), and let $u \in C(]\,0,T\,] \times \Omega)$. Then:

(i) *u is a viscosity subsolution of (2.16) if $\forall \phi \in C^2(]\,0,T\,] \times \Omega)$, $\forall (t_0, x_0)$ a local maximum of $(u - \phi)(t,x)$,*

$$\frac{\partial \phi}{\partial t}(t_0, x_0) + H(t_0, x_0, u(t_0, x_0), \nabla \phi(t_0, x_0), \nabla^2 \phi(t_0, x_0)) \leq 0.$$

(ii) *u is a viscosity supersolution of (2.16) if $\forall \phi \in C^2(]\,0,T\,] \times \Omega)$, $\forall (t_0, x_0)$ a local minimum of $(u - \phi)(t,x)$,*

$$\frac{\partial \phi}{\partial t}(t_0, x_0) + H(t_0, x_0, u(t_0, x_0), \nabla \phi(t_0, x_0), \nabla^2 \phi(t_0, x_0)) \geq 0.$$

(iii) *u is a viscosity solution of (2.16) if u is both a viscosity subsolution and a viscosity supersolution.*

2.3.3   About the Existence

We can wonder why the word "viscosity" is used in the former definition. This terminology has a historical basis. Initial work on Hamilton–Jacobi equations was first concerned with first-order PDEs like

$$H(x, u(x), \nabla u(x)) = 0. \tag{2.18}$$

The way to solve (2.18) was to introduce in (2.18) an additional regularizing (viscosity) term:

$$\boxed{-\varepsilon \Delta u(x)} + H(x, u(x), \nabla u(x)) = 0. \tag{2.19}$$

To prove existence, one can follow these two steps:

(i) Under suitable hypotheses, we show that (2.19) admits a unique regular solution u_ε, which we uniformly bound.

(ii) The second step is to study the behavior of u_ε as $\varepsilon \to 0$ and to pass to the limit in (2.19).

This method is well known in mechanics as the *vanishing viscosity method*. We will use it in Chapter 4 (for active contour models).
In this approach, the main concern is, in fact, the behavior of the solution as $\varepsilon \to 0$. We have the following stability result:

Lemma 2.3.1 (stability [39]) *Let $H_\varepsilon(x, u, p, M)$ be a sequence of continuous functions on $\Omega \times R \times R^N \times S^N$ ($\Omega \subset R^N$) satisfying the ellipticity* *condition*

$$H_\varepsilon(x, u, p, M_1) \leq H_\varepsilon(x, u, p, M_2) \quad \text{if} \quad M_1 - M_2 \geq 0,$$

and let $u_\varepsilon \in C(\Omega)$ *be a viscosity solution of* $H_\varepsilon(x, u_\varepsilon, \nabla u_\varepsilon, \nabla^2 u_\varepsilon) = 0$ *in* Ω. *If* $u_\varepsilon \to u$ *in* $C(\Omega)$ *and if* $H_\varepsilon \to H$ *in* $C(\Omega \times R \times R^N \times S^N)$, *then* u *is a viscosity solution of* $H(x, u, \nabla u, \nabla^2 u) = 0$ *in* Ω.

Another way to get a viscosity solution is to use *Perron's method*. Roughly speaking, the method runs as follows. Under appropriate assumptions:

(i) One proves that the set S of subsolutions is not empty.

(ii) Let $w(x) = \sup \{v(x); \ v \in S\}$. By stability $w(x)$ is a subsolution. It remains to show that $w(x)$ is a supersolution.

2.3.4 About the Uniqueness

Theorems concerning uniqueness are perhaps the strong point of this theory. But, as expected, it is also the most difficult part. For second-order Hamilton–Jacobi equations uniqueness results are built on a powerful technical lemma due to Crandall–Ishii. Before stating it, we need the definitions of superjet and subjet of u.

Definition 2.3.2 (superjet, subjet) *Let* $u \colon \Omega_t =]0, T] \times \Omega \to R$. *Then we call the superjet of* $u(s, z)$ *the set* $P_\Omega^+ u(s, z)$ *defined by* : $(a, p, X) \in R \times R^N \times S^N$ *lies in* $P_\Omega^+ u(s, z)$ *if* $(s, z) \in \Omega_t$ *and*

$$u(t, x) \leq u(s, z) + a(t - s) + p \cdot (x - z) + \frac{1}{2}(x - z)^t X(x - z) + o(|t - s| + |x - z|^2)$$

as $\Omega_t \supset (t, x) \to (s, z)$. *Similarly, the subjet of* $u(s, z)$ *is defined by*

$$P_\Omega^- u(s, z) = -P_\Omega^+ (-u)(s, z).$$

We can show [39, 118] that u is a subsolution (a supersolution) of (2.15) if and only if

$$a + H(t, x, u(t, x), p, X) \leq 0 \quad (\geq 0) \tag{2.20}$$

for all $(t, x) \in \Omega_t$ and $(a, p, X) \in P_{\Omega_t}^+ u(t, x)$ $(\in P_{\Omega_t}^- u(t, x))$. The interest of this equivalent definition of sub (super) solution is that the use of test functions ϕ is no longer required.

We can now state the Crandall–Ishii lemma.

Lemma 2.3.2 (Crandall–Ishii's Lemma [116, 118]) *Let* Ω_i *be locally compact subsets of* R^{N_i} *and let* $u_i \colon]0, T] \times \Omega_i \to R$ *be upper semicontinuous functions,* $i = 1, \ldots, k$. *Let* φ *be defined on an open neighborhood of* $(0, T) \times \Omega_1 \times \cdots \times \Omega_k$ *and such that* $(t, x_1, \ldots, x_k) \to \varphi(t, x_1, \ldots, x_k)$ *is once continuously differentiable and twice continuously differentiable in*

$(x_1, \ldots, x_k) \in \Omega_1 \times \cdots \times \Omega_k$. *Suppose that* $\bar{t} \in (0, T)$, $\overline{x_i} \in \Omega_i$, $i = 1, \ldots, k$
and

$$w(t, x_1, \ldots, x_k) \equiv u_1(t, x_1) + \cdots + u_k(t, x_k) - \varphi(t, x_1, \ldots, x_k) \leq w(\bar{t}, \overline{x_1}, \ldots, \overline{x_k})$$

for $0 < t < T$ *and* $x_i \in \Omega_i$. *Assume, moreover, that there is an* $r > 0$ *such that for* $M > 0$ *there is a constant* C *such that for* $(b_i, q_i, X_i) \in P_{\Omega_i}^+ u_i(t, x_i)$,
$|x_i - \overline{x_i}| + |t - \bar{t}| \leq r$, *and* $|u_i(t, x_i)| + |q_i| + |X_i| \leq M$ *we have*

$$b_i \leq C, \quad i = 1, \ldots, k. \tag{2.21}$$

Then for each $\varepsilon > 0$ *there exists* $X_i \in S^{N_i}$ *such that*

$$(b_i, \nabla_{x_i} \varphi(\bar{t}, \overline{x_1}, \ldots, \overline{x_k}), X_i) \in P_{\Omega_i}^+ u(\bar{t}, \overline{x_i}) \quad \text{for} \ i = 1, \ldots, k, \tag{2.22}$$

$$-\left(\frac{1}{\varepsilon} + |A|\right) I \leq \begin{pmatrix} X_1 & \cdots & 0 \\ \vdots & \ddots & \vdots \\ 0 & \cdots & X_k \end{pmatrix} \leq A + \varepsilon A^2, \tag{2.23}$$

$$b_1 + b_2 + \cdots + b_k = \frac{\partial \varphi}{\partial t}(\bar{t}, \overline{x_1}, \ldots, \overline{x_k}), \tag{2.24}$$

where $A = (\nabla_{x_i}^2 \varphi)(\bar{t}, \overline{x_1}, \ldots, \overline{x_k})$.

This lemma is technical, and it is rather difficult to understand its role in the proof of uniqueness. Let us describe it in a simple example to show how it is important.

Example Let Ω be bounded. We assume that u and v are two continuous functions, respectively sub- and supersolutions of

$$H(x, u, \nabla u, \nabla^2 u) = 0, \quad x \in \Omega, \tag{2.25}$$

and $u \leq v$ on $\partial \Omega$. We want to show that $u \leq v$ on $\overline{\Omega}$. This is a "maximum principle-like result," and it gives uniqueness as soon as (2.25) is associated with Dirichlet conditions.

Let us first assume that u and v are, in fact, classical sub- and super-solutions of (2.25), that is, twice differentiable. If the function $w = u - v$ admits a local maximum $\hat{x} \in \Omega$, then we have

$$\nabla u(\hat{x}) = \nabla v(\hat{x}), \tag{2.26}$$
$$\nabla^2 u(\hat{x}) \leq \nabla^2 v(\hat{x}), \tag{2.27}$$

and so

$$H(\hat{x}, u(\hat{x}), \nabla u(\hat{x}), \nabla^2 u(\hat{x})) \leq 0 \leq H(\hat{x}, v(\hat{x}), \boxed{\nabla v(\hat{x})}, \nabla^2 v(\hat{x})) \tag{2.28}$$

$$\leq H(\hat{x}, v(\hat{x}), \boxed{\nabla u(\hat{x})}, \nabla^2 u(\hat{x})). \tag{2.29}$$

Observe that we have chosen u and v as test functions, which was possible here because we assumed u and $v \in C^2(\Omega)$. Now, if we assume[3] that $H(x, r, p, X)$ is strictly increasing with respect to the variable r, then $u - v$ is negative at $\hat{x} \in \Omega$, and then $u \leq v$ on $\overline{\Omega}$ (since we have assumed that $u \leq v$ on $\partial\Omega$).

In the generic case, one can no longer choose u and v as test functions and write (2.26)–(2.27). Let us highlight what needs to be adapted from the simple proof to the general case:

- Choosing u and v as test functions and looking at the maximum point of $u(x) - v(x)$ allowed us to get the equality of the derivative at the maximum point $\hat{x}$ (2.26). This is used in (2.28)–(2.29) (see terms $\boxed{}$). In the general case, the classical idea is to duplicate the variables.

- In the same way, in (2.28)–(2.29) we used the comparison between the second derivatives (2.27) (see terms $\blacksquare$). This comparison will be given by the Crandall–Ishii lemma, more precisely by (2.23).

See the detailed proof of Theorem 4.3.2 in Section 4.3.3. $\blacksquare$

Remark The theory of viscosity solutions described in this section can be extended to the discontinuous case. For example, the case of discontinuous solutions is discussed in [10], and the case of a discontinuity in the second member of the PDE in [187, 307, 308]. We refer the reader to Section 4.3.4 where such notions are used. $\blacksquare$

2.4 Elements of Differential Geometry: Curvature

In this section we recall some basic definitions and properties of differential geometry and focus on the notion of curvature. Curvature has different meanings depending on the "object" that we are considering. We shall define the notion of curvatures for:

- Parametrized curves (for snakes, for instance),

- Images that can be represented by their isovalues,

- Images seen as surfaces where the height is the gray-scale intensity (see Figure 2.3).

Naturally, there are many other notions of curvature, especially for surfaces, and we refer to [218] for more details.

---

[3] Again, we recall that this example is just for educational purposes. In practice, such a strong assumption will not be necessary.

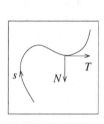 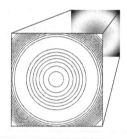 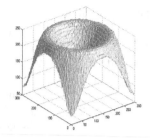

Figure 2.3. Different "objects" encountered in image analysis: parametrized curves, curves as isolevels of an image, image as a surface.

2.4.1   Parametrized Curves

Let $x(p) = (x_1(p), x_2(p))$ be a regular planar oriented curve in R^2, $0 \leq p \leq 1$. We note:

$T(p) = x'(p) = (x_1'(p), x_2'(p))$, the tangent vector at $x(p)$,

$N(p) = (-x_2'(p), x_1'(p))$, the normal vector at $x(p)$,

$s(p) = \int\limits_0^p \sqrt{(x_1'(r))^2 + (x_2'(r))^2} \, dr$, the curvilinear abscissa (or arc length).

If $x(p)$ is regular, we can parametrize it by s, and then $T(s) = \frac{dx}{ds}(s)$ is such that $|T(s)| = 1$. The curvature tensor is defined by $\frac{dT}{ds}(s) = \frac{d^2x}{ds^2}(s)$. One can show that the curvature tensor is collinear to $\frac{N(s)}{|N(s)|}$, i.e., $\frac{dT}{ds}(s) = \kappa(s) \frac{N(s)}{|N(s)|}$, where $\kappa(s)$ is the curvature and $\frac{1}{|\kappa(s)|}$ is the radius of curvature. For any parametrization we have

$$\kappa(p) = \frac{x_1'(p) \, x_2''(p) - x_2'(p) \, x_1''(p)}{((x_1'(p)^2 + x_2'(p)^2)^{3/2}} \tag{2.30}$$

and

$$\frac{1}{|x'(p)|} \frac{\partial}{\partial p}\left(\frac{x'(p)}{|x'(p)|}\right) = \kappa(p) \frac{N(p)}{|N(p)|}.$$

2.4.2   Curves as Isolevel of a Function u

Let us now consider the case where $x(s)$, parametrized by its curvilinear abscissa, is the k-level of a function $u : R^2 \to R$, that is,

$$x(s) = \{(x_1(s), x_2(s)) \; ; \; u((x_1(s), x_2(s)) = k\}.$$

By differentiating the equality $u((x_1(s), x_2(s)) = k$ with respect to s we obtain

$$x_1'(s) \, u_{x_1} + x_2'(s) \, u_{x_2} = 0, \tag{2.31}$$

where u_{x_i} stands for $\frac{\partial u}{\partial x_i}(x_1(s), x_2(s))$. Therefore, the vectors $(x'_1(s), x'_2(s))$ and $(-u_{x_2}, u_{x_1})$ are collinear. For some λ we have

$$\begin{cases} x'_1(s) = -\lambda u_{x_2}, \\ x'_2(s) = \lambda u_{x_1}, \end{cases} \qquad (2.32)$$

so the vectors (u_{x_1}, u_{x_2}) and $(-u_{x_2}, u_{x_1})$ are respectively normal and tangent to the curve $x(s)$. If we differentiate again (2.31) with respect to s, we obtain

$$(x'_1(s))^2 u_{x_1^2} + (x'_2(s))^2 u_{x_2^2} + 2\,x'_1(s)x'_2(s)\,u_{x_1 x_2} + x''_1(s)\,u_{x_1} + x''_2(s)\,u_{x_2} = 0.$$

hence with (2.32):

$$\lambda^2((u_{x_1})^2 u_{x_2^2} + (u_{x_2})^2 u_{x_1^2} - 2\,u_{x_1} u_{x_2}\,u_{x_1 x_2}) + \frac{1}{\lambda}(x''_1(s)x'_2(s) - x''_2(s)x'_1(s)) = 0.$$

But since $|x'(s)| = 1$, we get from (2.32) $\lambda^2 = \frac{1}{|\nabla u|^2}$. Therefore, with (2.30) we deduce the expression of the curvature (of course, we suppose $|\nabla u(x)| \neq 0$),

$$\kappa = \frac{(u_{x_1})^2 u_{x_2^2} + (u_{x_2})^2 u_{x_1^2} - 2\,u_{x_1} u_{x_2}\,u_{x_1 x_2}}{((u_{x_1})^2 + (u_{x_2})^2)^{3/2}}, \qquad (2.33)$$

and we leave it as an exercise to the reader to verify that

$$\kappa = \operatorname{div}\left(\frac{\nabla u}{|\nabla u|}\right). \qquad (2.34)$$

2.4.3 Images as Surfaces

Let us now examine quickly the case of 3-D surfaces. Denote by D an open set in R^2, $S : D \to R^3$, $(u, v) \to S(u, v)$ a regular parametrized surface. We assume that the vectors S_u and S_v are noncollinear for every $(u, v) \in D$, so that they form a basis of the tangent plane. The vector $N(u, v) = \frac{S_u \wedge S_v}{|S_u \wedge S_v|}$ is the unit normal vector to $S(u, v)$. Some classical notation is as follows:

$$\left.\begin{aligned} E(u, v) &= |S_u|^2, \\ F(u, v) &= S_u \cdot S_v, \\ G(u, v) &= |S_v|^2, \end{aligned}\right\} \quad \begin{array}{l} \text{coefficients of the first} \\ \text{quadratic fundamental form} \\ \text{of } S(u, v) \end{array}$$

$$\left.\begin{aligned} L(u, v) &= S_{uu} \cdot N = -S_u \cdot N_u, \\ M(u, v) &= S_{uv} \cdot N = -\frac{1}{2}(S_u \cdot N_v + S_v \cdot N_u), \\ P(u, v) &= S_{vv} \cdot N = S_v \cdot N_v. \end{aligned}\right\} \quad \begin{array}{l} \text{coefficients of the second} \\ \text{quadratic fundamental form} \\ \text{of } S(u, v) \end{array}$$

Using this notation, we have $|S_u \wedge S_v| = \sqrt{EG - F^2}$, the surface element is $ds = \sqrt{EG - F^2}\, du\, dv$, and the mean curvature H can be rewritten as

$$H = \frac{EP + GL - 2\,FM}{2\,(EG - F^2)}.$$

☞ For a general review on differential geometry: [218].

2.5   Other Classical Results Used in This Book

We now summarize some classical theorems, propositions, and inequalities.
The purpose is not to give a complete list of important results but only to
help the reader to find those that are used in this book. For a complete and
very clear overview, we refer the interested reader to the appendices of the
book by Evans [148] from which we have selected the relevant results.

☞ Functional analysis [148],

2.5.1   Inequalities

Cauchy's inequality with ε: $ab \leq \varepsilon a^2 + \dfrac{b^2}{4\varepsilon}$ $(a, b > 0, \varepsilon > 0)$.

Cauchy–Schwarz inequality: $|x \cdot y| \leq |x||y|$ $(x, y \in R^N)$.

Gronwall's inequality (differential form):

(i) *Let $\eta(.)$ be a nonnegative, absolutely continuous function on $[0, T]$
that satisfies for a.e. t the differential inequality*

$$\eta'(t) \leq \phi(t)\eta(t) + \Psi(t),$$

*where $\phi(t)$ and $\Psi(t)$ are nonnegative, integrable functions on $[0, T]$.
Then*

$$\eta(t) \leq e^{\int_0^t \phi(s)ds} \left[\eta(0) + \int_0^t \Psi(s)ds \right] \quad \text{for all } 0 \leq t \leq T.$$

(ii) *In particular, if $\eta' \leq \phi\eta$ on $[0, T]$ and $\eta(0) = 0$, then
$\eta \equiv 0$ on $[0, T]$.*

Gronwall's inequality (integral form):

(i) *Let $\xi(t)$ be a nonnegative, integrable function on $[0, T]$ that satisfies
for a.e. t the integral inequality*

$$\xi(t) \leq C_1 \int_0^t \xi(s)ds + C_2$$

for some constants C_1, $C_2 \geq 0$. Then

$$\xi(t) \leq C_2 \left(1 + C_1 t e^{C_1 t}\right) \quad \text{for a.e. } 0 \leq t \leq T.$$

(ii) *In particular, if $\xi(t) \leq C_1 \int_0^t \xi(s)ds$ for a.e. $0 \leq t \leq T$. Then*

$$\xi(t) = 0 \ a.e.$$

Hölder's inequality: *Assume $1 \leq p$, $q \leq \infty$, $\frac{1}{p} + \frac{1}{q} = 1$. Then, if $u \in L^p(\Omega)$, $v \in L^q(\Omega)$, we have*

$$\int_\Omega |uv| \ dx \leq |u|_{L^p(\Omega)} |v|_{L^q(\Omega)} .$$

Jensen's inequality: *Assume that $f : R \to R$ is convex, and $\Omega \subset R^N$ is open and bounded $(|\Omega| = 1)$. Let $u : \Omega \to R$ be integrable. Then*

$$f \left(\fint_\Omega u \ dx \right) \leq \left(\fint_\Omega f(u) \ dx \right),$$

where $\fint_\Omega u \ dx$ denotes the mean value of u over Ω.

Minkowski's inequality: *Assume $1 \leq p \leq \infty$ and u, $v \in L^p(\Omega)$. Then*

$$|u + v|_{L^p(\Omega)} \leq |u|_{L^p(\Omega)} + |v|_{L^p(\Omega)} .$$

Poincaré inequality: *Let Ω be an open bounded set of R^N and $u \in W_0^{1,p}(\Omega) = \{u \ / \ u \in W^{1,p}(\Omega); \ u|_{\partial\Omega} = 0\}$, $1 \leq p < n$. Then*

$$|u|_{L^q(\Omega)} \leq c_1 |\nabla u|_{L^p(\Omega)} , \quad q \in \left[1, \frac{np}{n-p} \right]$$

for some constant c_1 depending only on p, n, q, and Ω.

Poincaré–Wirtinger inequality: *Let Ω be open, bounded, and connected with a C^1 boundary. Then for all $u \in W^{1,p}(\Omega)$, $1 \leq p \leq +\infty$,*

$$\left| u - \fint_\Omega u \ dx \right|_{L^p(\Omega)} \leq c_2 |\nabla u|_{L^p(\Omega)}$$

for some constant c_2 depending only on p, n, and Ω. Observe that the same inequality holds for functions of bounded variation, where $|\nabla u|_{L^p(\Omega)}$ is replaced by the total variation $|Du|(\Omega)$.

Young's inequality: *Let $1 < p$, $q < \infty$, $\frac{1}{p} + \frac{1}{q} = 1$. Then:*

$$ab \leq \frac{a^p}{p} + \frac{b^q}{q} \quad (a, b > 0).$$

Young's inequality with ε: *Let $1 < p$, $q < \infty$, $\frac{1}{p} + \frac{1}{q} = 1$. Then:*

$$ab \leq \varepsilon a^p + C(\varepsilon) b^q \quad for \quad C(\varepsilon) = (\varepsilon p)^{-q/p} q^{-1}.$$

2.5.2 Calculus Facts

We suppose that Ω is a bounded, open subset of R^N, and $\partial\Omega$ is C^1.

Theorem 2.5.1 (Gauss–Green theorem) *Suppose $u \in C^1(\overline{\Omega})$. Then*

$$\int_\Omega u_{x_i} \ dx = \int_{\partial\Omega} u\nu^i \ ds \quad (i = 1,\dots,N),$$

where ν is the outward unit normal of $\partial\Omega$.

Theorem 2.5.2 (Integration by parts formula) *Let $u,\ v \in C^1(\overline{\Omega})$. Then*

$$\int_\Omega u_{x_i} v \ dx = -\int_\Omega u v_{x_i} \ dx + \int_{\partial\Omega} u v \nu^i \ ds \quad (i = 1,\dots,N).$$

Theorem 2.5.3 (Green's formulas) *Let $u,\ v \in C^2(\Omega)$. Then:*

(i) $\displaystyle \int_\Omega \triangle u \ dx = \int_{\partial\Omega} \frac{\partial u}{\partial\nu} \ ds.$

(ii) $\displaystyle \int_\Omega \nabla v \cdot \nabla u \ dx = -\int_\Omega u \triangle v \ dx + \int_{\partial\Omega} \frac{\partial v}{\partial\nu} u \ ds.$

(iii) $\displaystyle \int_\Omega u \triangle v - v \triangle u \ dx = \int_{\partial\Omega} u \frac{\partial v}{\partial\nu} - v \frac{\partial u}{\partial\nu} \ ds.$

Theorem 2.5.4 (Coarea formula) *Let $u : R^N \to R$ be Lipschitz continuous and assume that for a.e. $r \in R$, the level set*

$$\{x \in R^N \mid u(x) = r\}$$

is a smooth, $(n-1)$-dimensional hypersurface in R^N. Suppose also that $f : R^N \to R$ is continuous and integrable. Then

$$\int_{R^N} f|\nabla u| \ dx = \int_{-\infty}^{+\infty} \left(\int_{\{u=r\}} f \ ds \right) dr.$$

2.5.3 About Convolution and Smoothing

For $\Omega \subset R^N$, we set $\Omega_\varepsilon = \{x \in \Omega \mid \operatorname{dist}(x, \partial\Omega) > \varepsilon\}$.

Definition 2.5.1 (mollifier)

(i) *Define* $\eta \in C_c^\infty(R^N)$ *by*

$$\eta(x) = \begin{cases} C \exp\left(\dfrac{1}{|x|^2 - 1}\right) & \text{if } |x| < 1, \\ 0 & \text{if } |x| \geq 1, \end{cases}$$

where the constant $C > 0$ *is selected such that* $\int_{R^N} \eta \, dx = 1$. *Then* η *is called a standard mollifier.*

(ii) *For each* $\varepsilon > 0$, *set*

$$\eta_\varepsilon(x) = \frac{1}{\varepsilon^N} \eta\left(\frac{x}{\varepsilon}\right).$$

The functions η_ε *are* $C_c^\infty(R^N)$ *and satisfy*

$$\int_{R^N} \eta_\varepsilon \, dx = 1, \quad \mathrm{spt}(\eta_\varepsilon) \subset B(0, \varepsilon).$$

Definition 2.5.2 (mollification) *If* $f : \Omega \to R$ *is locally integrable, define its mollification by*

$$f^\varepsilon = \eta_\varepsilon * f \quad \text{in } \Omega_\varepsilon,$$

that is

$$f^\varepsilon(x) = \int_\Omega \eta_\varepsilon(x - y) f(y) dy = \int_{B(0,\varepsilon)} \eta_\varepsilon(y) f(x - y) dy \quad \text{for } x \in \Omega_\varepsilon.$$

Theorem 2.5.5 (properties of mollifiers)

(i) $f^\varepsilon \in C_c^\infty(\Omega_\varepsilon)$.

(ii) $f^\varepsilon \to f$ *a.e. as* $\varepsilon \to 0$.

(iii) *If* $f \in C(\Omega)$, *then* $f^\varepsilon \to f$ *uniformly on compact subsets of* Ω.

(iv) *If* $1 \leq p < \infty$ *and* $f \in L^p_{\mathrm{loc}}(\Omega)$, *then* $f^\varepsilon \to f$ *in* $L^p_{\mathrm{loc}}(\Omega)$.

2.5.4   Uniform Convergence

Theorem 2.5.6 (Arzelà–Ascoli compactness criterion) *Suppose that* $\{f_k\}_{k=1}^\infty$ *is a sequence of real-valued functions defined on* R^N *such that*

$$|f_k(x)| \leq M \quad (k = 1, \ldots, x \in R^N)$$

for some constant M, *and the* $\{f_k\}_{k=1}^\infty$ *are uniformly equicontinuous. Then there exist a subsequence* $\{f_{k_j}\}_{j=1}^\infty \subseteq \{f_k\}_{k=1}^\infty$ *and a continuous function* f *such that*

$$f_{k_j} \to f \text{ uniformly on compact subsets of } R^N.$$

We recall that saying that $\{f_k\}_{k=1}^{\infty}$ are uniformly equicontinuous means that for each $\varepsilon > 0$ there exists $\delta > 0$ such that $|x - y| < \delta$ implies $|f_k(x) - f_k(y)| < \varepsilon$, for $x, y \in R^N$, $k = 1, \ldots$.

2.5.5   Dominated Convergence Theorem

Theorem 2.5.7 (dominated convergence theorem) *Assume that the functions $\{f_k\}_{k=1}^{\infty}$ are Lebesgue integrable and*

$$f_k \to f \quad a.e.$$

Suppose also that

$$|f_k| \leq g \quad a.e.,$$

for some integrable function g. Then

$$\int\limits_{R^N} f_k \, dx \to \int\limits_{R^N} f \, dx.$$

This theorem, which is fundamental in the Lebesgue theory of integration, will be used very often in this book.

2.5.6   Well-Posed Problems

Finally, recall the classical definition concerning the well-posedness of a minimization problem or a PDE.

Definition 2.5.3 (well-posed) *When a minimization problem or a PDE admits a unique solution that depends continuously on the data, we say that the minimization problem or the PDE is well posed in the sense of Hadamard.*

If existence, uniqueness, or continuity fails, we say that the minimization problem or the PDE is ill posed.

3
Image Restoration

How to Read This Chapter

Image restoration is historically one of the oldest concerns in image processing and is still a necessary preprocessing step for many applications. So we start with a precise study of this problem, which will give the reader a broad overview of the variational and PDE-based approaches as applied to image analysis.

- We first give in Section 3.1 some precise information about what we mean by degradation or noise. This is actually a difficult question, and we focus on a simple model with additive noise and convolution by a linear operator for the blur.

- Section 3.2 presents restoration through the minimization of a functional involving two terms: a fidelity term to the data (based on the model of noise) plus a regularization term. We discuss in Section 3.2.2 some qualitative properties we would like for the restored image. This leads to a certain functional that we study in detail in Section 3.2.3 (existence and uniqueness of a solution). This subsection is rather mathematical, and presents an example of relaxation in the BV-framework. Section 3.2.4 is concerned with the the numerical computation of the solution found previously. Two algorithms and some experimental results are presented. We finally mention in Section 3.2.5 some scale-invariance properties and conclude in Section 3.2.6 by considering the nonconvex case.

- Section 3.3 is a survey of some PDE-based models proposed in the literature over the last decade for restoration and enhancement. Three types are distinguished:
  - Smoothing or forward–parabolic PDEs (Section 3.3.1), used mainly in pure restoration.
  - Smoothing–enhancing or backward–parabolic PDEs (Section 3.3.2) concerning restoration-enhancement processes.
  - Hyperbolic PDEs (Section 3.3.3) for enhancing blurred images, focusing on shock filters.

For each case we develop in detail the model and its mathematical justification (when it is possible and instructive). The mathematical background involves the theory of maximal operators, the notion of viscosity solutions (Section 3.3.1), and fixed-point techniques (Section 3.3.2). As for Section 3.3.3, no rigorous results are available. We mention only a conjecture by Osher and Rudin regarding the existence of a weak solution for shock filters.

Finally, we present in Section 3.3.4 how to extend the notion of Gaussian filtering: In [77, 78], Buades et al. revisit the notion of neighborhood filtering. Here the notion of neighborhood must be understood in a wide meaning, each of these meanings corresponding to specific filter. Interestingly, the authors also proved the link between these filters and well known PDEs such as the heat equation or the Perona–Malik equation.

3.1 Image Degradation

It is well known that during formation, transmission, and recording processes images deteriorate. Classically, this degradation is the result of two phenomena. The first one is deterministic and is related to the image acquisition modality, to possible defects of the imaging system (for example blur created by an incorrect lens adjustment or by motion). The second phenomenon is random and corresponds to the *noise* coming from any signal transmission. It is important to choose a degradation model as close as possible to reality. The random noise is usually modeled by a probabilistic distribution. In many cases, a Gaussian distribution is assumed. However, some applications require more specific ones, like the gamma distribution for radar images (speckle noise) or the Poisson distribution for tomography. We show in Figures (3.1)–(3.2) few examples of possible degradation.

Our aim in this chapter is to explain the methods that allow one to remove or diminish the effects of such degradation. To fix the terminology, we will call this processing *restoration*.

Unfortunately, it is usually impossible to identify the kind of noise involved for a given real image. If no model of degradation is available some

Figure 3.1. "Borel building" image (building from INRIA Sophia-Antipolis).

Figure 3.2. Examples of degradation in the top left-hand side corner of the "Borel building" image.

assumptions have to be made. A commonly used model is the following. Let $u : \Omega \subset R^2 \to R$ be an original image describing a real scene, and let u_0 be the observed image of the same scene (i.e., a degradation of u). We assume that

$$u_0 = R\,u + \eta, \qquad (3.1)$$

where η stands for a white additive Gaussian noise and where R is a linear operator representing the blur (usually a convolution). Given u_0, the problem is then to reconstruct u knowing (3.1). As we will see, the problem is *ill posed*, and we are able to carry out only an approximation of u.

3.2   The Energy Method

3.2.1   An Inverse Problem

Let u be the original image describing a real scene (the unknown), and let u_0 be the observed image (the data). Let us assume that the model of degradation (3.1) is valid. Recovering u from u_0 knowing (3.1) is a typical example of an inverse problem.

☞   For a general presentation on inverse problems: [197, 52].

This is not an easy task, since we know little about the noise η. We know only some statistics, such as its mean and its variance. Of course, since η is a random variable, a natural way to interpret equation (3.1) is to use probabilities (see, for instance, [68, 109, 131, 160]). It is not our goal to develop such a theory. Let us mention only that by supposing that η is a white Gaussian noise, and according to the maximum likelihood principle, we can find an approximation of u by solving the least-square problem

$$\inf_u \int_\Omega |u_0 - Ru|^2 \, dx, \qquad (3.2)$$

where Ω is the domain of the image. To fix ideas, let us imagine for a moment that u_0 and u are discrete variables in R^M, that R is an $M \times M$ matrix, and that $|\cdot|$ stands for the Euclidean norm. If a minimum u of (3.2) exists, then it necessarily satisfies the following equation:

$$R^* u_0 - R^* Ru = 0, \qquad (3.3)$$

where R^* is the adjoint of R. Solving (3.3) is in general an ill-posed problem, since $R^* R$ is not always one-to-one, and even if $R^* R$ were one-to-one, its eigenvalues may be small, causing numerical instability. Therefore, the idea is to regularize the problem (3.2), that is to consider a related problem that admits a unique solution.

☞ *From now on, we suppose that $u_0 \in L^\infty(\Omega)$ ($\Omega \subset R^2$, bounded) and that R is a linear operator of $L^2(\Omega)$. We do not use any probabilistic arguments. The noise η is regarded as a perturbation causing spurious oscillations in the image. One of the goals of the restoration is to remove these oscillations while preserving salient features such as edges.*

3.2.2 Regularization of the Problem

A classical way to overcome ill-posed minimization problems is to add a regularization term to the energy. This idea was introduced in 1977 by Tikhonov and Arsenin [317]. The authors proposed to consider the following minimization problem:

$$F(u) = \int_\Omega |u_0 - Ru|^2 \, dx + \lambda \int_\Omega |\nabla u|^2 \, dx. \qquad (3.4)$$

The first term in $F(u)$ measures the fidelity to the data. The second is a smoothing term. In other words, we search for a u that best fits the data so that its gradient is low (so that noise will be removed). The parameter λ is a positive weighting constant.

In the study of this problem, the functional space for which both terms are well-defined is

$$W^{1,2}(\Omega) = \left\{ u \in L^2(\Omega); \nabla u \in L^2(\Omega)^2 \right\}.$$

Under suitable assumptions on R (we will return later to these assumptions) the problem inf $\left\{ F(u), u \in W^{1,2}(\Omega) \right\}$ admits a unique solution characterized by the Euler–Lagrange equation

$$R^*Ru - R^*u_0 - \lambda \Delta u = 0 \qquad (3.5)$$

with the Neumann boundary condition

$$\frac{\partial u}{\partial N} = 0 \text{ on } \partial\Omega \text{ (N is the outward normal to $\partial\Omega$).}$$

Is the solution u of (3.5) a good candidate for our original restoration problem? The answer is no, since it is well known that the Laplacian operator has very strong isotropic smoothing properties and does not preserve edges (see Figure 3.3). Equivalently, this oversmoothing can be explained by looking at the energy (3.4). The L^p norm with $p = 2$ of the gradient allows us to remove the noise but unfortunately penalizes too much the gradients corresponding to edges. One should then decrease p in order to preserve the edges as much as possible. Some of the first work in this direction was done by Rudin, Osher, and Fatemi [284, 283], who proposed to use the L^1

original noisy image   result

Figure 3.3. Restoration of the noisy "Borel building" image (additive Gaussian noise) by minimizing (3.4): edges are lost (in this case R =Id).

norm of the gradient of u in (3.5), also called the total variation, instead of the L^2 norm.

In order to study more precisely the influence of the smoothing term, let us consider the following energy [28, 331]:

$$E(u) = \frac{1}{2} \int_\Omega |u_0 - Ru|^2 \, dx + \lambda \int_\Omega \phi(|\nabla u|) \, dx. \qquad (3.6)$$

☞ *We need to find the properties of ϕ so that the solution of the minimization problem is close to a piecewise constant image that is formed by homogeneous regions separated by sharp edges.*

Let us suppose that $E(u)$ has a minimum point u. Then it formally satisfies the Euler–Lagrange equation

$$R^*Ru - \lambda \, \mathrm{div}\left(\frac{\phi'(|\nabla u|)}{|\nabla u|}\nabla u\right) = R^*u_0. \qquad (3.7)$$

Equation (3.7) can be written in an expanded form by formally developing the divergence term.

We are going to show that it can be decomposed using the local image structures, that is, the tangent and normal directions to the isophote lines (lines along which the intensity is constant). More precisely, for each point x where $|\nabla u(x)| \neq 0$ we can define the vectors $N(x) = \frac{\nabla u(x)}{|\nabla u(x)|}$ and $T(x)$, $|T(x)| = 1$, $T(x)$ orthogonal to $N(x)$. With the usual notation $u_{x_1}, u_{x_2}, u_{x_1 x_1}, \dots$ for the first and second partial derivatives of u, we can

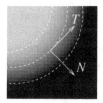

rewrite (3.7) as

$$R^*Ru - \lambda\left(\frac{\phi'(|\nabla u|)}{|\nabla u|} u_{TT} + \phi''(|\nabla u|) u_{NN} \right) = R^*u_0, \qquad (3.8)$$

where we denote by u_{TT} and u_{NN} the second derivatives of u in the T-direction and N-direction, respectively:

$$u_{TT} = {}^tT \nabla^2 u\, T = \frac{1}{|\nabla u|^2} \left(u_{x_1}^2 u_{x_2 x_2} + u_{x_2}^2 u_{x_1 x_1} - 2u_{x_1} u_{x_2} u_{x_1 x_2} \right),$$

$$u_{NN} = {}^tN \nabla^2 u\, N = \frac{1}{|\nabla u|^2} \left(u_{x_1}^2 u_{x_1 x_1} + u_{x_2}^2 u_{x_2 x_2} + 2u_{x_1} u_{x_2} u_{x_1 x_2} \right).$$

In fact, decomposing the divergence term as a weighted sum of the two directional derivatives along T and N can be done for most classical diffusion operators [205]. This allows us to see clearly the action of the operators in the directions T and N.

In our case, this is also useful for determining how the function ϕ should be chosen:

- At locations where the variations of the intensity are weak (low gradients), we would like to encourage smoothing, the same in all directions. Assuming that the function ϕ is regular, this isotropic smoothing condition may be achieved by imposing

$$\phi'(0) = 0, \quad \lim_{s\to 0^+} \frac{\phi'(s)}{s} = \lim_{s\to 0^+} \phi''(s) = \phi''(0) > 0. \qquad (3.9)$$

Therefore, at points where $|\nabla u|$ is small, (3.8) becomes

$$R^*Ru - \lambda\phi''(0)(u_{TT} + u_{NN}) = R^*u_0;$$

i.e., since $u_{TT} + u_{NN} = \Delta u$,

$$R^*Ru - \lambda\phi''(0)\Delta u = R^*u_0. \qquad (3.10)$$

So, at these points, u locally satisfies (3.10), which is a uniformly elliptic equation having strong regularizing properties in all directions.

- In a neighborhood of an edge C, the image presents a strong gradient. If we want to preserve this edge, it is preferable to diffuse along C (in the T-direction) and not across it. To do this, it is sufficient in (3.7) to annihilate, for strong gradients, the coefficient of u_{NN} and to assume that the coefficient of u_{TT} does not vanish:

$$\lim_{s\to +\infty} \phi''(s) = 0, \quad \lim_{s\to +\infty} \frac{\phi'(s)}{s} = \beta > 0. \qquad (3.11)$$

Unfortunately, these two conditions are incompatible. One must find a compromise. For example, $\phi''(s)$ and $\phi'(s)/s$ both converge to zero as $s \to +\infty$, but at different rates:

$$\lim_{s \to +\infty} \phi''(s) = \lim_{s \to +\infty} \frac{\phi'(s)}{s} = 0 \text{ and } \lim_{s \to +\infty} \frac{\phi''(s)}{\phi'(s)/s} = 0. \quad (3.12)$$

Notice that many functions ϕ satisfying the conditions (3.9)–(3.12) can be found, for example, the function

$$\phi(s) = \sqrt{1 + s^2}, \quad (3.13)$$

which is usually called the *hypersurface minimal function*.

Remark The assumptions (3.9) and (3.12) on ϕ are qualitative. They have been imposed in order to describe the regularization conditions. Naturally, they are not sufficient to ensure that the model is mathematically well posed. Other hypotheses such as convexity, and linear growth are necessary. This is developed in the coming section.   ■

3.2.3   Existence and Uniqueness of a Solution for the Minimization Problem

This section is devoted to the mathematical study of

$$\inf \left\{ E(u) = \frac{1}{2} \int_\Omega |u_0 - Ru|^2 \ dx + \lambda \int_\Omega \phi(|\nabla u|) \ dx \right\}. \quad (3.14)$$

In order to use the direct method of the calculus of variations, we have to assume some minimal hypotheses on ϕ:

ϕ is a strictly convex, nondecreasing function from R^+ to R^+, with $\phi(0) = 0$ (without a loss of generality);   (3.15)

$$\lim_{s \to +\infty} \phi(s) = +\infty. \quad (3.16)$$

This latter growth condition must not be too strong, because it must not penalize strong gradients, i.e., the formation of edges (see what happened with $\phi(s) = s^2$). Hence we assume that ϕ grows at most linearly:

There exist two constants $c > 0$ and $b \geq 0$ such that $cs - b \leq \phi(s) \leq cs + b \ \forall s \geq 0$.   (3.17)

Remark To recover and preserve edges in an image, it would certainly be preferable to impose a growth condition of the type $\lim_{s \to +\infty} \phi(s) = \beta > 0$. In this case the contribution of the term $\phi(|\nabla u|)$ in $E(u)$ would not penalize the formation of strong gradients, since "it would cost nothing." Unfortunately, since we also want ϕ to have a quadratic behavior near zero, then necessarily ϕ should have a nonconvex shape, which is an undesirable property (see, for instance, the Bolza problem discussed in Sections 2.1.2 and

2.1.3). ∎

According to (3.17), the natural space on which we would be able to seek a solution is the space

$$V = \left\{ u \in L^2(\Omega); \; \nabla u \in L^1(\Omega)^2 \right\}.$$

Unfortunately, this space is not reflexive. In particular, we cannot say anything about minimizing sequences that are bounded in V. However, an interesting remark is that sequences bounded in V are also bounded in $BV(\Omega)$ (Section 2.2). Therefore, they are compact for the BV$-$w* topology. Still, the energy E extended by $+\infty$ if u belongs to $BV(\Omega) - V$ is not lower semicontinuous for this topology. In this case, it is classical to compute the relaxed energy. Moreover, note that the space $BV(\Omega)$ is also interesting a priori to model images, since it allows discontinuities across hypersurfaces, i.e. across edges for images.

Theorem 3.2.1 *The relaxed functional of* (3.14) *for the* BV$-$w* *topology is defined by:*

$$\overline{E}(u) = \frac{1}{2} \int_\Omega |u_0 - Ru|^2 \, dx \tag{3.18}$$

$$+ \lambda \int_\Omega \phi(|\nabla u|) \, dx + \lambda c \int_{S_u} (u^+ - u^-) d\mathcal{H}^1 + \lambda c \int_{\Omega - S_u} |C_u|,$$

where $c = \lim\limits_{s \to +\infty} \dfrac{\phi(s)}{s}$.

Proof Let us define

$$e(u) = \begin{cases} \dfrac{1}{2} \int_\Omega |u_0 - Ru|^2 \, dx + \lambda \int_\Omega \phi(|\nabla u|) \, dx & \text{if } u \in V, \\ +\infty & \text{if } u \in BV(\Omega) - V. \end{cases}$$

We note that $e(u) = E(u)$ if $u \in V$. Since $e(u)$ is not l.s.c. for the BV$-$w* topology, we need to compute its l.s.c. envelope (for the BV$-$w* topology), i.e., the greatest l.s.c. functional $\overline{e}(u)$ less than or equal to $e(u)$. Since $\overline{E}(u)$ is l.s.c. (see Section 2.2.3), we have $\overline{e}(u) \geq \overline{E}(u)$. Thus, we have to show that $\overline{e}(u) \leq \overline{E}(u)$.

Thanks to [130], for each $u \in BV(\Omega)$ there exists a sequence $u_n \in C^\infty(\Omega) \cap V$ such that $u_n \xrightarrow[\text{BV}-\text{w}^*]{} u$ and $\overline{E}(u) = \lim\limits_{n \to +\infty} e(u_n)$. Therefore

$$\overline{E}(u) = \lim\limits_{n \to +\infty} e(u_n) \geq \inf_{\substack{u_n \in BV(\Omega) \\ u_n \xrightarrow[\text{BV}-\text{w}^*]{} u}} \left\{ \underline{\lim} \, e(u_n) \right\} = \overline{e}(u),$$

which concludes the proof.   ∎

In the sequel, we also assume that

$R : L^2(\Omega) \to L^2(\Omega)$ is a linear continuous operator, and $R.1 \neq 0$.   (3.19)

The second assumption of (3.19) means that R does not annihilate the constants, which guarantees the BV-coercivity of $\overline{E}(u)$.

Theorem 3.2.2 *Under assumptions (3.15)–(3.17) and (3.19), the minimization problem*

$$\inf_{u \in BV(\Omega)} \overline{E}(u), \qquad (3.20)$$

where $\overline{E}$ is defined by (3.18), admits a unique solution $u \in BV(\Omega)$.

Proof The proof follows [331, 332].
Step 1: Existence
Let u_n be a minimizing sequence for (3.20). Thanks to (3.16), we have

$$\begin{cases} |Du_n|(\Omega) = \displaystyle\int_\Omega |\nabla u_n| \; dx + \int_{S_{u_n}} |u_n^+ - u_n^-| \, d\mathcal{H}^1 + \int_{\Omega - S_{u_n}} |C_{u_n}| \le M, \\ \displaystyle\int_\Omega |Ru_n - u_0|^2 \; dx \le M, \end{cases}$$

where M denotes a universal strictly positive constant that may differ from line to line. The first inequality above says that the total variation of Du_n is uniformly bounded. It remains to prove that $|u_n|_{L^1(\Omega)}$ is bounded. Let $w_n = \frac{1}{|\Omega|} \int_\Omega u_n \; dx$ and $v_n = u_n - w_n$. Then $\int_\Omega v_n \; dx = 0$ and $Dv_n = Du_n$. Hence $|Dv_n| \le M$. Using the generalized Poincaré–Wirtinger inequality we obtain

$$|v_n|_{L^2(\Omega)} \le K |Dv_n|(\Omega) \le M, \qquad (3.21)$$

where K is a constant. Now, from the inequality $\int_\Omega |Ru_n - u_0|^2 \; dx \le M$, we deduce

$$|Ru_n - u_0|^2_{L^2(\Omega)} \le M,$$

which is equivalent to:

$$|Rv_n + Rw_n - u_0|^2_{L^2(\Omega)} \le M.$$

Rewriting Rw_n by

$$Rw_n = (Rv_n + Rw_n - u_0) - (Rv_n - u_0),$$

we get

$$|Rw_n|_{L^2(\Omega)} \le M + |Rv_n|_{L^2(\Omega)} + |u_0|_{L^2(\Omega)} \le M.$$

So we have

$$|Rw_n|_{L^2(\Omega)} = \left|\frac{1}{|\Omega|}\int_\Omega u_n \ dx\right| |R.1|_{L^2(\Omega)} \leq M,$$

and thanks to (3.19), we obtain that $\left|\int_\Omega u_n \ dx\right|$ is uniformly bounded. From (3.21) we obtain:

$$|u_n|_{L^2(\Omega)} = \left|v_n + \frac{1}{|\Omega|}\int_\Omega u_n dx\right|_{L^2(\Omega)} \leq |v_n|_{L^2(\Omega)} + \left|\int_\Omega u_n dx\right| \leq M.$$

Hence u_n is bounded in $L^2(\Omega)$ and in $L^1(\Omega)$ (Ω is bounded). Since $|Du_n|(\Omega)$ is also bounded, we get that u_n is bounded in $BV(\Omega)$.

Thus, up to a subsequence, there exists u in $BV(\Omega)$ such that $u_n \xrightarrow[BV-w^*]{} u$ and $Ru_n \xrightarrow[L^2(\Omega)]{} Ru$.

Finally, from the weak semicontinuity property of the convex function of measures and the weak semicontinuity of the L^2-norm, we get

$$\int_\Omega |Ru - u_0|^2 \ dx \leq \lim_{n\to+\infty}\int_\Omega |Ru_n - u_0|^2 \ dx,$$

$$\int_\Omega \phi(Du) \leq \lim_{n\to+\infty}\int_\Omega \phi(Du_n),$$

that is,

$$\overline{E}(u) \leq \lim_{n\to+\infty} \overline{E}(u_n) = \inf_{v\in BV(\Omega)} \overline{E}(v);$$

i.e., u is minimum point of $\overline{E}(u)$.

Step **2**: Uniqueness
Let u and v be two minima of $\overline{E}(u)$. From the strict convexity of ϕ we easily get that $Du = Dv$, which implies that $u = v + c$. But since the function $u \to \int_\Omega |Ru - u_0|^2 \ dx$ is also strictly convex, we deduce that $Ru = Rv$, and therefore $Rc = 0$, and from (3.19) we conclude that $c = 0$ and $u = v$. ∎

Remark In fact, the space $BV(\Omega)$ may not be the most suitable space for images. In [254], Nikolova showed that the minimization of the total variation leads to the staircase effect. Also, in [171], Gousseau and Morel proved that natural images are only partially described by $BV(\Omega)$ functions. These two results suggest that in some situations, more adapted functional spaces should be proposed. This remark is closely related to Section 5.2, where the reader will find some answers on this subject. ∎

3.2.4   Toward the Numerical Approximation

The next question is to characterize the solution of the problem (3.20) in order to get a numerical approximation. If we try to write directly the Euler–Lagrange equations, the difficulty is to define variations on $BV(\Omega)$ because of the presence of measures. Some interesting results can still be obtained. For instance, it can be proved that the solution of (3.20) satisfies the following conditions:

(i) $R^*Ru - R^*u_0 - \lambda \operatorname{div}\left(\dfrac{\phi'(|\nabla u|)}{|\nabla u|}\nabla u\right) = 0$ in $L^2(\Omega)$.

(ii) $\dfrac{\phi'(|\nabla u|)}{|\nabla u|}\dfrac{\partial u}{\partial N} = 0$ on $\partial\Omega$.

(iii) For all $w \in BV(\Omega)$ with $Dw = \nabla w\, dx + D_s w$ and $D_s w = \rho D_s u + \mu'$,

$$\int_\Omega (Ru - u_0)Rw\, dx + \int_\Omega \frac{\phi'(|\nabla u|)}{|\nabla u|}\nabla u \cdot \nabla w\, dx + \int_\Omega \rho|D_s u| + \int_\Omega |\mu'|$$

$$\geq -\int_\Omega \operatorname{div}\left(\frac{\phi'(|\nabla u|)}{|\nabla u|}\nabla u\right) w\, dx,$$

where ∇u is the approximate derivative of u. This is fully detailed in Proposition 3.3.4 (Section 3.3.1), which involves similar operators. So, it is mathematically incorrect to use only (i) and (ii) to find a numerical solution, which is usually done. However, condition (iii) remains difficult to handle.

To circumvent this difficulty, two methods are presented here, namely a projection algorithm proposed by Chambolle [91], and the half-quadratic approach introduced Geman and Reynolds [159], and Charbonnier et al. [108].

THE PROJECTION APPROACH

When $\phi(t) = t$ and R is the identity operator, Chambolle in [91] has remarked that the minimization of the total variation can be viewed as a projection problem on a suitable convex set. Moreover, he gave a very nice algorithm to compute this projection. To explain the method we follow his paper [91]. The presentation is given here in the discrete setting but everything is the same in the continuous case.

Let us define some notation. We will denote by $u_{i,j}, i, j = 1, \ldots, N$, a discrete image and by $X = \mathbb{R}^{N^2}$ the set of all discrete images of size N^2. In order to define the total variation (TV) of a discrete image, we introduce the gradient $\nabla : X \to X \times X$:

$$(\nabla u)^1_{i,j} = \begin{cases} u_{i+1,j} - u_{i,j} & \text{if } i < N, \\ 0 & \text{if } i = N, \end{cases} \qquad (\nabla u)^2_{i,j} = \begin{cases} u_{i,j+1} - u_{i,j} & \text{if } j < N, \\ 0 & \text{if } j = N. \end{cases}$$

We also introduce a discrete version of the divergence operator defined by analogy with the continuous case by div $= -\nabla^*$, where ∇^* is the adjoint of ∇. Thus

$$(\operatorname{div} p)_{i,j} = (\operatorname{div} p)_{i,j}^1 + (\operatorname{div} p)_{i,j}^2,$$

with

$$(\operatorname{div} p)_{i,j}^1 = \begin{cases} p_{i,j}^1 - p_{i-1,j}^1 & \text{if } 1 < i < N, \\ p_{i,j}^1 & \text{if } i = 1, \\ -p_{i-1,j}^1 & \text{if } i = N, \end{cases}$$

$$(\operatorname{div} p)_{i,j}^2 = \begin{cases} p_{i,j}^2 - p_{i,j-1}^2 & \text{if } 1 < j < N, \\ p_{i,j}^2 & \text{if } j = 1, \\ -p_{i,j-1}^2 & \text{if } j = N. \end{cases}$$

Then the discrete TV denoted by J_{TV} is defined as the l^1-norm of the vector ∇u by

$$J_{\text{TV}}(u) = \sum_{i,j=1}^{N} |(\nabla u)_{i,j}|,$$

where $|y| = \sqrt{y_1^2 + y_2^2}$ for every $y = (y_1, y_2) \in \mathbb{R}^2$. We can observe that J_{TV} is a discretization of the total variation defined in the continuous setting for a function $u \in L^1(\Omega)$ by

$$J(u) = \sup \left\{ \int_{\Omega} u(x) \operatorname{div} \xi(x)\, dx;\ \xi \in C_c^1(\Omega; \mathbb{R}^2), |\xi(x)| \leqslant 1, \forall x \in \Omega \right\}.$$

So, this notation being fixed, the problem we want to solve is

$$\inf_{u \in X} \left\{ J_{\text{TV}}(u) + \frac{1}{2\lambda} |f - u|_{X^2} \right\}, \tag{3.22}$$

which is the discrete version of (3.20) and corresponds to the Rudin, Osher, and Fatemi model [284]. Notice that we have replaced the parameter λ by $\frac{1}{2\lambda}$ to match the original notation from [91].

(A) Writing the Solution as a Projection

Let us introduce the set G, which we will use also in Section 5.2.

Definition 3.2.1 *We define the set G by*

$$G = \{v \in X;\ \exists p \in X, |p_{i,j}| \leqslant 1, \forall i, j,\ \text{such that } v = \operatorname{div} p\}. \tag{3.23}$$

Proposition 3.2.1 *The unique minimizer of (3.22) is given by*

$$u = f - P_{\lambda G}(f),$$

where $P_{\lambda G}(f)$ is the L^2-orthogonal projection of f on the set λG.

Proof   The proof relies on convex and duality principles. Since J_{TV} is homogeneous and of degree one (i.e., $J_{\mathrm{TV}}(\lambda u) = \lambda J_{\mathrm{TV}}(u)$ for all $u \in X$ and $\lambda > 0$), it is then standard from convex analysis [279] that the polar transform of J_{TV} (also called the Legendre–Fenchel transform), defined by

$$J_{\mathrm{TV}}^*(v) = \sup_u \left\{ \langle u, v \rangle_X - J_{\mathrm{TV}}(u) \right\}, \tag{3.24}$$

is the indicator function of a closed convex set G, i.e.,

$$J_{\mathrm{TV}}^*(v) = \chi_G(v) = \begin{cases} 0 & \text{if } v \in G, \\ +\infty & \text{otherwise.} \end{cases} \tag{3.25}$$

It easy to check that the set G is given by (3.23). Now if u is a minimizer of (3.22), necessarily u satisfies

$$0 \in \frac{u - f}{\lambda} + \partial J_{\mathrm{TV}}(u), \tag{3.26}$$

where $\partial J_{\mathrm{TV}}(u)$ denotes the subdifferential[1] of J_{TV} at u. Equation (3.26) is equivalent to

$$\frac{f - u}{\lambda} \in \partial J_{\mathrm{TV}}(u),$$

which, in turn, is equivalent to (see [279])

$$u \in \partial J_{\mathrm{TV}}^* \left(\frac{f - u}{\lambda} \right), \tag{3.27}$$

or, since $\lambda > 0$,

$$0 \in \frac{f - u}{\lambda} - \frac{f}{\lambda} + \frac{1}{\lambda} \partial J_{\mathrm{TV}}^* \left(\frac{f - u}{\lambda} \right). \tag{3.28}$$

Therefore we get that $w = \frac{f-u}{\lambda}$ is a minimizer of

$$\frac{1}{2} \left| w - \frac{f}{\lambda} \right|_{X^2} + \frac{1}{\lambda} J_{\mathrm{TV}}^*(w).$$

But since J_{TV}^* is the indicator function of the set G, it is then clear that w is the solution of the problem

$$\min_{w \in G} \frac{1}{2} \left| w - \frac{f}{\lambda} \right|_{X^2},$$

which is given by an orthogonal projection of $\frac{f}{\lambda}$ on the convex set G or equivalently by a projection of f on λG.   ∎

---

[1]We recall that $w \in \partial J_{\mathrm{TV}}(u)$ means $J_{\mathrm{TV}}(v) \geqslant J_{\mathrm{TV}}(u) + \langle v - u, w \rangle_X$ for all v in X.

Now it remains to effectively compute a projection on λG, which is the most original part of the work done by Chambolle in [91].

(B) Projection Algorithm on λG

We first remark that computing the nonlinear projection $P_{\lambda G}(f)$ amounts to solving the following problem:

$$\min\left\{|\lambda \operatorname{div} p - f|^2_{X \times X} \, ; \, p \in X \times X, |p_{i,j}| \leq 1 \,\, \forall i,j = 1, \ldots, N\right\}. \quad (3.29)$$

Let $\alpha_{i,j} \geqslant 0$ be the Lagrange multipliers associated with each constraint in problem (3.29). We have for each i, j,

$$-(\nabla(\lambda \operatorname{div} p - f))_{i,j} + \alpha_{i,j} p_{i,j} = 0 \quad (3.30)$$

with either $\alpha_{i,j} > 0$ and $|p_{i,j}| = 1$, or $|p_{i,j}| < 1$ and $\alpha_{i,j} = 0$. In the latter case we also have $(\nabla(\lambda \operatorname{div} p - f))_{i,j} = 0$. Therefore in any case we obtain

$$\alpha_{i,j} = |(\nabla(\lambda \operatorname{div} p - f))_{i,j}|.$$

From this observation, we can deduce the following fixed-point algorithm. Let $\tau > 0$ be given and let $p^0 = 0$ be an initial guess. We compute $p_{i,j}^{n+1}$ as

$$p_{i,j}^{n+1} = p_{i,j}^n + \tau((\nabla(\operatorname{div} p^n - f/\lambda))_{i,j} - |(\nabla(\operatorname{div} p^n - f/\lambda))_{i,j}| p_{i,j}^{n+1}). \quad (3.31)$$

Rewriting (3.31), the final algorithm is described in Table 3.1.

_____ **For p^0 given** _____

$$p_{i,j}^{n+1} = \frac{p_{i,j}^n + \tau((\nabla(\operatorname{div} p^n - f/\lambda)))_{i,j}}{1 + \tau |(\nabla(\operatorname{div} p^n - f/\lambda)))_{i,j}|}.$$

_____**The limit (p^∞) is the solution (see Proposition 3.2.2)**_____

Table 3.1. Presentation of Chambolle's algorithm [91].

Chambolle has proved the following convergence result. We refer the reader to [91] for a detailed proof.

Proposition 3.2.2 *Let us assume that* $0 < \tau \leqslant \frac{1}{8}$. *Then* $\lambda \operatorname{div} p^n$ *converges to* $P_{\lambda G}(f)$ *as* $n \to +\infty$.

Remark Chambolle's algorithm still works when the operator R is not the identity. For extensions of this work, we refer to [44]. ∎

THE HALF-QUADRATIC MINIMIZATION APPROACH

In this section, we consider a general operator R and functions $\phi(t)$ satisfying assumptions (3.15)–(3.17) and (3.19). We will define a "close" energy

for which the Euler–Lagrange equation will be easier to implement. This
can be done through Γ-convergence. The general idea is the following (see
also Figure 3.4):

- We construct a sequence of energies E_ε such that for each $\varepsilon > 0$,
  the associated minimization problem admits a unique minimum u_ε
  in the Sobolev space $W^{1,2}(\Omega)$. Then we prove, via Γ-convergence
  theory, that u_ε converges in $L^1(\Omega)$-strong to the minimum of $\overline{E}(u)$.

- Then, for $\varepsilon > 0$ fixed, we propose a suitable numerical scheme called
  the *half-quadratic algorithm*, for which we give a convergence result.
  The idea is to introduce an additional variable, also called a *dual
  variable*, such that the extended functional $J_\varepsilon(u, b)$ has the same
  minimum value as $E_\varepsilon(u)$. More precisely, we show that (u_n, b_n), the
  minimizing sequence of J_ε, is convergent and that u_n converges for
  the $L^1(\Omega)$-strong topology to u_ε.

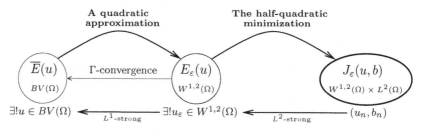

Figure 3.4. Overview of the approach.

This methodology was originally proposed with the total variation ($\phi(s) = s$) by Chambolle and Lions [94]. It was then extended to convex functions
of measures and applied to several problems (see, for instance, Sections
5.3.2 and 5.3.3). It is general in the sense that it is a way to approximate
the smoothing term, which is usually the difficult point. Let us present in
detail this two-step approach.

(A) Quadratic Approximation

For a function ϕ satisfying (3.15)–(3.17), let us define ϕ_ε by

$$
\phi_\varepsilon(s) = \begin{cases}
\dfrac{\phi'(\varepsilon)}{2\varepsilon}s^2 + \phi(\varepsilon) - \dfrac{\varepsilon\phi'(\varepsilon)}{2} & \text{if } 0 \le s \le \varepsilon, \\[2mm]
\phi(s) & \text{if } \varepsilon \le s \le 1/\varepsilon, \\[2mm]
\dfrac{\varepsilon\phi'(1/\varepsilon)}{2}s^2 + \phi(1/\varepsilon) - \dfrac{\phi'(1/\varepsilon)}{2\varepsilon} & \text{if } s \ge \varepsilon.
\end{cases}
$$

We have $\forall \varepsilon$, $\phi_\varepsilon \geq 0$ and $\forall s$, $\lim\limits_{\varepsilon \to 0} \phi_\varepsilon(s) = \phi(s)$. Now let us define the functional E_ε by

$$E_\varepsilon(u) = \begin{cases} \dfrac{1}{2} \displaystyle\int_\Omega |Ru - u_0|^2 \, dx + \lambda \int_\Omega \phi_\varepsilon(|\nabla u|) \, dx & \text{if } u \in W^{1,2}(\Omega), \\ +\infty & \text{otherwise.} \end{cases}$$

$$(3.32)$$

From now on, we suppose that ϕ and R satisfy hypotheses (3.15)–(3.17) and (3.19). The existence and uniqueness of u_ε are quite obvious and derive from classical arguments.

Proposition 3.2.3 *For each $\varepsilon > 0$, the functional E_ε has a unique minimum u_ε in $W^{1,2}(\Omega)$.*

Now, to show that u_ε converges in $L^1(\Omega)$-strong to the unique minimum of $\overline{E}(u)$, we are going to use the notion of Γ-convergence. In particular, we use the two following results, which we recall for the convenience of the reader (see Chapter 2, Theorems 2.1.7 and 2.1.8):

Let X be a topological space, endowed with a τ-topology, and let F_h, $F : X \to \overline{R}$.
Theorem 2.1.7 *Let us assume that (F_h) is equicoercive and Γ-converges to F. Let us suppose that F has a unique minimum x_0 in X. If (x_h) is a sequence in X such that x_h is a minimum for F_h, then (x_h) converges to x_0 in X and $(F_h(x_h))$ converges to $F(x_0)$.*
Theorem 2.1.8 *If (F_h) is a decreasing sequence converging to F pointwise, then (F_h) Γ-converges to the lower semicontinuous envelope of F in X, denoted by $R_\tau F$.*

Then, we have the following result:

Theorem 3.2.3 *The sequence u_ε from Proposition 3.2.3 converges in $L^1(\Omega)$-strong to the unique minimum u of $\overline{E}(u)$ and $E_\varepsilon(u_\varepsilon)$ converges to $\overline{E}(u)$.*

Proof In our case $X = L^1(\Omega)$. Let us denote by $\widetilde{E}(u) : BV(\Omega) \to \overline{R}$ the functional defined by

$$\widetilde{E}(u) = \begin{cases} \overline{E}(u) & \text{if } u \in W^{1,2}(\Omega), \\ +\infty & \text{otherwise.} \end{cases}$$

By construction, we observe that $E_\varepsilon(u)$ is a decreasing sequence converging pointwise to $\widetilde{E}(u)$. Therefore, E_ε Γ-converges to the lower semicontinuous envelope $R_\tau \widetilde{E}$ of $\widetilde{E}$. In order to apply Theorem 2.1.7, we need to check that $\overline{E} = R_\tau \widetilde{E}$.

Step 1: $\overline{E}$ is l.s.c. in $L^1(\Omega)$ with respect to the $L^1(\Omega)$-strong topology. Indeed, let $u_h \in L^1(\Omega)$ be such that $u_h \xrightarrow[L^1(\Omega)]{} u$ as $h \to +\infty$ and

$\varlimsup\limits_{h\to+\infty} \overline{E}(u_h) < \infty$. Then, since $\overline{E}(u_h)$ is bounded, we deduce that u_h is uniformly bounded in $BV(\Omega)$. Thus, up to a subsequence, $u_h \xrightarrow[BV-w^*]{} u$ and $\varliminf\limits_{h\to+\infty} \overline{E}(u_h) \geq \overline{E}(u)$, i.e., $\overline{E}$ is l.s.c. with respect to the $L^1(\Omega)$ topology.

Step 2: Let us show that $\overline{E} = R_\tau \widetilde{E}$.

From Step 1, it suffices to prove that for u in $BV(\Omega)$ there exists a sequence $u_h \in W^{1,2}(\Omega)$ such that $u_h \xrightarrow[L^1(\Omega)]{} u$ and $\overline{E}(u) = \lim\limits_{h\to+\infty} \widetilde{E}(u_h)$. Such a sequence can be constructed using classical approximation arguments [149, 130].

Finally, applying the Γ-convergence result from Theorem 2.1.7, we conclude that u_ε, the unique minimum of E_ε, converges in $L^1(\Omega)$-strong to the unique minimum u of $\overline{E}$. ∎

(B) Changing the problem using a "duality" result

Now it remains to compute u_ε numerically. To do this, we can use the Euler–Lagrange equation satisfied by u_ε:

$$R^* R u_\varepsilon - \operatorname{div}\left(\frac{\phi'_\varepsilon(|\nabla u_\varepsilon|)}{|\nabla u_\varepsilon|}\nabla u_\varepsilon\right) = R^* u_0. \tag{3.33}$$

Equation (3.33) is a highly nonlinear equation. To overcome this difficulty we propose the half-quadratic algorithm based on the following "duality" result [108, 159].

Proposition 3.2.4 *Let $\phi : [0,+\infty[\to [0,+\infty[$ be such that $\phi(\sqrt{s})$ is concave on $]0,+\infty[$ and $\phi(s)$ is non-decreasing. Let L and M be defined by $L = \lim\limits_{s\to+\infty} \phi'(s)/2s$ and $M = \lim\limits_{s\to 0} \phi'(s)/2s$. Then there exists a convex and decreasing function $\psi :]L, M] \to [\beta_1, \beta_2]$ such that*

$$\phi(s) = \inf_{L \leq b \leq M} (bs^2 + \psi(b)), \tag{3.34}$$

where $\beta_1 = \lim\limits_{s\to 0^+} \phi(s)$ and $\beta_2 = \lim\limits_{s\to+\infty} (\phi(s) - s\,\phi'(s)/2)$. Moreover, for every $s \geq 0$, the value b for which the minimum is reached is given by $b = \dfrac{\phi'(s)}{2s}$. The additional variable b is usually called the dual variable.

Proof Let $\theta(s) = -\phi(\sqrt{s})$. By construction $\theta(s)$ is convex. Thus, $\theta(s)$ identifies with its convex envelope, i.e.,

$$\theta(s) = \theta^{**}(s) = \sup_{s^*}(ss^* - \theta^*(s^*)),$$

where $\theta^*(s^*)$ is the polar function of $\theta(s)$ defined as

$$\theta^*(s^*) = \sup_s(ss^* - \theta(s)).$$

Therefore,

$$\phi(\sqrt{s}) = \inf_{s^*}(-ss^* + \theta^*(s^*)).$$

Let $b = -s^*$ and $s = \sqrt{s}$. Then ϕ can be written as

$$\phi(s) = \inf_b(bs^2 + \theta^*(-b)), \tag{3.35}$$

which gives the first part of the theorem with $\psi(b) = \theta^*(-b)$. Now,

$$\theta^*(-b) = \sup_s(-sb - \theta(s)),$$

and since the map $s \to -sb - \theta(s)$ is concave, the supremum is given by the zero of its derivative:

$$-b - \theta'(s) = 0.$$

That is, $b = \phi'(\sqrt{s})/2s$. Since the map $s \to \phi'(s)/2s$ is nonincreasing, it is easy to see that the infimum in (3.34) is achieved for $b \in [L, M]$. The expressions of β_1 and β_2 follow immediately. ∎

Remark It is interesting to observe that Proposition 3.2.4 applies to convex and nonconvex functions ϕ. But in all cases the function ψ appearing in (3.34) is always convex. We list in Table 3.2 three examples of functions ϕ and their corresponding functions ψ.

	$\phi(s)$	convex?	$\psi(b)$	$\dfrac{\phi'(s)}{2s}$	Scaled $\dfrac{\phi'(s)}{2s}$
1	$2\sqrt{1+s^2} - 2$	Yes	$b + \dfrac{1}{b}$	$\dfrac{1}{\sqrt{1+s^2}}$	$\dfrac{1}{\sqrt{1+(10s)^2}}$
2	$\log(1+s^2)$	No	$b - \log(b) - 1$	$\dfrac{1}{1+s^2}$	$\dfrac{1}{1+(3s)^2}$
3	$\dfrac{s^2}{1+s^2}$	No	$b - 2\sqrt{b} + 1$	$\dfrac{1}{(1+s^2)^2}$	$\dfrac{1}{(1+(3s/2)^2)^2}$

Table 3.2. Examples of ϕ functions.

We mention that the first one is often called the hypersurface minimal function. The last column presents scaled versions of $\phi'(s)/s$ that are close to 0.1 for $s = 1$ (see also Figure 3.5). This permits a better comparison from a numerical point of view. ∎

Now let us look at how we may apply Proposition 3.2.4 to solving the problem

$$\inf_u \left\{ E_\varepsilon(u) = \frac{1}{2}\int_\Omega |Ru - u_0|^2 \, dx + \lambda \int_\Omega \phi_\varepsilon(|\nabla u|) \, dx, \quad u \in W^{1,2}(\Omega) \right\}.$$

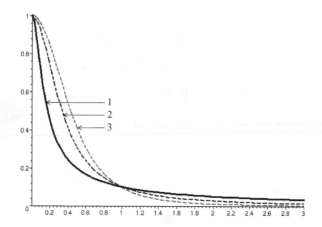

Figure 3.5. Functions $\phi'(s)/2s$ with different choices of ϕ (see Table 3.2).

Let us assume that ϕ_ε satisfies the hypotheses of Proposition 3.2.4. Then there exist L_ε, M_ε, and ψ_ε such that

$$E_\varepsilon(u) = \frac{1}{2} \int_\Omega |Ru - u_0|^2 \; dx + \lambda \int_\Omega \inf_{L_\varepsilon \leq b \leq M_\varepsilon} (b |\nabla u|^2 + \psi_\varepsilon(b)) \; dx.$$

Supposing that we can invert the infimum with respect to b and the integral (this can be justified), we have

$$\inf_u E_\varepsilon(u) = \inf_u \inf_b \left[\frac{1}{2} \int_\Omega |Ru - u_0|^2 \; dx + \lambda \int_\Omega (b |\nabla u|^2 + \psi_\varepsilon(b)) \; dx \right]$$

$$= \inf_b \inf_u \left[\frac{1}{2} \int_\Omega |Ru - u_0|^2 \; dx + \lambda \int_\Omega (b |\nabla u|^2 + \psi_\varepsilon(b)) \; dx \right].$$

If we introduce the functional

$$J_\varepsilon(u, b) = \frac{1}{2} \int_\Omega |Ru - u_0|^2 \; dx + \lambda \int_\Omega (b |\nabla u|^2 + \psi_\varepsilon(b)) \; dx,$$

then

☞ J_ε is convex in u, and for each u fixed in $W^{1,2}(\Omega)$ it is convex in b.

Of course, $J_\varepsilon(u, b)$ is not convex in the pair (u, b). So this leads to the alternate semiquadratic algorithm described in Table 3.3.

C++

To illustrate this, we display in Figure 3.6 the result obtained on the "Borel building" image.

_____For (u^0, b^0) given_____

- Step 1: $u_\varepsilon^{n+1} = \underset{u}{argmin}\ J_\varepsilon(u, b^n)$. Since the problem is convex (and quadratic), this is equivalent to solving

$$\begin{cases} R^*Ru - R^*u_0 - \lambda\mathrm{div}(b^n\nabla u) = 0 \ \text{in}\ \ \Omega, \\ b^n\dfrac{\partial u}{\partial N} = 0 \ \text{on}\ \partial\Omega. \end{cases} \qquad (3.36)$$

Once discretized, the linear system can be solved with a Gauss--Seidel iterative method, for example.

- Step 2: $b^{n+1} = \underset{b}{argmin}\ J_\varepsilon(u_\varepsilon^{n+1}, b)$. According to Proposition 3.2.4, the minimum in b is reached for

$$b^{n+1} = \frac{\phi_\varepsilon'\left(\left|\nabla u_\varepsilon^{n+1}\right|\right)}{2\left|\nabla u_\varepsilon^{n+1}\right|}. \qquad (3.37)$$

- Go back to the first step until there is convergence.

_____The limit $(u_\varepsilon^\infty, b^\infty)$ is the solution (see Theorem 3.2.4)._____

Table 3.3. Presentation of the half-quadratic algorithm, also called ARTUR (see [106, 108]). From a numerical point of view, the only difficulty is the discretization of the term $\mathrm{div}(b\nabla u)$ in (3.36), where b is given and defined by (3.37). Several possibilities can be considered. This is detailed in Section A.3.2 of the Appendix.

The sequence of functions $b^n(x)$ can be seen as an indicator of contours. If ϕ satisfies the edge-preserving hypotheses $\lim_{s\to+\infty} \phi'(s)/2s = 0$ and $\lim_{s\to 0^+} \phi'(s)/2s = 1$, then the following conditions are satisfied:

- If $b^n(x) = 0$, then x belongs to a contour.

- If $b^n(x) = 1$, then x belongs to a homogeneous region.

So the property of this iterative algorithm is to detect and take into account progressively the discontinuities of the image. This is illustrated in Figure 3.7 where the initial condition was such that $(u^0, b^0) \equiv (0, 1)$.

Now, as far as the convergence of this algorithm is concerned, we have the following theorem

Theorem 3.2.4 *If ϕ_ε and R satisfy (3.15)–(3.17) and (3.19), and if ϕ_ε satisfies the assumptions of Proposition 3.2.4, then the sequence (u^n, b^n) is convergent in $L^2(\Omega)$-strong$\times L^\infty(\Omega)$-weak*. Moreover, u^n converges strongly in $L^2(\Omega)$ (and weakly in $W^{1,2}(\Omega)$) to the unique solution u_ε of E_ε.*

We do not reproduce here the proof of Theorem 3.2.4. It is rather long and technical. For more details, we refer the reader to [94, 24].

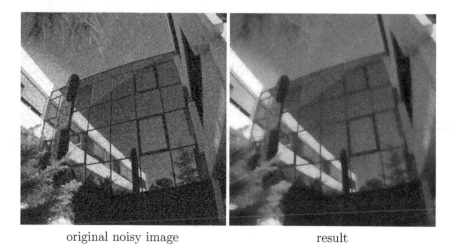

original noisy image   result

Figure 3.6. Result with half-quadratic minimization (compare with Figure 3.3: noise is removed, while discontinuities are retained).

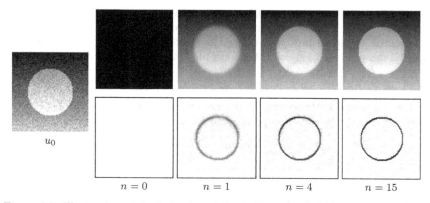

u_0

$n = 0$ $n = 1$ $n = 4$ $n = 15$

Figure 3.7. Illustration of the behavior of the half-quadratic algorithm on a synthetic image u_0. On the right-hand side, four iterations of the algorithm are displayed ($n = 0, 1, 4, 15$). The first row is u^n, while the second one is b^n. Initialization is $u^0 \equiv 0$ and $b^0 \equiv 1$. This shows the interpretation of b as an edge detector that becomes more precise as time evolves.

Remarks

- The only assumption ensuring that the semiquadratic algorithm works is that $\phi(\sqrt{s})$ is concave. That is, we may apply this algorithm even to nonconvex functions ϕ. Of course, we will be able to give a convergence result only for convex functions.

- The half-quadratic approach can also be realized with another duality result based on the Legendre transform. This has been studied in [23, 107].

■

3.2.5 Some Invariances and the Role of λ

In this section we set some elementary properties to highlight the invariance and the "scale" nature of the parameter λ in front of the regularization term in $E(u)$:

$$E(u) = \frac{1}{2} \int_\Omega |u - u_0|^2 \, dx + \lambda \int_\Omega \phi(|\nabla u|) \, dx.$$

Here we assume that $u_0 \in L^2(\Omega)$ and that ϕ satisfy hypotheses (3.15)–(3.17). Let $u(x, \lambda)$ be the unique minimizer of $E(u)$. To simplify, we suppose that

$$u(., \lambda) \in W^{1,1}(\Omega) \cap L^2(\Omega).$$

Let us define the operator $T_\lambda : L^2(\Omega) \to L^2(\Omega)$ by $T_\lambda u_0 = u(\lambda)$, where $u(\lambda) = u(x, \lambda)$. By definition, we have for all $v \in W^{1,1}(\Omega) \cap L^2(\Omega)$,

$$\frac{1}{2} \int_\Omega |u(x, \lambda) - u_0(x)|^2 \, dx + \lambda \int_\Omega \phi(|\nabla u(x, \lambda)|) \, dx \qquad (3.38)$$

$$\leq \frac{1}{2} \int_\Omega |v(x) - u_0(x)|^2 \, dx + \lambda \int_\Omega \phi(|\nabla v(x)|) \, dx.$$

Moreover, $u(x, \lambda)$ necessarily satisfies the Euler–Lagrange equation:

$$\begin{cases} u(x, \lambda) - u_0(x) = \lambda \operatorname{div} \left(\dfrac{\phi'(|\nabla u(x, \lambda)|)}{|\nabla u(x, \lambda)|} \nabla u(x, \lambda) \right) \text{ in } \Omega, \\ \dfrac{\phi'(|\nabla u(x, \lambda)|)}{|\nabla u(x, \lambda)|} \dfrac{\partial u}{\partial N}(x, \lambda) = 0 \text{ on } \partial\Omega. \end{cases} \qquad (3.39)$$

We begin by carrying out some invariance properties, which can be easily proved. These invariances with respect to some image transformation Q express the fact that T_λ and Q can commute.

(A1) *Gray-level invariance:*
$T_\lambda 0 = 0$ and $T_\lambda(u_0 + c) = T_\lambda u_0 + c$, for every constant c.

(A2) *Translation invariance:*
Define the translation τ_h by $(\tau_h)(f)(x) = f(x + h)$.
Then $T_\lambda(\tau_h u_0) = \tau_h(T_\lambda u_0)$.

(A3) *Isometry invariance:*
Let us set $(Rf)(x) = f(Rx)$ for any isometry of R^2 and function f

from R^2 to R. Then $T_\lambda(Ru_0) = R(T_\lambda u_0)$. Of course, this invariance is true because the regularization function depends on the norm of ∇u.

Now let us examine some properties of the correspondence

$$\lambda \to T_\lambda u_0(x) = u(x, \lambda).$$

Before proving them, it is interesting to do some quantitative tests on the "Borel building" image. From these experiments, we can observe the following:

- The SNR reaches a maximum rapidly and decreases rapidly (see Figure 3.8).

- $|u(., \lambda)|_{L^2(\Omega)}$ seems to be constant.

- $\overline{u}(., \lambda)$, the mean of u, is constant.

- $\int_\Omega |u(x, \lambda) - \overline{u_0}|\ dx$ tends to zero, which means that u converges in the $L^1(\Omega)$-strong topology to the average of the initial data.

Let us see whether we can prove some of these empirical properties.

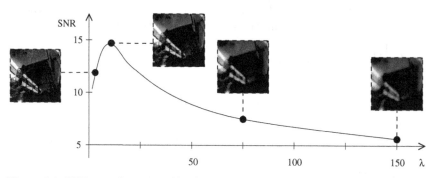

Figure 3.8. SNR as a function of λ. After being optimal, high values of λ smooth the image too much.

Property 1 *The L^2 norm of $u(., \lambda)$ is bounded by a constant independent of λ.*

Proof By letting $v = 0$ in (3.38) and remembering that $\phi(0) = 0$, we obtain

$$\int_\Omega |u(x, \lambda) - u_0|^2\ dx \le \int_\Omega |u_0|^2\ dx,$$

from which we deduce

$$\int_\Omega |u(x, \lambda)|^2\ dx \le 2 \int_\Omega |u_0|^2\ dx,$$

i.e., the L^2-norm of u is bounded by a constant independent of λ.   ∎

Property 2 *For every λ, we have $\int_\Omega u(x,\lambda)\,dx = \int_\Omega u_0(x)\,dx$.*

Proof From (3.39), we get

$$\int_\Omega u(x,\lambda)\,dx = \int_\Omega u_0(x)\,dx + \lambda \int_\Omega \operatorname{div}\left(\frac{\phi'(|\nabla u(x,\lambda)|)}{|\nabla u(x,\lambda)|}\nabla u(x,\lambda)\right)\,dx.$$

But thanks to the Green formula and the boundary condition in (3.39),

$$\int_\Omega \operatorname{div}\left(\frac{\phi'(|\nabla u(x,\lambda)|)}{|\nabla u(x,\lambda)|}\nabla u(x,\lambda)\right)\,dx = \int_{\partial\Omega}\frac{\phi'(|\nabla u(x,\lambda)|)}{|\nabla u(x,\lambda)|}\frac{\partial u}{\partial N}d\Gamma = 0,$$

which concludes the proof.   ∎

Property 3 $u(.,\lambda)$ *converges in $L^1(\Omega)$-strong to the average of the initial data.*

Proof Again from (3.38), letting $v = 0$, we obtain

$$0 \le \lambda \int_\Omega \phi(|\nabla u(x,\lambda)|\,dx \le \frac{1}{2}\int_\Omega u_0^2(x)\,dx.$$

Therefore,

$$0 \le \lim_{\lambda\to+\infty}\int_\Omega \phi(|\nabla u(x,\lambda)|\,dx \le \lim_{\lambda\to+\infty}\frac{1}{2\lambda}\int_\Omega u_0^2(x)\,dx = 0,$$

i.e., $\lim_{\lambda\to+\infty}\int_\Omega \phi(|\nabla u(x,\lambda)|)\,dx = 0$. Since ϕ is strictly convex with linear growth, we easily deduce from the above equality that

$$\lim_{\lambda\to+\infty}|\nabla u(.,\lambda)|_{L^1(\Omega)} = 0. \tag{3.40}$$

On the other hand, thanks to the Poincaré–Wirtinger inequality,

$$\int_\Omega |u(x,\lambda) - \overline{u_0}|\,dx \le cte\,|\nabla u(x,\lambda)|_{L^1(\Omega)} \quad \text{with } \overline{u_0} = \frac{1}{|\Omega|}\int_\Omega u_0(x,\lambda)\,dx.$$

Thus, with (3.40), $\lim_{\lambda\to+\infty}\int_\Omega |u(x,\lambda) - \overline{u_0}|\,dx = 0$, i.e., $u(.,\lambda)$ converges in $L^1(\Omega)$-strong to the average of the initial data.   ∎

This latter property shows that λ can be interpreted as a *scale parameter*. Starting from the initial image u_0 we construct a family of images $\{u(x,\lambda)\}_{\lambda>0}$ of gradually simplified (smoothed) versions of it. This scale notion, also called *scale space theory*, plays a central role in image analysis and has been investigated by many authors [43, 245, 5, 198, 338]. As we will see in Section 3.3, "scale space ideas" are strongly present in PDE theory.

3.2.6 Some Remarks on the Nonconvex Case

As noted at the beginning of Section 3.2.3, a "good" edge-preserving behavior for ϕ would be such that

$$\phi(s) \approx cs^2 \text{ as } s \to 0^+, \qquad (3.41)$$

$$\lim_{s \to +\infty} \phi(s) \approx \gamma > 0. \qquad (3.42)$$

Unfortunately, conditions (3.41) and (3.42) imply that ϕ is nonconvex. Of course, there is no longer the existence of a solution to the minimization problem, and one cannot prove any convergence result. Nevertheless, potentials satisfying (3.41)–(3.42), like

$$\phi(s) = \frac{s^2}{1 + s^2},$$

seem to provide better (sharper) results than convex potentials with linear growth. This is illustrated in Figures 3.9 and 3.10. So what can be said in this case?

First of all, we have just mentioned that using a nonconvex potential yields an ill-posed problem. The existence may indeed not be straightforward. Let us prove it for the following energy:

$$E(u) = \int_\Omega |u - u_0|^2 \, dx + \lambda \int_\Omega \frac{|\nabla u|^2}{1 + |\nabla u|^2} \, dx.$$

The function $\phi(s) = s^2/(1 + s^2)$ satisfies (3.41) and (3.42) with $\gamma = 1$. We are going to show that $E(u)$ has no minimizer. We assume that $u_0 \in L^\infty(\Omega)$.

Proposition 3.2.5 *If $u_0(x)$ is not a constant, the functional $E(u)$ has no minimizer in $W^{1,2}(\Omega)$ and $\inf \{E(u); \ u \in W^{1,2}(\Omega)\} = 0$.*

Proof For clarity's sake, we prove the proposition in the one-dimensional case $\Omega =]a, b[$. The same proof goes for $N \geq 2$. We follow the proof given by Chipot et al. [111].

By density, we always may find a sequence of step functions $\overline{u_n}$ such that

$$|\overline{u_n}| \leq |u_0|_{L^\infty}, \quad \lim_{n \to +\infty} |\overline{u_n} - u_0|_{L^2(\Omega)} = 0.$$

In fact, we can find a partition $a = x_0 < x_1 < \cdots < x_n = b$ such that $\overline{u_n}$ is constant on each interval (x_{i-1}, x_i), $h_n = \max_i(x_i - x_{i-1}) < 1$ with $\lim_{n \to +\infty} h_n = 0$. Let us set $\sigma_i = x_i - x_{i-1}$. Next, we define a sequence of continuous functions u_n by

$$u_n(x) = \begin{cases} \overline{u_n}(x) & \text{if } x \in [x_{i-1}, x_i - \sigma_i^2], \\ \dfrac{(\overline{u_{n,i+1}} - \overline{u_{n,i}})}{\sigma_i^2}(x - x_i) + \overline{u_{n,i+1}} & \text{if } x \in [x_i - \sigma_i^2, x_i], \end{cases}$$

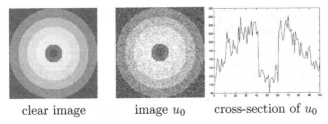

clear image   image u_0 cross-section of u_0

Figure 3.9. Clear image (without noise), original noisy image (obtained by adding a Gaussian noise of variance 20), and a cross-section of it passing through the center of the image.

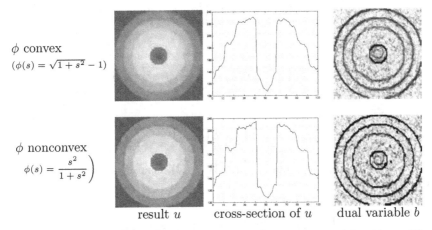

ϕ convex

$(\phi(s) = \sqrt{1+s^2} - 1)$

ϕ nonconvex

$\phi(s) = \dfrac{s^2}{1+s^2}$

result u cross-section of u dual variable b

Figure 3.10. Results obtained using a convex and a nonconvex ϕ function, with the half-quadratic algorithm (see Section 3.2.4). We display the result u, a cross-section passing through the center of the image, and the dual variable b. One can observe that nonconvex ϕ functions allow the reconstruction of sharper images.

where $\overline{u_{n,i}} = \overline{u_n}(x)_{/]x_{i-1},x_i[}$. It is easy to check that

$$|\overline{u_n} - u_n|_{L^2(\Omega)} \leq 2\,|u_0|_{L^\infty}^2 \sum_{i=1}^{n} \sigma_i^2 \leq 2\,|u_0|_{L^\infty}^2\, h_n \sum_{i=1}^{n} \sigma_i = 2\,|u_0|_{L^\infty}^2\,(b-a)h_n.$$

Therefore, $\displaystyle\lim_{n\to+\infty} |\overline{u_n} - u_n|_{L^2(\Omega)} = 0.$ Since

$$\lim_{n\to+\infty} |u_n - u_0|_{L^2(\Omega)} \leq \lim_{n\to+\infty} |u_n - \overline{u_n}|_{L^2(\Omega)} + \lim_{n\to+\infty} |\overline{u_n} - u_0|_{L^2(\Omega)},$$

we also deduce that $\lim\limits_{n\to+\infty} |u_n - u_0|_{L^2(\Omega)} = 0$. Moreover

$$\int_a^b \frac{u_n'^2(x)}{1 + u_n'^2(x)}\, dx = \sum_i \int_{x_i - \sigma_i^2}^{x_i} \frac{(\overline{u}_{n,i+1} - \overline{u}_{n,i})^2}{\sigma_i^4 + (\overline{u}_{n,i+1} - \overline{u}_{n,i})^2}\, dx$$

$$\leq \sum_i \sigma_i^2 \leq h_n \sum_i \sigma_i = h_n(b - a).$$

Thus

$$\lim_{n\to+\infty} \int_a^b \frac{u_n'^2(x)}{1 + u_n'^2(x)}\, dx = 0,$$

and finally,

$$0 \leq \inf_{u\in W^{1,2}(\Omega)} E(u) \leq \lim_{n\to+\infty} E(u_n) = 0,$$

i.e.,

$$\inf_{u\in W^{1,2}(\Omega)} E(u) = 0.$$

Now, if there exists a minimizer $u \in W^{1,2}(\Omega)$, then necessarily $E(u) = 0$, which implies

$$\begin{cases} \int_a^b |u - u_0|^2\, dx = 0 & \Leftrightarrow \quad u = u_0 \text{ a.e.} \\ \int_a^b \frac{u'^2}{1 + u'^2}\, dx = 0 & \Leftrightarrow \quad u' = 0 \text{ a.e.} \end{cases}$$

The first equality is possible only if $u_0 \in W^{1,2}(\Omega)$ (since $u \in W^{1,2}(\Omega)$), and in this case the second equality implies $u_0' = 0$, which is possible only if u_0 is a constant. Therefore, excluding this trivial case, $E(u)$ has no minimizer in $W^{1,2}(\Omega)$.   ∎

☛ *By density arguments, there is no hope of obtaining the existence of a minimizer for $E(u)$ in any reasonable space.*

Then what can we do? A possibility is to regularize the functional $E(u)$ either by adding constraints, or by adding a supplementary term. This latter idea has been investigated by Chipot et al. [111]. They introduced the following energy:

$$E_\varepsilon(u) = \int_\Omega |u - u_0|^2\, dx + \lambda \int_\Omega \frac{|\nabla u|^2}{1 + |\nabla u|^2}\, dx + \varepsilon \int_\Omega |\nabla u|^2\, dx.$$

$E_\varepsilon(u)$ is convex for $\varepsilon \geq \lambda/4$ and nonconvex for $\varepsilon < \lambda/4$. The former case is not interesting, since it is too regularizing and not edge-preserving. Though $E_\varepsilon(u)$ is nonconvex for $\varepsilon < \lambda/4$, it has quadratic growth at infinity. This fact allows us to use convexification tools that permit us to obtain the existence of a minimizer for $E_\varepsilon(u)$ in the one-dimensional case [111].

❋ *For dimensions greater than one, the problem is quite open. The behavior of the minimizing sequences is also a challenging problem, which is closely related to Perona–Malik anisotropic diffusion, as we shall see further.*

Another attempt would be to work directly with the discrete version of $E(u)$. For example, we mention a recent paper by Rosati [280, 281], who studies the asymptotic behavior of a discrete model related to $E(u)$. In his paper, Rosati chooses $\phi(s) = s^2(1 + \mu s^2)$, and with the condition that μ is proportional to the mesh size h, he proves that the discrete model Γ-converges to a modified Mumford–Shah functional as h tends to 0. This is surely a wise approach, since if we make such efforts to introduce nonconvex potentials, it is because they give very good numerical results in restoration problems. Discrete problems generally have solutions even in the nonconvex case. It should be very interesting to investigate more thoroughly the relationships between discrete and continuous models.

We also mention that choosing $\phi(s) = \log(1 + s^2)$ also leads to very good results and is often used in experiments. This is again another situation, since ϕ is nonconvex but with sublinear growth at infinity. It would also be very interesting to understand the problem from a theoretical point of view.

3.3   PDE-Based Methods

In the previous section we considered a class of approach that consists in setting the best energy according to our needs. The equations that were to be solved numerically were the Euler–Lagrange equations associated with the minimization problems. Another possibility is to work directly on the equations, without thinking of any energy. It is the aim of this section to present some classical PDE-based methods for restoration, trying to follow the chronological order in which they appeared in the literature. These models can be formally written in the following general form:

$$
\begin{cases}
\dfrac{\partial u}{\partial t}(t,x) + F(x, u(t,x), \nabla u(t,x), \nabla^2 u(t,x)) = 0 \quad \text{in } (0,T) \times \Omega, \\
\qquad \text{(a second-order differential operator)} \\
\dfrac{\partial u}{\partial N}(t,x) = 0 \quad \text{on } (0,T) \times \partial\Omega \quad \text{(Neumann boundary condition)}, \\
u(0,x) = u_0(x) \quad \text{(initial condition)},
\end{cases}
\tag{3.43}
$$

where $u(t,x)$ is the restored version of the initial degraded image $u_0(x)$. As usual, ∇u and $\nabla^2 u$ stand respectively for the gradient and the Hessian matrix of u with respect to the space variable x. Let us comment.

One of the main difference, with the equations encountered up to now is the presence of the parameter t. Starting from the initial image $u_0(x)$ and by running (3.43) we construct a family of functions (i.e., images) $\{u(t,x)\}_{t>0}$ representing successive versions of $u_0(x)$. As t increases we expect that $u(t,x)$ changes into a more and more simplified image, or in other words, structures for large t constitute simplifications of corresponding structures at small t. Moreover, no new structure must be created. For these reasons t is called a *scale variable*.

As we will see further, the choice of F in (3.43) is determining, since we would like to attain two goals that may seem a priori contradictory. The first is that $u(t,x)$ should represent a smooth version of $u_0(x)$ where the noise has been removed. The second is to be able to preserve some features such as edges, corners, and T-junctions, which may be viewed as singularities.

Finally, a natural question is how to classify PDE-based models. The answer is inevitably subjective. Perhaps, the simplest way is to choose the classical PDE classification, namely forward parabolic PDE, backward parabolic PDE and hyperbolic PDEs corresponding respectively to smoothing, smoothing–enhancing, and enhancing processes. Let us follow this classification.

3.3.1 Smoothing PDEs

THE HEAT EQUATION

The oldest and most investigated equation in image processing is probably
the parabolic linear heat equation[43, 5, 198]:   C++

$$\begin{cases} \frac{\partial u}{\partial t}(t,x) - \Delta u(t,x) = 0, & t \geq 0, \quad x \in R^2, \\ u(0,x) = u_0(x). \end{cases} \quad (3.44)$$

Notice that we have here $x \in R^2$. In fact, we consider that $u_0(x)$ is primarily
defined on the square $[0,1]^2$. By symmetry we extend it to $C = [-1,+1]^2$
and then in all of R^2 by periodicity (see Figure 3.11). This way of extending $u_0(x)$ is classical in image processing. The motivation will become
clearer in the sequel. If $u_0(x)$ extended in this way satisfies in addition
$\int_C |u_0(x)|\, dx < +\infty$, we will say that $u_0 \in L^1_\#(C)$.

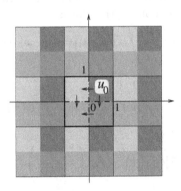

Figure 3.11. Extension of u_0 primarily defined on $[0,1]^2$ to R^2 by symmetry and
periodicity.

The motivation to introduce such an equation came from the following
remark: Solving (3.44) is equivalent to carrying out a Gaussian linear filtering, which was widely used in signal processing. More precisely, let u_0
be in $L^1_\#(C)$. Then the explicit solution of (3.44) is given by

$$u(t,x) = \int_{R^2} G_{\sqrt{2t}}(x-y)\, u_0(y)\, dy = (G_{\sqrt{2t}} * u_0)(x), \quad (3.45)$$

where $G_\sigma(x)$ denotes the two-dimensional Gaussian kernel

$$G_\sigma(x) = \frac{1}{2\pi\,\sigma^2}\, \exp\left(-\frac{|x|^2}{2\,\sigma^2}\right). \quad (3.46)$$

Convolution by a positive kernel is the basic operation in linear image
filtering. It corresponds to low-pass filtering (see Figure 3.12). This formula

gives the correspondence between the time t and the scale parameter σ of the Gaussian kernel.

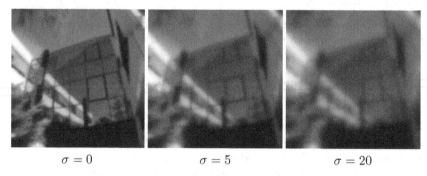

$\sigma = 0$ $\sigma = 5$ $\sigma = 20$

Figure 3.12. Examples of the test image at different scales.

The action of the Gaussian kernel can also be interpreted in the frequency domain. Let us define the Fourier transform

$$F[f](w) = \int_{R^2} f(x) \exp\left(-i\,w \cdot x\right)\,dx,$$

where $w \in R^2$. It is well known that

$$F[G_\sigma * f](w) = F[G_\sigma](w)F[f](w),$$

and since

$$F[G_\sigma](w) = \exp\left(-\frac{|w|^2}{2/\sigma^2}\right),$$

it follows that

$$F[G_\sigma * f](w) = \exp\left(-\frac{|w|^2}{2/\sigma^2}\right) F[f](w),$$

i.e., the convolution by a Gaussian is a low-pass filter that inhibits high frequencies (oscillations in the space domain).

Remark As we can observe in Figure 3.12, the smoothing is isotropic: It does not depend on the image, and it is the same in all directions. In particular, edges are not preserved.

In fact, if we introduce two arbitrary orthonormal directions D_1 and D_2, we have $\Delta u = u_{D_1 D_1} + u_{D_2 D_2}$. If we rewrite this equality with the directions $D_1 = N = \nabla u / |\nabla u|$ and $D_2 = T$ with $T \cdot N = 0$, $|T| = 1$, then $\Delta u = u_{NN} + u_{TT}$. The isotropy means that the diffusion is equivalent in the two directions.

As will be shown in the sequel, most of the diffusion operators can be decomposed as a weighted sum of u_{NN} and u_{TT} [205].   ∎

To set the properties satisfied by $u(t, x)$, we either can use the fact that u is a convolution product or we can deduce these properties from the general theory of uniformly parabolic equations. Choosing the latter approach, we summarize below some of the main properties of $u(t, x)$ (see [176] for the proofs).

Proposition 3.3.1 *Let u_0 be in $L^1_\#(C)$ and define $u(t, x)$ by (3.45). Then for all $t > 0$ and $x \in R^2$, $u(t, x)$ satisfies the heat equation with initial value u_0:*

$$\frac{\partial u}{\partial t}(t, x) = \Delta u(t, x) \quad and \quad \lim_{t \to 0} \int_C |u(t, x) - u_0(x)| \ dx = 0,$$

$$u(t, .) \in L^1_\#(C) \quad and \quad u \in C^\infty((0, T) \times R^2) \quad for \ all \ T > 0.$$

Moreover, if t_1 is any positive real number, there exists a constant $c(t_1)$ such that for $t \in [t_1, +\infty[$,

$$\sup_{x \in R^2} |u(t, x)| \leq c(t_1) |u_0|_{L^1_\#(C)}. \tag{3.47}$$

If $u_0 \in L^\infty_\#(C)$, then we have a maximum principle

$$\inf_{x \in R^2} u_0(x) \leq u(t, x) \leq \sup_{x \in R^2} u_0(x). \tag{3.48}$$

Here $u(t, x)$ given by (3.45) is the unique solution of the heat equation satisfying conditions (3.47) and (3.48).

Remark The uniqueness of u in Proposition 3.3.1 has been obtained in the class of periodic functions. If we drop this assumption, we have to replace it by another one. It is well known that the heat equation $\frac{\partial u}{\partial t} - \Delta u = 0$ in $(0, T) \times R^2$, $u(0, x) = 0$ has infinitely many solutions. Each of the solutions besides $u \equiv 0$ grows very rapidly as $|x| \to +\infty$. To get a uniqueness result in this case it suffices to impose that u satisfy the growth estimate

$$|u(t, x)| \leq A e^{a |x|^2},$$

for some constants A and $a > 0$. We refer to [148] for more details.   ∎

Now let T_t, $t > 0$, be the family of scale operators from $L^1_\#(C)$ into $L^1_\#(C)$ defined by $(T_t u_0)(x) = u(t, x)$, where $u(t, x)$ is the unique solution of (3.44) given by (3.45). T_t is a family of linear operators. We list below some invariance properties of T_t that can be easily proved by observing that $G_\sigma(x) > 0$, $\int_{R^2} G_\sigma(x) \ dx = 1$, and by using the periodicity of u_0 and classical theorems of integration theory such as Fubini's theorem:

(A1) *Gray-level shift invariance:*
$T_t(0) = 0$ and $T_t(u_0 + c) = T_t\,u_0 + c$, for any constant c.

(A2) *Translation invariance:*
$T_t(\tau_h\,u_0) = \tau_h\,(T_t\,u_0)$, where τ_h is the translation $\tau_h(f)(x) = f(x+h)$.

(A3) *Scale invariance:*
$T_t(H_\lambda\,u_0) = H_\lambda(T_{t'}\,u_0)$ with $t' = t\,\lambda^2$, where $(H_\lambda f)(x) = f(\lambda\,x)$.

(A4) *Isometry invariance:*
$T_t(R\,u_0) = R(T_t\,u_0)$, for any orthogonal transformation R of R^2 where $(Rf)(x) = f(Rx)$.

(A5) *Conservation of average value:*
$T_t(M\,u_0) = M(T_t\,u_0)$, where $Mf = \displaystyle\fint_C f(x)\,dx$.

(A6) *Semigroup property:*
$T_{t+s}\,u_0 = T_t\,(T_s\,u_0)$.

(A7) *Comparison principle:*
If $u_0 \le v_0$ then $(T_t u_0) \le (T_t v_0)$.

These invariance properties are quite natural from an image analysis point of view. For example, the gray-level shift invariance means that the analysis must be independent of the range of the brightness of the initial image. The other geometric properties express the invariance of image analysis under the respective positions of percipiens and perceptum.

Are these properties sufficient to ensure correct qualitative properties for $T_t u$? The answer is no. Though the heat equation has been (and is) successfully applied in image processing, it has some drawbacks: It is too smoothing. In fact, whatever the regularity of the initial data, $u(t, x)$ is C^∞ in x, $\forall t > 0$: Edges are lost. We sometimes say that the heat equation has infinite speed of propagation. Of course, this instantaneous regularity is not a desirable property, since in particular, edges can be lost or severely blurred.

NONLINEAR DIFFUSION

We are going to describe models that are generalizations of the heat equation. What we would like to do is to find models (if possible, well-posed models) for removing the noise while preserving the edges at best. For now, the domain image will be a bounded open set Ω of R^2. Let us consider the following equation, initially proposed by Perona and Malik [275]:

$$\begin{cases} \dfrac{\partial u}{\partial t} = \operatorname{div}\left(c(|\nabla u|^2)\,\nabla u\right) & \text{in } \Omega \times (0, \mathrm{T}), \\[2mm] \dfrac{\partial u}{\partial N} = 0 & \text{on } \partial\Omega \times (0, T), \\[2mm] u(0, x) = u_0(x) & \text{in } \Omega, \end{cases} \qquad (3.49)$$

with $c(s): [0, +\infty[\rightarrow]0, +\infty[$. Before going further, we can remark that if we choose $c \equiv 1$, then we recover the heat equation. Now, imagine that $c(s)$ is a decreasing function satisfying $c(0) = 1$ and $\lim_{s \to +\infty} c(s) = 0$. With this choice:

- Inside the regions where the magnitude of the gradient of u is weak, equation (3.49) acts like the heat equation, resulting in isotropic smoothing.

- Near the region's boundaries where the magnitude of the gradient is large, the regularization is "stopped" and the edges are preserved.

Indeed, we can be more precise if we interpret this divergence operator using the directions T, N associated to the image (as for (3.7)–(3.8)). By developing formally the divergence operator, we get (with the usual notation $u_x, u_{xx}, \ldots$)

$$\text{div}\left(c(|\nabla u|^2)\, \nabla u \right)$$
$$= 2\left(u_x^2 u_{xx} + u_y^2 u_{yy} + 2\, u_x u_y u_{xy} \right) c'(|\nabla u|^2) + c(|\nabla u|^2)\, (u_{xx} + u_{yy}).$$

If we define $b(s) = c(s) + 2sc'(s)$, then (3.49) reads as

$$\frac{\partial u}{\partial t}(t, x) = c(|\nabla u|^2)\, u_{TT} + b(|\nabla u|^2)\, u_{NN}. \tag{3.50}$$

Therefore, (3.50) may be interpreted as a sum of a diffusion in the T-direction plus a diffusion in the N-direction, the functions c and b acting as weighting coefficients. Of course, since N is normal to the edges, it would be preferable to smooth more in the tangential direction T than in the normal direction N. Thus, we impose $\lim_{s \to +\infty} b(s)/c(s) = 0$, or equivalently, according to the definition of b,

$$\lim_{s \to +\infty} \frac{s\, c'(s)}{c(s)} = -\frac{1}{2}. \tag{3.51}$$

If we restrict ourselves to functions $c(s) > 0$ with power growth, then the above limit implies that $c(s) \approx 1/\sqrt{s}$ as $s \to +\infty$. The question now is to know whether (3.49) is well posed or not.

Firstly, we have to examine the parabolicity of equation (3.49). To do this, we observe that (3.49) can be written as

$$\frac{\partial u}{\partial t} = a_{11}(|\nabla u|^2)\, u_{xx} + 2\, a_{12}(|\nabla u|^2)\, u_{xy} + a_{22}(|\nabla u|^2)\, u_{yy} \tag{3.52}$$

with

$$a_{11}(|\nabla u|^2) = 2\, u_x^2\, c'(|\nabla u|^2) + c(|\nabla u|^2),$$
$$a_{12}(|\nabla u|^2) = 2\, u_x\, u_y\, c'(|\nabla u|^2),$$
$$a_{22}(|\nabla u|^2) = 2\, u_y^2\, c'(|\nabla u|^2) + c(|\nabla u|^2),$$

and (3.52) is parabolic if and only if $\sum_{i=1,2} a_{ij}(|\nabla u|^2)\xi_i\xi_j \geq 0$, $\forall \xi \in R^2$. An easy algebraic calculation shows that this condition reduces to the single inequality

$$b(s) > 0.$$

To summarize, the assumptions imposed on $c(s)$ are

$$\begin{cases} c : [0, +\infty[\rightarrow]0, +\infty[\text{ decreasing,} \\ c(0) = 1, \quad c(s) \approx \dfrac{1}{\sqrt{s}} \quad \text{as} \quad s \to +\infty, \\ b(s) = c(s) + 2\,s\,c'(s) > 0. \end{cases} \qquad (3.53)$$

A canonical example of a function $c(s)$ satisfying (3.53) is $c(s) = 1/\sqrt{1+s}$.

☛ *With the assumptions* (3.53) *the nonlinear diffusion model* (3.49) *acts as a forward parabolic equation smoothing homogeneous regions while preserving edges.*

Remark If we remove the condition $b(s) > 0$, by supposing, for example, that for some s_0, $b(s) > 0$ for $s \leq s_0$, and $b(s) < 0$ for $s > s_0$, then (3.49) changes into a backward parabolic equation for $|\nabla u|^2 > s_0$, or equivalently into a smoothing–enhancing model. We will return later to this model (see Section 3.3.2). ∎

What can be said concerning the existence of a solution for (3.49)? Unfortunately, with the assumption (3.53) we cannot directly apply general results for parabolic equations. The difficulty comes from the highly degenerate behavior of (3.49) due to the vanishing condition $c(s) \approx 1/\sqrt{s}$ as s tends to infinity. As a matter of fact, one can find some classical results for equations of the form

$$\frac{\partial u}{\partial t} - \text{div}\, a(t, x, u, \nabla u) = 0,$$

where the function a satisfies the structural conditions

$$a(t, x, u, \nabla u)\,.\,\nabla u \geq \alpha_0\,|\nabla u|^p - \beta_0(t, x), \qquad (3.54)$$

$$|\,a(t, x, u, \nabla u)\,| \leq \alpha_1\,|\nabla u|^{p-1} + \beta_1(t, x) \qquad (3.55)$$

a.e. (t, x) with $p > 1$; $(\alpha_i)_{i=0,1}$ are given constants and $(\beta_i)_{i=0,1}$ are given nonnegative functions satisfying some integrability conditions [139]. Here we have $p = 1$.

☛ *In fact, the difficulties in obtaining an existence result for* (3.49) *have to be compared to those encountered in Section 3.1.2 for variational problems. We saw that a linear growth assumption on the potential required enlarging the problem and working in the space* $BV(\Omega)$.

A well-adapted framework for solving (3.49) with assumptions (3.53) is *nonlinear semigroup theory* and the notion of *maximal operator*. We recall only some basic definitions, referring the reader to [73, 89] for the complete theory. Let $(H, \langle . \rangle)$ be a Hilbert space and $A : H \to \mathcal{P}(H)$ an operator where $\mathcal{P}(H)$ is the set of subsets of H. The domain of A is the set $D(A) = \{x \in H \, ; \, Ax \neq \emptyset\}$ and the range of A is $R(A) = \bigcup_{x \in H} Ax$. If for any $x \in H$ the set Ax contains more than one element, we say that A is multivalued. The graph of A is the set $G(A) = \{(x, y) \in D(A) \times H; y \in Ax\}$.

Definition 3.3.1 (monotone operator) $A : H \to \mathcal{P}(H)$ *is said to be* monotone *if*

$$\forall x_1, x_2 \in D(A) : \langle Ax_1 - Ax_2, x_1 - x_2 \rangle_{H \times H} \geq 0,$$

or $\forall y_1 \in Ax_1$, $\forall y_2 \in Ax_2 :$ $\langle y_1 - y_2, x_1 - x_2 \rangle_{H \times H} \geq 0$ *if A is multivalued.*

The set $\mathcal{A}$ of monotone operators can be ordered by graph inclusion. We will say that $A_1 \leq A_2$ if $G(A_1) \subset G(A_2)$, which is equivalent to $A_1(x) \subset A_2(x) \; \forall x \in H$. It can be shown that every totally ordered subset of $\mathcal{A}$ has an upper bound. Then, thanks to the well-known Zorn's lemma [285], $\mathcal{A}$ contains at least one maximal element, which is called a maximal monotone operator. Accordingly, a monotone operator $A : H \to \mathcal{P}(H)$ is maximal monotone if and only if $G(A) \subset G(B)$ implies $A = B$, where $B : H \to \mathcal{P}(H)$ is an arbitrary monotone operator.

In practice, to show that a monotone operator is maximal, it is easier to use the following characterizing property:

Proposition 3.3.2 *Let* $A : H \to \mathcal{P}(H)$, A *monotone then A is maximal* monotone *if and only if one of the following conditions holds:*

(i) *The operator* $(A+I)$ *is surjective, i.e.,* $R(A+I) = H$ *(I is the identity operator).*

(ii) $\forall \lambda > 0$ $(I + \lambda A)^{-1}$ *is a contraction on all of H.*

Example Let $\varphi : H \to] - \infty, +\infty]$ be convex and proper ($\varphi \not\equiv +\infty$). Then for any $x \in H$, the subdifferential of φ at x defined as

$$\partial \varphi(x) = \{y \in H \, ; \, \forall \xi \in H, \; \varphi(\xi) \geq \varphi(x) + \langle y, \, \xi - x \rangle_{H \times H}\}$$

is monotone. It can be proved [73] that if φ is a lower semicontinuous proper convex function, then $\partial \varphi(x)$ is maximal monotone. ∎

The main interest of this notion, in our context, is that it permits us to solve certain nonlinear evolution PDEs:

Proposition 3.3.3 [73] *Let* $A : H \to \mathcal{P}(H)$, A *maximal monotone, and* $u_0 \in D(A)$ *then there exists a unique function* $u(t) : [0, +\infty[\to H$ *such that*

$$\begin{cases} 0 \in \dfrac{du}{dt} + Au(t), \\ u(0) = u_0. \end{cases} \qquad (3.56)$$

So, if we want to solve equations like (3.56) we have only to check, according to Proposition 3.3.2, that $(A + I)$ is surjective. Thus, the study of an evolution equation reduces to the study of a stationary one, which represents a big advantage. Let us apply these results to (3.49). Our aim is to show that the divergence operator in (3.49),

$$Au = -\operatorname{div}\left(c\left(|\nabla u|^2\right)\nabla u\right),$$

is maximal monotone. As suggested before, a classical and convenient way is to identify A with the subdifferential of a convex l.s.c. functional. Let $\Phi(t)$ be the function defined by

$$\Phi(s) = \int_0^s \tau\, c(\tau^2)\, d\tau + 1.$$

We have $\Phi(0) = 1$, $\Phi'(s) = s\, c(s^2)$ and if $c(s)$ satisfies (3.53), then $\Phi(s)$ is strictly convex. Let us set

$$J(u) = \begin{cases} \displaystyle\int_\Omega \Phi(|\nabla u(x)|)\, dx & \text{if } u \in W^{1,1}(\Omega), \\ +\infty & \text{if } u \in L^2(\Omega) - W^{1,1}(\Omega). \end{cases} \qquad (3.57)$$

We can easily verify that $Au = \operatorname{div}\left(\dfrac{\Phi'(|\nabla u|)}{|\nabla u|}\nabla u\right)$ identifies with $\partial J(u)$, the subdifferential of J at u, with

$$\mathrm{Dom}(A) = \Big\{ u \in W^{1,1}(\Omega),\ \operatorname{div}\left(\dfrac{\Phi'(|\nabla u|)}{|\nabla u|}\nabla u\right) \in L^2(\Omega),$$

$$\dfrac{\Phi'(|\nabla u|)}{|\nabla u|}\dfrac{\partial u}{\partial N} = 0 \text{ on } \partial\Omega \Big\}.$$

Unfortunately, $J(u)$ is not lower semicontinuous on $L^2(\Omega)$, and then A is not maximal monotone. To overcome this difficulty, as in the variational case, we introduce the relaxed functional

$$\overline{J}(u) = \begin{cases} \displaystyle\int_\Omega \Phi(|\nabla u(x)|)\, dx + \alpha\, |D_s u| & \text{if } u \in BV(\Omega), \\ +\infty & \text{if } u \in L^2(\Omega) - BV(\Omega), \end{cases} \qquad (3.58)$$

where $\nabla u\, dx + D_s u$ is the Lebesgue decomposition of the measure Du and $\alpha = \lim_{s \to +\infty} \Phi(s)/s$; $\overline{J}(u)$ is convex and l.s.c. on $L^2(\Omega)$. Then, we associate to $\overline{J}(u)$ the evolution problem on $L^2(\Omega)$:

$$\begin{cases} 0 \in \dfrac{du}{dt} + \partial \overline{J}(u) \quad \text{on} \ ]0, +\infty[, \\ u(0, x) = u_0(x). \end{cases} \tag{3.59}$$

We can check that $\partial \overline{J}$ is maximal monotone, and from general results concerning evolution equations governed by a maximal monotone operator, in [331] the following theorem is proved:

Theorem 3.3.1 [331] *Let Ω be an open, bounded, and connected subset of R^2, with Lipschitz boundary $\Gamma = \partial \Omega$. Let $u_0 \in \text{Dom}(\partial \overline{J}) \cap L^\infty(\Omega)$. Then there exists a unique function $u(t) : [0, +\infty[\rightarrow L^2(\Omega)$ such that*

$$u(t) \in \text{Dom}(\partial \overline{J}), \ \forall t > 0, \ \frac{du}{dt} \in L^\infty((0, +\infty); L^2(\Omega)), \tag{3.60}$$

$$-\frac{du}{dt} \in \partial \overline{J}(u(t)), \quad a.e. \ \ t > 0, u(0) = u_0. \tag{3.61}$$

If $\widehat{u}$ is a solution with $\widehat{u_0}$ instead of u_0, then

$$|u(t) - \widehat{u}(t)|_{L^2(\Omega)} \le |u_0 - \widehat{u_0}|_{L^2(\Omega)} \quad \text{for all} \ \ t \ge 0. \tag{3.62}$$

Remarks

- For each $t > 0$, the map $u_0 \rightarrow u(t)$ is a contraction of $\text{Dom}(\partial \overline{J})$ in $\overline{\text{Dom}(\partial \overline{J})}$. We denote by $S(t)$ its unique extension by continuity to $\overline{\text{Dom}(\partial \overline{J})} = \text{Dom}\,\overline{J} = BV(\Omega)$. If u_0 is in $BV(\Omega)$, then $u(t) = S(t)\,u_0$ is called the generalized solution of (3.49).

- We observe that if $u(t, x)$ is a solution of (3.49), then a.e. t

$$\int_\Omega u(t, x)\, dx = \int_\Omega u_0(x)\, dx.$$

To prove this, we differentiate $I(t) = \int_\Omega u(t, x)\, dx$ and apply the Green formula

$$I'(t) = \int_\Omega \frac{du}{dt}(t, x)\, dx = \int_\Omega \text{div}\left(c(|\nabla u|^2\,(t, x))\nabla u(t, x)\right)\, dx$$

$$= \int_\Gamma c(|\nabla u|^2\,(t, x))\frac{\partial u}{\partial N}(t, x)\, ds = 0.$$

Thus $I(t) = I(0) = \int_\Omega u_0(x)\, dx$. ∎

Although this theorem ensures the existence of a solution, it remains difficult to understand. In particular, if we look at result (3.60) or (3.61), we would like to know more about $\partial \overline{J}$. Let us characterize $\partial \overline{J}(u)$ for

$u \in BV(\Omega)$. Let us assume that $\partial \bar{J}(u) \neq \emptyset$ and $\xi \in L^2(\Omega)$ is such that $\xi \in \partial \bar{J}(u)$. By definition we have

$$\bar{J}(u + sw) \geq \bar{J}(u) + s < \xi, w >_{L^2(\Omega) \times L^2(\Omega)} \quad \forall s, \forall w \in L^2(\Omega). \quad (3.63)$$

From (3.63), we can deduce some conditions by choosing successively functions $w \in \mathcal{C}_0^\infty(\Omega)$, $\mathcal{C}_0^\infty(\overline{\Omega})$, and $BV(\Omega)$. So we have the following result.

Proposition 3.3.4 *If $\partial \bar{J}(u) \neq \emptyset$, then:*

(i) *$\partial \bar{J}(u)$ has only one element, given by*

$$\xi = \mathrm{div} \left(\frac{\Phi'(|\nabla u|)}{|\nabla u|} \nabla u \right) \in L^2(\Omega).$$

(ii) $\dfrac{\Phi'(|\nabla u|)}{|\nabla u|} \dfrac{\partial u}{\partial N} = 0$ *on* $\partial \Omega$.

(iii) *For all $w \in BV(\Omega)$ with $Dw = \nabla w dx + D_s w$ and $D_s w = \rho D_s u + \mu'$,*

$$\int_\Omega \frac{\Phi'(|\nabla u|)}{|\nabla u|} \nabla u \cdot \nabla w \, dx + \int_\Omega \rho |D_s u| + \int_\Omega |\mu'| \geq - \int_\Omega \mathrm{div} \left(\frac{\Phi'(|\nabla u|)}{|\nabla u|} \nabla u \right) w \, dx.$$

Proof Starting from the definition (3.63), we look for necessary conditions on ξ. Several choices of w are made:

Step 1: $w \in \mathcal{C}_0^\infty(\Omega)$
Since $u \in BV(\Omega)$ we have the decomposition $Du = \nabla u \, dx + D_s u$. Then $Du + sDw = (\nabla u + s\nabla w) dx + D_s u$, and (3.63) can be rewritten as:

$$\frac{\bar{J}(u + sw) - \bar{J}(u)}{s} = \int_\Omega \frac{\Phi(|\nabla u + s\nabla w|) - \Phi(|\nabla u|)}{s} \, dx$$

$$\geq \langle \xi, w \rangle_{L^2(\Omega) \times L^2(\Omega)} \quad \forall s, \forall w \in \mathcal{C}_0^\infty(\Omega).$$

When $s \to 0^+$, we have, after integrating by parts,

$$- \int_\Omega \mathrm{div} \left(\frac{\Phi'(|\nabla u|)}{|\nabla u|} \nabla u \right) w \, dx \geq \langle \xi, w \rangle_{L^2(\Omega) \times L^2(\Omega)} \quad \forall w \in \mathcal{C}_0^\infty(\Omega).$$

By changing w into $-w$, we obtain

$$\xi = -\mathrm{div} \left(\frac{\Phi'(|\nabla u|)}{|\nabla u|} \nabla u \right) \quad \text{in the distributional sense.} \quad (3.64)$$

Since $\xi \in L^2(\Omega)$, this equality is also true in $L^2(\Omega)$.

Step 2: $w \in C^\infty(\overline{\Omega})$
We also have $Du + sDw = (\nabla u + s\nabla w) dx + D_s u$, and by (3.63)

$$\int_\Omega \frac{\Phi'(|\nabla u|)}{|\nabla u|} \nabla u \cdot \nabla w \, dx \geq \langle \xi, w \rangle_{L^2(\Omega) \times L^2(\Omega)} \quad \forall w \in C^\infty(\overline{\Omega}).$$

After integrating by parts we have

$$-\int_{\Omega} \text{div}\left(\frac{\Phi'(|\nabla u|)}{|\nabla u|}\nabla u\right)w\,dx + \int_{\partial\Omega}\frac{\Phi'(|\nabla u|)}{|\nabla u|}\nabla u \cdot N\,w\,ds$$

$$\geq \langle \xi, w\rangle_{L^2(\Omega)\times L^2(\Omega)} \quad \forall w \in C^\infty(\overline{\Omega}).$$

Notice that the term on $\partial\Omega$ is well-defined, since $\sigma = \frac{\Phi'(|\nabla u|)}{|\nabla u|}\nabla u \in L^2(\Omega)^2$ and $\text{div}(\sigma) \in L^2(\Omega)$, thanks to (3.64) (see [216]). From (3.64), we easily deduce from the previous inequality the condition (ii).

Step 3: $w \in BV(\Omega)$
Let $D_s w = \rho D_s u + \mu'$ be the Lebesgue decomposition of the measure $D_s w$ with respect to $D_s u$. So, we have

$$Du + sDw = (\nabla u + s\nabla w)dx + D_s u(1 + s\rho) + s\mu'.$$

Thus

$$\frac{\overline{J}(u + sw) - \overline{J}(u)}{s}$$

$$= \int_{\Omega}\frac{\Phi(|\nabla u + s\nabla w|) - \Phi(|\nabla u|)}{s}\,dx + \int_{\Omega}\frac{|1 + s\rho| - 1}{s}|D_s u| + \int_{\Omega}|\mu'|,$$

and by letting $s \to 0$, with (3.63), we obtain the inequality (iii). ∎

Remarks

- It is useless to search for further characterization by choosing $w \in L^2(\Omega)$ (and not in $BV(\Omega)$), since in this case $\overline{J}(u + sw) = +\infty$.

- In fact, since Φ is convex, it is easy to show that if $u \in BV(\Omega)$ satisfies (i), (ii), and (iii), then $\partial\overline{J}(u) \neq \emptyset$. Therefore, (i), (ii), and (iii) are necessary and sufficient conditions for ensuring that $\partial\overline{J}(u) \neq \emptyset$.

- The existence and uniqueness theorem from Brezis tells us that $-\frac{du}{dt} \in \partial\overline{J}(u(t))$. We just showed that this means $\frac{du}{dt} = \frac{\Phi'(|\nabla u|)}{|\nabla u|}\nabla u$ in $L^2(\Omega)$.

∎

Let us now interpret the last inequality (iii) in a particular case which is representative in image analysis. Before that, we need the following lemma:

Lemma 3.3.1 *Let Ω be regular, $\sigma \in C(\Omega)^M$ bounded, $\text{div}\,\sigma \in L^2(\Omega)$ and $w \in BV(\Omega)\cap L^2(\Omega)$. Then we have the following Stokes formula:*

$$\int_{\Omega}\text{div}\,\sigma\,w\,dx = \int_{\partial\Omega}\sigma \cdot N\,w\,ds - \int_{\Omega}\sigma \cdot Dw,$$

where $\displaystyle\int_\Omega \sigma \cdot Dw = \int_\Omega \sigma \cdot \nabla w \ dx + \int_\Omega \sigma \cdot D_s w.$

Proof By regularization, we can find a sequence $w_n \in \mathcal{C}^\infty(\Omega)$ such that

$$w_n \xrightarrow[L^2(\Omega)]{} w,$$

$$|Dw_n| \longrightarrow |Dw|.$$

Since w_n is regular, we have

$$\int_\Omega \operatorname{div} \sigma \ w_n \ dx = \int_{\partial\Omega} \sigma \cdot N \ w_n \ ds - \int_\Omega \sigma \cdot \nabla w_n \ dx.$$

Moreover the convergence for the strong topology of $BV(\Omega)$ induces the BV–w* convergence and the convergence of the trace operator. The result is obtained as n tends to infinity.   ∎

As announced previously, let us explain the equality (iii) from Proposition 3.3.4, where we suppose that u admits a discontinuity along a single curve Σ and $\Omega = \Omega_0 \cup \Sigma \cup \Omega_1$. So $u \in C^1(\Omega_0 \cup \Omega_1)$, and we have

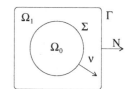

$$D_s u = [u]\nu \ d\mathcal{H}^1|_\Sigma = (u_1 - u_0)\nu \ d\mathcal{H}^1|_\Sigma.$$

We can still decompose $D_s w$ in the following way: $D_s w = \rho[u] \ d\mathcal{H}^1|_\Sigma + \mu'$. Now, the idea is to look at the term

$$A = \int_\Omega \operatorname{div}\left(\frac{\Phi'(|\nabla u|)}{|\nabla u|}\nabla u\right) w \ dx \equiv \int_\Omega \sigma w \ dx \quad \text{with} \quad \sigma = \frac{\Phi'(|\nabla u|)}{|\nabla u|}\nabla u.$$

By decomposing A on Ω_0 and Ω_1 and applying Lemma 3.3.1, we have

$$A = -\int_{\Omega_0} \sigma \cdot \nabla w \ dx - \int_{\Omega_0} \rho\sigma \cdot D_s u - \int_{\Omega_0} \sigma\mu' + \int_\Sigma \sigma \cdot \nu w \ ds$$

$$- \int_{\Omega_1} \sigma \cdot \nabla w \ dx - \int_{\Omega_1} \rho\sigma \cdot D_s u - \int_{\Omega_1} \sigma\mu' - \int_\Sigma \sigma \cdot \nu w \ ds + \underbrace{\int_{\partial\Omega} \sigma \cdot N \ w \ ds}_{=0}$$

$$= -\int_\Omega \sigma \cdot \nabla w \ dx - \int_\Omega \rho\sigma \cdot D_s u - \int_\Omega \sigma\mu'.$$

By replacing this expression in (iii) with $D_s u = [u]\nu \ d\mathcal{H}^1|_\Sigma$, we obtain after some computation

$$\int_\Sigma \rho\left\{|[u]| - \frac{\Phi'(|\nabla u|)}{|\nabla u|}\nabla u \cdot \nu[u]\right\} ds + \int_\Omega \left\{|\mu'| - \frac{\Phi'(|\nabla u|)}{|\nabla u|}\nabla u\mu'\right\} dx \geq 0.$$

The second integral is always positive (if $|\Phi'| \leq 1$), and in the first integral ρ is arbitrary. So we have

$$|[u]| - \frac{\Phi'(|\nabla u|)}{|\nabla u|}\nabla u \cdot \nu[u] = 0 \quad \text{a.e. on } \Sigma,$$

which we can rewrite (if $[u] \neq 0$) as

$$\text{sign}[u] = \frac{\Phi'(|\nabla u|)}{|\nabla u|}\nabla u \cdot \nu \quad \text{a.e. on } \Sigma. \tag{3.65}$$

This condition is an interpretation of (iii). Naturally, this could be generalized to the case where u admits discontinuities on a finite or countable set of curves. It would be interesting to see whether this condition could be used numerically.

To conclude this section, we would like to mention an interesting book by Andreu et al. [18] which is an extensive study of several aspects related to the total variation.

THE ALVAREZ–GUICHARD–LIONS–MOREL SCALE SPACE THEORY

In this section we examine the remarkable work of Alvarez et al. [5]. In this paper the connection between scale space analysis and PDEs is rigorously established. Starting from a set of very natural filtering axioms (based on desired image properties) they prove that the resulting filtered image must necessarily be the viscosity solution of a PDE. In addition, they completely describe these PDEs. Most of their results have been collected in a very recent monograph by Guichard and Morel [176]. It is not our intention to set out all the details and refinements of their axioms. Our purpose is to review some aspects of this very nice theory. We follow the presentation given in the original Alvarez et al. paper [5, 6]. Note that Brockett and Maragos [75] have independently developed a theory generating multiscale morphological erosion-dilatation of a more general type.

As said previously, we define a multiscale analysis as a family of operators $\{T_t\}_{t \geq 0}$, which applied to the original image $u_0(x)$ yield a sequence of images $u(t, x) = (T_t u_0)(x)$. We are going to list below a series of axioms to be satisfied by $\{T_t\}_{t \geq 0}$. These formal properties are very natural from an image analysis point of view. For simplicity, we suppose for all $t \geq 0$, $T_t : C_b^\infty(R^2) \to C_b(R^2)$, where $C_b^\infty(R^2)$ (respectively $C_b(R^2)$) is the space of bounded functions having derivatives of any order (respectively bounded continuous functions). This is not restrictive, since as is usual in mathematical analysis, once properties are proved for regular functions we can extend them to nonregular functions by density arguments.

List of axioms and invariance properties (X denotes the space $C_b^\infty(R^2)$)

(A1) *Recursivity:*
   $T_0(u) = u$, $T_s \circ T_t(u) = T_{s+t}(u)$ for all $s, t \geq 0$ and all $u \in X$.

(A2) *Regularity:*
$|T_t(u + h\,v) - (T_t(u) + h\,v)|_{L^\infty} \le c\,h\,t$ for all h and t in $[0,1]$ and all $u,\,v \in X$.

(A3) *Locality:*
$(T_t(u) - T_t(v))(x) = o(t),\ t \to 0^+$ for all u and $v \in X$ such that $\nabla^\alpha u(x) = \nabla^\alpha v(x)$ for all $|\alpha| \ge 0$ and all x ($\nabla^\alpha u$ stands for the derivative of order α).

(A4) *Comparison principle:*
$T_t(u) \le T_t(v)$ on R^2, for all $t \ge 0$ and $u, v \in X$ such that $u \le v$ on R^2.

(I1) *Gray-level shift invariance:*
$T_t(0) = 0,\ T_t(u + c) = T_t(u) + c$ for all u in X and all constant c.

(I2) *Translation invariance:*
$T_t(\tau_h.u) = \tau_h.(T_t u)$ for all h in R^2, $t \ge 0$, where $(\tau_h.u)(x) = u(x+h)$.

We emphasize that these axioms and invariance properties are quite natural from an image analysis point of view. A1 means that a coarser analysis of the original image can be deduced from a finer one without any dependence upon the original picture. A2 states a continuity assumption of T_t. A3 means that $(T_t u)(x)$ is determined by the behavior of u near x. A4 expresses the idea that if an image v is brighter than another image u, this ordering is preserved across scale. Finally, I1 and I2 state respectively that no a priori assumption is made on the range of the brightness and that all points are equivalent.

We are now in position to give the main result. We denote by S^2 the space of all 2×2 symmetric matrices endowed with its natural ordering.

Theorem 3.3.2 [6] *Under assumptions A1, A2, A3, A4, I1, and I2:*

(i) *There exists a continuous function $F : R^2 \times S^2 \to R$ satisfying $F(p, A) \ge F(p, B)$ for all $p \in R^2$, A and B in S^2 with $A \ge B$ such that*

$$\delta_t(u) = \frac{T_t(u) - u}{t} \to F(\nabla u, \nabla^2 u),\ t \to 0^+$$

uniformly for $x \in R^2$, uniformly for $u \in X$.

(ii) *If $u_0 \in C_b(R^2)$, then $u(t, x) = (T_t u_0)(x)$ is the unique viscosity solution of*

$$\begin{cases} \dfrac{\partial u}{\partial t} = F(\nabla u, \nabla^2 u), \\ u(0, x) = u_0(x), \end{cases}$$

and $u(t, x)$ is bounded, uniformly continuous on R^2.

Proof We only give only the main steps, sometimes omitting technical details and referring for the original proofs to [5, 6].

Step **1**: Existence of an infinitesimal generator.
There exists an operator $S : X \to C_b(R^2)$ such that

$$\delta_t(u) = \frac{T_t(u) - u}{t} \to S[u], \ t \to 0^+$$

uniformly on R^2 and for all $u \in X$. See [5, 6] for this very technical proof.

Step **2**: A general lemma
Let X, Y, Z be three sets, $A : X \to Y$ and $G : X \to Z$. Let us suppose that the equality $G(x) = G(x')$ implies $A(x) = A(x')$. Then there exists $F : G(X) \subset Z \to Y$ such that for all x in X, $A(x) = F(G(x))$; i.e., A is a function only of G.

Step **3**: Let us show that if u and v satisfy $u(0) = v(0)$, $\nabla u(0) = \nabla v(0) = p \in R^2$, and $\nabla^2 u(0) = \nabla^2 v(0) = A \in S^2$, then $S[u](0) = S[v](0)$.
Let $z(x)$ be a function in $C_b^\infty(R^2)$ such that $z(x) \geq 0$ and $z(x) = |x|^2$ for x near 0. Set $u^\varepsilon(x) = u(x) + \varepsilon z(x)$. We claim that $u^\varepsilon(x) \geq v(x)$ for $|x|$ small enough. Indeed, thanks to Taylor's formula,

$$u^\varepsilon(x) = u(0) + x \cdot \nabla u(0) + \frac{1}{2} \nabla^2 u(0)\, x \cdot x + o(|x|^2) + \varepsilon\, z(x)$$

$$= v(x) + o(|x|^2) + \varepsilon\, z(x).$$

But since u, $v \in C_b^\infty(R^2)$ and $z(x) = |x|^2$ in a neighborhood of zero, it is clear that there exists a constant $c > 0$ such that $o(|x|^2) + \varepsilon\, z(x) \geq 0$ for $|x| \leq c\varepsilon$. Therefore, $u^\varepsilon(x) \geq v(x)$ for $|x| \leq c\varepsilon$. Then let us set $w^\varepsilon(x) = w(x/\varepsilon)$, where $w \in C_b^\infty(R^2)$, $0 \leq w \leq 1$, $w(x) = 1$ if $|x| \leq c/2$, and $w(x) = 0$ if $|x| \geq c$. Finally, let us define $\overline{u}^\varepsilon(x) = w^\varepsilon(x)\, u^\varepsilon(x) + (1 - w^\varepsilon(x))\, v(x)$. The function $\overline{u}^\varepsilon(x)$ satisfies:

$$\nabla^\alpha \overline{u}^\varepsilon(0) = \nabla^\alpha u^\varepsilon(0), \ \forall \alpha,$$

$$\overline{u}^\varepsilon(x) \geq v(x) \ \text{ on } \ R^2.$$

Thus, from A4 we get $T_t(\overline{u}^\varepsilon) \geq T_t(v)$ on R^2, and (since $\overline{u}^\varepsilon(0) = u^\varepsilon(0) = u(0) = v(0)$),

$$\frac{T_t(\overline{u}^\varepsilon)(0) - \overline{u}^\varepsilon(0)}{t} \geq \frac{T_t(v)(0) - v(0)}{t}.$$

If $t \to 0^+$, we get $S[\overline{u}^\varepsilon](0) \geq S[v](0)$. But from A3, we have $(T_t \overline{u}^\varepsilon)(0) - (T_t u^\varepsilon)(0) = o(t)$, which implies $S[u^\varepsilon](0) \geq S[v](0)$. Now, letting $\varepsilon \to 0$, and using the uniform convergence in Step 1, we deduce $S[u](0) \geq S[v](0)$ and by symmetry (changing $z(x)$ into $-z(x)$) we finally obtain

$$S[u](0) = S[v](0).$$

Step 4: Let us prove the same property as in Step 3 for any $x_0 \in R^2$. Let u and v satisfy: $u(x_0) = v(x_0)$, $\nabla u(x_0) = \nabla v(x_0) = p \in R^2$, and $\nabla^2 u(x_0) = \nabla^2 v(x_0) = A \in S^2$. We have to prove $S[u](x_0) = S[v](x_0)$. Without loss of generality, we can suppose that $u(x_0) = v(x_0) = 0$. Let us define

$$u_{x_0}(x) = (\tau_{x_0}.u)(x) = u(x_0 + x),$$
$$v_{x_0}(x) = (\tau_{x_0}.v)(x) = v(x_0 + x).$$

We have $u_{x_0}(0) = v_{x_0}(0) = 0$, $\nabla u_{x_0}(0) = \nabla v_{x_0}(0)$, $\nabla^2 u_{x_0}(0) = \nabla^2 v_{x_0}(0)$. Step 3 implies $S[u_{x_0}](0) = S[v_{x_0}](0)$. But, from I2,

$$S[u_{x_0}](0) = S[\tau_{x_0}.u](0) = \lim_{t \to 0^+} \frac{T_t(\tau_{x_0}.u)(0) - (\tau_{x_0}.u)(0)}{t}$$

$$\underset{\text{from I2}}{=} \lim_{t \to 0^+} \frac{\tau_{x_0}.(T_t u - u)(0)}{t} = (\tau_{x_0}.S[u])(0) = S[u](x_0),$$

and the same holds for $S[v](x_0)$. Therefore,

$$S[u](x_0) = S[v](x_0).$$

To summarize, we have for all $x \in R^2$ the following property: If $(u(x), \nabla u(x), \nabla^2 u(x)) = (v(x), \nabla v(x), \nabla^2 v(x))$, Then $S[u](x) = S[v](x)$. So we apply the general lemma of Step 2 with

$$X = \left\{ u(x);\ u \in C_b^\infty(R^2) \right\},$$
$$Y = \left\{ S[u](x);\ u \in C_b^\infty(R^2) \right\},$$
$$Z = \left\{ (u(x), \nabla u(x), \nabla^2 u(x));\ u \in C_b^\infty(R^2) \right\},$$

and $G : u(x) \to (u(x), \nabla u(x), \nabla^2 u(x))$ $(G : X \to Z)$. We have

$$G(u(x)) = G(v(x)) \text{ implies } S[u](x) = S[v](x).$$

Therefore, there exists a function $F : G(X) \subset Z \to Y$ such that

$$S[u](x) = F(x, u(x), \nabla u(x), \nabla^2 u(x)).$$

Step 5: F does not depend on x and u.
This a direct consequence of I1 and I2. From I2, we have for all h

$$\tau_h.S[u](x) = S[\tau_h.u](x),$$

i.e., for all h,

$$F(x + h, u(x + h), \nabla u(x + h), \nabla^2 u(x + h))$$
$$= F(x, u(x + h), \nabla u(x + h), \nabla^2 u(x + h)).$$

Thus F does not depend on x. Now, from I1, if c is any constant, then

$$S[u + c] = \lim_{t \to 0^+} \frac{T_t(u + c) - (u + c)}{t} = \lim_{t \to 0^+} \frac{T_t(u) - u}{t} = S[u],$$

i.e., for c,

$$F(u(x) + c, \nabla u(x), \nabla^2(x)) = F(u(x), \nabla u(x), \nabla^2(x)).$$

Thus F does not depend on u.

Step **6**: F is continuous and nondecreasing with respect to its second argument.

The continuity of F follows from Step 1. Let us show that for all $p \in R^2$, $A, B \in S^2$ such that $A \geq B$, then $F(p, A) \geq F(p, B)$. Let us define

$$u(x) = \left(p \cdot x + \frac{1}{2} A x \cdot x \right) w(x) \quad \text{and} \quad v(x) = \left(p \cdot x + \frac{1}{2} B x \cdot x \right) w(x),$$

where $w(x)$ is the function defined in Step 3. We have

$$u(0) = v(0) = 0, \ \nabla u(0) = \nabla v(0) = p, \ \nabla^2 u(0) = A, \ \nabla^2 v(0) = B.$$

Moreover,

$$A - B \geq 0 \ \Rightarrow \ u(x) \geq v(x) \ \Rightarrow \ (T_t u)(x) \geq (T_t v)(x) \quad \text{(from A4)},$$

and the last inequality at $x = 0$ implies

$$S[u](0) = F(p, A) \geq S[v](0) = F(p, B).$$

Step **7**: $u(t, x) = (T_t u_0)(x)$ is the unique viscosity solution of

$$\begin{cases} \dfrac{\partial u}{\partial t} = F(\nabla u, \nabla^2 u), \\ u(0, x) = u_0(x). \end{cases} \tag{3.66}$$

Let us first prove that $u(t, x)$ is a subsolution of (3.66). Let $\phi(t, x)$ be a test function and (t_0, x_0) a global maximum of $(u - \phi)(t, x)$. Without loss of generality, we may suppose

$$u(t_0, x_0) - \phi(t_0, x_0) = \max_{(t,x)} (u - \phi)(t, x) = 0$$

and that $\phi(t, x)$ is of the form $\phi(t, x) = f(x) + g(t)$ with $g(t_0) = 0$. These two simplifications imply

$$u(t, x) \leq \phi(t, x) \quad \text{for all} \quad (t, x),$$
$$u(t_0, x_0) = \phi(t_0, x_0) = f(x_0).$$

Now let h be in $]0, t_0[$. From the recursivity axiom A1, we get

$$T_h(u(t_0 - h, x_0)) = (T_h \circ T_{t_0 - h})(u_0(x_0)) = T_{t_0}(u_0(x_0)) = u(t_0, x_0) = f(x_0),$$

but since $u \leq \phi$,

$$f(x_0) = u(t_0, x_0) = T_h(u(t_0 - h, x_0))$$
$$\leq T_h(\phi(t_0 - h, x_0)) = T_h(f(x_0) + g(t_0 - h)) = T_h(f(x_0)) + g(t_0 - h).$$

Thus, since $g(t_0) = 0$,

$$\frac{1}{h}(g(t_0) - g(t_0 - h)) + \frac{1}{h}(f(x_0) - T_h(f(x_0))) \leq 0,$$

and if $h \to 0^+$,

$$g'(t_0) - F(\nabla f(x_0), \nabla^2 f(x_0)) \leq 0; \text{ i.e., } \frac{\partial \phi}{\partial t}(t_0, x_0) - F(\nabla f(x_0), \nabla^2 f(x_0)) \leq 0,$$

which means that $u(t, x)$ is a subsolution of (3.66). It can be proved using similar arguments that $u(t, x)$ is a supersolution. So $u(t, x)$ is a viscosity solution of (3.66).

This last step concludes the proof of Theorem 3.3.2.   ∎

If the multiscale analysis satisfies additional invariance properties, then the function F can be written in an explicit form. We state below two important cases.

Theorem 3.3.3 [5, 6] *Let us suppose that T_t satisfy the assumptions of Theorem 3.3.2 and*

(I3) *Isometry invariance:*
   $T_t(R.u)(x) = R.(T_t u)(x)$ *for all orthogonal transformation R on R^2, where $(R.u)(x) = u(Rx)$.*

We also assume that $u \to T_t u$ is linear. Then $u(t, x) = (T_t u_0)(x)$ is the solution of the heat equation $\frac{\partial u}{\partial t} = c \Delta u$, $u(0, x) = u_0(x)$, where c is a positive constant.

Theorem 3.3.4 [5, 6] *Let us suppose that T_t satisfy the assumptions of Theorem 3.3.2 and*

(I4) *Gray-scale invariance:*
   $T_t(\varphi(u)) = \varphi(T_t(u))$ *for all nondecreasing real functions φ.*

(I5) *Scale invariance:*
   $\forall \lambda, t > 0$, *there exists $t'(t, \lambda) > 0$ such that $H_\lambda.(T_{t'} u) = T_t(H_\lambda.u)$, where $(H_\lambda.u)(x) = u(\lambda x)$. Moreover, we suppose that $t'(t, \lambda)$ is differentiable with respect to λ at $\lambda = 1$ and that the function $g(t) = \frac{\partial t'}{\partial \lambda}(t, 1)$ is continuous and positive for $t > 0$.*

(I6) *Projection invariance:*
   For all $A : R^2 \to R^2$ linear, for all $t > 0$, there exists $t'(t, A) > 0$ such that $A.(T_{t'} u) = T_t(A.u)$.

Then $u(t, x) = (T_t u_0)(x)$ is the solution of

$$\begin{cases} \dfrac{\partial u}{\partial t} = |\nabla u| \, (t \operatorname{curv} u)^{1/3}, \\ u(0, x) = u_0(x), \end{cases}$$

where

$$\operatorname{curv} u = \frac{u_{xx} u_y^2 + u_{yy} u_x^2 - 2 u_{xy} u_x u_y}{|\nabla u|^3}.$$

☛ *These two last theorems are very interesting, since they express that the Alvarez et al. theory is a very natural extension of the linear theory (Theorem 3.3.3) and also because the multi-scale framework leads to new nonlinear filters (Theorem 3.3.4).*

Remarks

- If in Theorem 3.3.4 we suppose that T_t satisfy I4, I5, and I3 instead of I6, we get only the PDE

$$\frac{\partial u}{\partial t} = |\nabla u| \beta \ (t \ \text{curv} \ u),$$

where β is a continuous nondecreasing function.

- The previous scale-space theory can be extended to the analysis of movies [6, 174, 175, 241, 242, 240].

- There are strong connections between PDEs described in this section and morphological operators (i.e., monotone, translation, and contrast invariant operators). In fact, let $\mathcal{F}$ be a set of functions containing continuous functions and characteristic functions of level sets of elements of $\mathcal{F}$. Then it can be proven [225] that any morphological operator on $\mathcal{F}$ is of the form

$$(Tu)(x) = \inf_{B \in \mathcal{B}} \sup_{y \in B} u(x + y),$$

where $\mathcal{B}$ is a family of structuring elements. As a very interesting result, one can prove that if we adequately scale morphological operators, then by iterating the resulting operators we retrieve all the equations given in this section. The interested reader can consult [226, 83, 176, 232] and the references therein.   ■

WEICKERT'S APPROACH

In order to take into account local variations of the gradient orientation, we need to define a more general descriptor than the magnitude of the gradient only. Let us start with simple remarks.

As seen in previous sections, it is a natural idea to say that the preferred smoothing direction is the one that minimizes gray-value fluctuations. Let $d(\theta)$ be the vector $(\cos \theta, \sin \theta)$. An elementary calculation shows that the function $F(\theta) = (d(\theta) \cdot \nabla u(x))^2$ is maximal if d is parallel to ∇u, and is minimal if d is orthogonal to ∇u. We can also remark that maximizing (respectively minimizing) $F(\theta)$ is equivalent to maximizing (respectively minimizing) the quadratic form $d^t \nabla u \nabla u^t d$. The matrix

$$\nabla u \nabla u^t = \begin{pmatrix} u_{x_1}^2 & u_{x_1} u_{x_2} \\ u_{x_1} u_{x_2} & u_{x_2}^2 \end{pmatrix} \tag{3.67}$$

is positive semidefinite, its eigenvalues are $\lambda_1 = |\nabla u|^2$, and $\lambda_2 = 0$ and there exists an orthonormal basis of eigenvectors v_1 parallel to ∇u and v_2 orthogonal to ∇u.

So, it would be tempting to define at x an orientation descriptor as a function of $\nabla u \nabla u^t(x)$. But by proceeding like this, we do not take into account possible information contained in a neighborhood of x. To this end, the idea proposed by Weickert is to introduce smoothing kernels at different scales. We only sketch the main ideas, since Weickert himself has written a monograph [336] based on his work.

To avoid false detections due to noise, $u(x)$ is first convolved with a Gaussian kernel k_σ : $u_\sigma(x) = (k_\sigma * u)(x)$. The local information is averaged by convolving componentwise $\nabla u_\sigma \nabla u_\sigma^t$ with a Gaussian kernel k_ρ. The result is a symmetric, positive semidefinite matrix

$$J_\rho(\nabla u_\sigma) = k_\rho * \nabla u_\sigma \nabla u_\sigma^t. \tag{3.68}$$

The matrix $J_\rho(\nabla u_\sigma)$ has orthonormal eigenvectors v_1, v_2 with v_1 parallel to

$$\begin{pmatrix} 2\,j_{12} \\ j_{22} - j_{11} + \sqrt{(j_{22} - j_{11})^2 + 4\,j_{12}^2} \end{pmatrix},$$

where j_{lk} are the elements of the matrix $J_\rho(\nabla u_\sigma)$. The corresponding eigenvalues are given by

$$\mu_1 = \frac{1}{2}\left[j_{11} + j_{22} + \sqrt{(j_{11} - j_{22})^2 + 4j_{12}^2} \right]$$

and

$$\mu_2 = \frac{1}{2}\left[j_{11} + j_{22} - \sqrt{(j_{11} - j_{22})^2 + 4j_{12}^2} \right].$$

They describe average contrast in the eigendirections within a neighborhood of size $O(\rho)$. The noise parameter σ makes the descriptor insensible to details of scale smaller than $O(\sigma)$. The vector v_1 indicates the orientation maximizing the gray-value fluctuations, while v_2 gives the preferred local direction of smoothing. The eigenvalues μ_1 and μ_2 convey shape information. Isotropic structures are characterized by $\mu_1 \cong \mu_2$, linelike structure by $\mu_1 \gg \mu_2 \approx 0$, corners by $\mu_1 \geq \mu_2 \gg 0$.

Now, the nonlinear diffusion process is governed by a parabolic equation that can be viewed as an extension of (3.49):

$$\begin{cases} \dfrac{\partial u}{\partial t} = \mathrm{div}(D(J_\rho(\nabla u_\rho))\nabla u) & \text{in }]0,T] \times \Omega, \\ u(0,x) = u_0(x) & \text{on } \Omega, \\ \langle D(J_\rho(\nabla u_\rho))\nabla u, N \rangle = 0 & \text{on }]0,T] \times \partial\Omega, \end{cases} \tag{3.69}$$

where D is an operator to be defined next and N is the unit outward normal to $\partial\Omega$. Notice the boundary condition, which is the natural condition associated with the divergence operator.[2] We have the following result:

Theorem 3.3.5 [336] *Let us assume that:*

(i) *The diffusion tensor $D = (d_{ij})$ belongs to $C^\infty(S^2, S^2)$, where S^2 denotes the set of symmetric matrices.*

(ii) *Uniform positive definiteness: for all $w \in L^2(\Omega, R^2)$ with $|w(x)| \le k$ on $\overline{\Omega}$, there exists a positive lower bound $\nu(k)$ for the eigenvalues of $D(J_\rho(w))$.*

Then for all $u_0 \in L^\infty(\Omega)$ equation (3.69) has a unique solution $u(t, x)$ satisfying

$$u \in C([0,T]; \, L^2(\Omega)) \cap L^2([0,T]; W^{1,2}(\Omega)),$$

$$\frac{\partial u}{\partial t} \in L^2((0,T); W^{1,2}(\Omega)).$$

Moreover, $u \in C^\infty(]0, T[\times\overline{\Omega})$. This solution depends continuously on u_0 with respect to the L^2-norm, and it satisfies the extremum principle:

$$\inf_\Omega u_0(x) \le u(t,x) \le \sup_\Omega u_0(x).$$

Related results have been proved for semidiscrete and fully discrete versions of the model. For the proofs as well as further properties (invariances, image simplification properties, behavior as t tends to infinity), we refer to Weickert [336].

Let us now describe two possibilities of how to choose the diffusion tensor $D(J_\rho)$. Since the eigenvectors of D should reflect the local image structure, one should choose the same orthonormal basis of eigenvectors as one gets from J_ρ. The choice of the corresponding eigenvalues λ_1 and λ_2 of D depends on the desired goal:

- **Edge-enhancing anisotropic diffusion** [335]. If one wants to smooth preferably within each region and aims to preserve edges, then one should reduce the diffusivity λ_1 perpendicular to edges all the more if the contrast μ_1 is large. This behavior may be accomplished by the following choice:

$$\lambda_1 = \begin{cases} 1 & \text{if } \mu_1 = 0, \\ 1 - \exp\left(\frac{-3.315}{\mu_1^4}\right) & \text{otherwise}, \end{cases} \quad (3.70)$$
$$\lambda_2 = 1.$$

  Figure 3.13 illustrates such a process.

- **Coherence-enhancing anisotropic diffusion** [337]. If one wants   C++

---

[2]This can be compared with the boundary condition that was associated with the divergence operator in (3.39)

original image   result

Figure 3.13. Example of Weickert's edge-enhancing (3.70) approach applied to the noisy "Borel building" image. It combines isotropic smoothing within flat regions with diffusion along edges. Diffusion across edges is reduced.

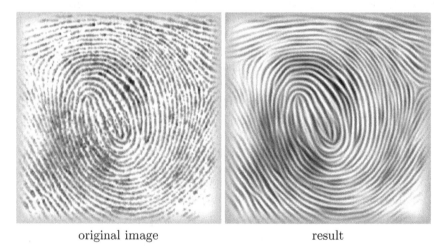

original image   result

Figure 3.14. Example of Weickert's coherence-enhancing approach (3.71), from [337]. Interrupted lines are closed, and the semantically important singularities are not destroyed. A typical application is for fingerprint enhancement, where structure is especially important. The right-hand side image presents the result.

to enhance flowlike structures and close interrupted lines, one should smooth preferably along the coherence direction v_2 with a diffusivity λ_2 that increases with respect to the coherence $(\mu_1 - \mu_2)^2$. This may be achieved by the following choice of the eigenvalues of $D(J_\rho)$:

$$
\begin{aligned}
\lambda_1 &= \alpha, \\
\lambda_2 &= \begin{cases} \alpha & \text{if } \mu_1 = \mu_2, \\ \alpha + (1-\alpha)\exp\left(\frac{-1}{(\mu_1-\mu_2)^2}\right) & \text{otherwise,} \end{cases}
\end{aligned}
\tag{3.71}
$$

where the small positive parameter $\alpha \in (0,1)$ keeps the diffusion tensor uniformly positive definite. Figure 3.14 shows the restoration properties of this diffusion filter as applied to a degraded fingerprint image.

SURFACE BASED APPROACHES

El-Fallah and Ford [145], Sochen, Kimmel, and Malladi [306, 305] and Yezzi [345] introduced the concept of images as embedded maps and minimal surfaces, and applied it to processing movies, color images, texture, and volumetric medical images (see [195]). According to their geometric framework for image processing, intensity images are considered as surfaces in the spatial-feature space. The image is thereby a two-dimensional surface in three-dimensional space (see Figure 3.15).

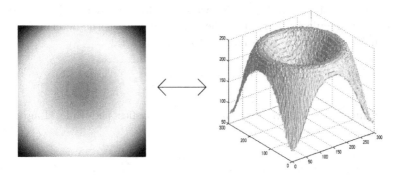

Figure 3.15. Interpretation of an image as a surface. The height is equal to the gray-level value.

Let us briefly explain the main ideas for gray-level images. As mentioned, an image is not considered as a function $u(x)$ from a domain Ω into R, but as an embedded surface $\mathcal{M}$ in R^3 defined by

$$
\begin{aligned}
(\sigma_1, \sigma_2) &\to X(\sigma_1, \sigma_2) = (X_1(\sigma_1, \sigma_2), X_2(\sigma_1, \sigma_2), X_3(\sigma_1, \sigma_2)), \\
\Sigma &\to \mathcal{M},
\end{aligned}
$$

where (σ_1, σ_2) denote the local coordinates of the surface.[3] The main point is that Σ and $\mathcal{M}$ are viewed as Riemannian manifolds equipped with suitable (Riemannian) metrics. To better understand, let us consider the following example corresponding to a particular choice of Σ and X:

$$\Sigma = \Omega, \text{ the image domain,}$$

$$\sigma_1 = x_1, \quad \sigma_2 = x_2 \text{ the classical Cartesian coordinates,}$$

$$X(x) = (x, u(x)), \text{ where } u(x) \text{ is the gray-level intensity.}$$

The metric on Σ is the usual one, $dx_1^2 + dx_2^2$, and the induced metric on $\mathcal{M}$ is $ds^2 = dx_1^2 + dx_2^2 + du^2$, that is, from elementary calculus,

$$ds^2 = dx_1^2 + dx_2^2 + (u_{x_1} dx_1 + u_{x_2} dx_2)^2$$
$$= (1 + u_{x_1}^2) dx_1^2 + 2 u_{x_1} u_{x_2} dx_1 dx_2 + (1 + u_{x_2}^2) dx_2^2.$$

This can be rewritten as

$$ds^2 = (dx_1, dx_2) \begin{pmatrix} 1 + u_{x_1}^2 & u_{x_1} u_{x_2} \\ u_{x_1} u_{x_2} & 1 + u_{x_2}^2 \end{pmatrix} \begin{pmatrix} dx_1 \\ dx_2 \end{pmatrix},$$

i.e., the metric is given by the symmetric positive definite matrix

$$G = \begin{pmatrix} 1 + u_{x_1}^2 & u_{x_1} u_{x_2} \\ u_{x_1} u_{x_2} & 1 + u_{x_2}^2 \end{pmatrix}.$$

How can these concepts be useful from an image analysis point of view? We know that most images are noisy or deteriorated. To obtain a restored approximation of a degraded image, we search for $\mathcal{M}$ having minimal area. In this way, singularities are smoothed. So, if g denotes the determinant of G, we have to minimize with respect to u the integral

$$S(\mathcal{M}) = \iint_\Omega \sqrt{g} \, dx_1 \, dx_2 = \iint_\Omega \sqrt{1 + u_{x_1}^2 + u_{x_2}^2} \, dx_1 \, dx_2.$$

If a minimizer $u(x)$ exists, it necessarily satisfies the Euler–Lagrange equation

$$\frac{\partial}{\partial x_1} \left(\frac{u_{x_1}}{\sqrt{1 + u_{x_1}^2 + u_{x_2}^2}} \right) + \frac{\partial}{\partial x_2} \left(\frac{u_{x_2}}{\sqrt{1 + u_{x_1}^2 + u_{x_2}^2}} \right) = 0,$$

i.e.,

$$\frac{u_{x_1 x_1}(1 + u_{x_2}^2) + u_{x_2 x_2}(1 + u_{x_1}^2) - 2 u_{x_1} u_{x_2} u_{x_1 x_2}}{(1 + u_{x_1}^2 + u_{x_2}^2)^{3/2}} = 0, \qquad (3.72)$$

which is equivalent to saying that the mean curvature H of $\mathcal{M}$ is zero. Surfaces of zero mean curvature are known as minimal surfaces.

---

[3] Σ is called the image manifold, and $\mathcal{M}$ the space feature manifold.

For computing a solution numerically, we embed equation (3.72) into a dynamical scheme

$$\frac{dX}{dt}(t) = F, \tag{3.73}$$

where F is an arbitrary flow field defined on $\mathcal{M}(t)$. If $X(t)$ is of the form

$$X(t) = (x_1, x_2, u(t, x_1, x_2))^T,$$

then we have

$$\frac{dX}{dt}(t) = \left(0, 0, \frac{\partial u}{\partial t}(t, x_1, x_2)\right)^T.$$

Therefore, the motion is necessarily in the z direction. If we choose $F = (0, 0, \alpha H)$, we obtain the scalar equation

$$\frac{\partial u}{\partial t} = \alpha H = \alpha \frac{u_{x_1 x_1}(1 + u_{x_2}^2) + u_{x_2 x_2}(1 + u_{x_1}^2) - 2u_{x_1} u_{x_2} u_{x_1 x_2}}{(1 + u_{x_1}^2 + u_{x_2}^2)^{3/2}}. \tag{3.74}$$

The coefficient $\alpha \in R$ can be interpreted as a weighting parameter. If $\alpha = 1/\sqrt{1 + u_{x_1}^2 + u_{x_2}^2}$, then (3.74) can be rewritten as

$$\frac{\partial u}{\partial t} = \frac{u_{x_1 x_1}(1 + u_{x_2}^2) + u_{x_2 x_2}(1 + u_{x_1}^2) - 2u_{x_1} u_{x_2} u_{x_1 x_2}}{(1 + u_{x_1}^2 + u_{x_2}^2)^2}. \tag{3.75}$$

The right-hand side term in (3.75) is known as the Laplace–Beltrami operator, and equation (3.75) can be viewed as the projection on the z-axis of the flow $\frac{dX}{dt} = H N$ where $N = 1/\sqrt{1 + u_{x_1}^2 + u_{x_2}^2}(u_{x_2}, -u_{x_1}, 1)^T$ is the unit normal to $\mathcal{M}(t)$ (see Figure 3.16).

The quantity $1/\sqrt{1 + u_{x_1}^2 + u_{x_2}^2} = 1/\sqrt{g}$ has, in fact, a remarkable interpretation. Let us consider the ratio

$$r = \frac{A^{\text{domain}}}{A^{\text{surface}}},$$

where A^{domain} is the area of an infinitesimal surface in the image domain (x_1, x_2), and A^{surface} is the corresponding area on the surface $\mathcal{M}$ (see Figure 3.16).

This ratio can be interpreted as an indicator of the height variation on the surface. For flat surfaces, r is equal to 1, and it is close to 0 near edges. In fact, r is related to the metric of the surface, since

$$r = \frac{dx_1 \, dx_2}{\sqrt{g} \, dx_1 \, dx_2} = \frac{1}{\sqrt{g}} = \frac{1}{\sqrt{1 + u_{x_1}^2 + u_{x_2}^2}}.$$

Hence, from a restoration point of view, it would be desirable to incorporate r in the model. For example, in (3.74), we can choose $\alpha = r^\gamma$, so the flow becomes

$$\frac{\partial u}{\partial t} = r^{\gamma+3}(u_{x_1 x_1}(1 + u_{x_2}^2) + u_{x_2 x_2}(1 + u_{x_1}^2) - 2u_{x_1} u_{x_2} u_{x_1 x_2}). \tag{3.76}$$

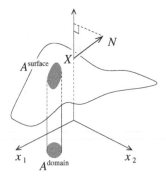

Figure 3.16. Representation of the normal N, A^{domain} and A^{surface}.

By selecting different γ we recover some flows already proposed:

- For $\gamma = -1$, (3.76) is the mean curvature flow projected onto the normal [306, 305].

- For $\gamma = 0$, (3.76) is the flow proposed by [145].

- For $\gamma = 1$, (3.76) is the Laplace–Beltrami flow. A result using this operator is shown in Figure 3.17.

original image   result

Figure 3.17. Example of the Laplace–Beltrami equation on the "Borel building" image.

Observing the result in Figure 3.17, we can remark that it is quite similar to the half-quadratic minimization and the total variation model described in Section 3.2. In fact, as shown in [306, 305], the Laplace–Beltrami equation has a direct relation with the total variation-based regularization methods. If we choose on Σ the metric $\varepsilon^2 dx_1^2 + \varepsilon^2 dx_2^2$ ($\varepsilon > 0$) and on $\mathcal{M}$ the metric

$ds^2 = \varepsilon^2 dx_1^2 + \varepsilon^2 dx_2^2 + du^2$, then the associated symmetric positive definite G is given by

$$G = \begin{pmatrix} \varepsilon^2 + u_{x_1}^2 & u_{x_1} u_{x_2} \\ u_{x_1} u_{x_2} & \varepsilon^2 + u_{x_2}^2 \end{pmatrix},$$

and a minimal surface is obtained by minimizing the functional

$$\mathcal{S}_\varepsilon(\mathcal{M}) = \varepsilon \iint_\Omega \sqrt{\varepsilon^2 + |\nabla u|^2} \; dx,$$

which is obviously a regularization of the total variation energy.

In fact, the real interest of this approach from a numerical point of view lies in the vectorial case. This Riemannian formalism can be developed in a wide variety of cases: textures, color, etc. For example, for color images, the space feature manifold $\mathcal{M}$ is defined by

$$X(\sigma_1, \sigma_2) = (X_1(\sigma_1, \sigma_2), X_2(\sigma_1, \sigma_2), X_r(\sigma_1, \sigma_2), X_g(\sigma_1, \sigma_2), X_b(\sigma_1, \sigma_2)),$$

where X is an embedded surface in R^5, and X_r, X_g, X_b are the three brightness components in the (Red, Green, Blue) system. We leave it as an exercise to the reader to write the associated metric and refer to [306, 305] for more details and experiments.

3.3.2 Smoothing–Enhancing PDEs

THE PERONA AND MALIK MODEL

Smoothing PDEs can be viewed as low-pass filters, and they can act as forward diffusion processes. What about introducing locally high-pass filters? We show in this section how a suitable choice of the weighting coefficient $c(.)$ in the Perona and Malik model (3.49) permits us to obtain this result.[4] $\boxed{\text{C++}}$

Let us first consider the 1-D case $(x \in R)$

$$\begin{cases} \dfrac{\partial u}{\partial t}(t, x) = [\, c(u_x^2(t, x))\, u_x(t, x)\,]_x, \\ u(0, x) = u_0(x), \end{cases} \tag{3.77}$$

and give some formal definitions:

Definition 3.3.2 (edge) *For a fixed time t, we say that $\bar{x}$ is an edge of a function $u(t, x)$ if $u_x(\bar{x}) = \max_x u_x(t, x)$. If u is smooth enough, we necessarily have at $\bar{x}$: $u_{xx}(t, \bar{x}) = 0$ and $u_{xxx}(t, \bar{x}) \le 0$. We will say that an edge $\bar{x}$ is blurred by a PDE if in a neighborhood of $\bar{x}$, $u_x(t, x)$ decreases as t increases, or in other words, if $\frac{\partial}{\partial t}(u_x(t, x)) \le 0$.*

We will say that an edge $\bar{x}$ is enhanced by a PDE if in a neighborhood of $\bar{x}$, $u_x(t, x)$ increases as t increases, i.e. if $\frac{\partial}{\partial t}(u_x(t, x)) \ge 0$.

---

[4]We recall that in Section 3.3.1, it was assumed $b(s) = c(s) + 2sc'(s) > 0$.

Let us examine the relationship between the coefficient $c(.)$ in the PDE (3.77) and the blurring/enhancing of an edge. From (3.77), we have formally

$$\frac{\partial}{\partial t}(u_x) = \left(\frac{\partial u}{\partial t}\right)_x = [c(u_x^2)\, u_x]_{xx} = u_{xxx}\, b(u_x^2) + 2\, u_{xx}^2\, b'(u_x^2),$$

where $b(s) = 2sc'(s) + c(s)$. If x is an edge at time t, then $u_{xx}(t,x) = 0$ and $u_{xxx}(t,x) \leq 0$. Thus

$$\text{sign}\left(\frac{\partial u}{\partial t}(t,x)\right)_x = \text{sign}(-b(u_x^2)(t,x)).$$

Therefore, we see that the blurring/enhancing process is governed by the sign of $b(u_x^2)$:

- If $b(u_x^2) > 0$, which means that (3.77) is a forward parabolic equation, the edge is blurred.

- If $b(u_x^2) < 0$, which means that (3.77) is a backward parabolic equation, the edge is sharpened.

Now, let us return to the general 2-D Perona and Malik model [275]:

$$\begin{cases} \dfrac{\partial u}{\partial t}(t,x) = \text{div}(\, c(\, |\nabla u(t,x)|^2)\, \nabla u(t,x)), \\ u(0,x) = u_0(x), \end{cases} \tag{3.78}$$

where $c : [0,+\infty[\rightarrow]0,+\infty[$ is a smooth decreasing function. As we saw in Section 3.3.1, equation (3.78) can be rewritten as

$$\begin{cases} \dfrac{\partial u}{\partial t}(t,x) = c(|\nabla u(t,x)|^2)\, u_{TT} + b(|\nabla u(t,x)|^2)\, u_{NN}, \\ u(0,x) = u_0(x). \end{cases}$$

Following our intuition from the 1-D case, if we want to sharpen edges, we need to impose that (3.78) is backward in the normal direction N, i.e.,

$$b(s) = 2s\, c'(s) + c(s) < 0 \quad \text{for large} \quad s \geq K, \tag{3.79}$$

where K is a given threshold. If we want to smooth homogeneous regions, we can impose

$$c(0) = b(0) = 1,$$

which implies that (3.78) acts like the heat equation for small gradients. Of course, there exist several possible choices for $c(.)$. A typical example is

$$c(s) = \frac{1}{1 + s/K}.$$

Now, what can be said about the existence of a solution for (3.78) with b satisfying (3.79)? The response is quite clear: hardly anything. To better understand the difficulty let us examine the 1-D "backward" heat equation:

$$\begin{cases} \dfrac{\partial u}{\partial t}(t,x) = -u_{xx}(t,x) & \text{on }]0,T[\times R, \\ u(0,x) = u_0(x). \end{cases} \tag{3.80}$$

By making the change of variable $\tau = T - t$, it is easy to see that whenever $u(t,x)$ is a solution of (3.80), then $v(\tau,x) = u(T-\tau,x)$ is a solution of

$$\begin{cases} \dfrac{\partial v}{\partial \tau}(\tau,x) = v_{xx}(\tau,x) & \text{on }]0,T[\times R, \\ v(T,x) = u_0(x), \end{cases} \tag{3.81}$$

which is exactly the heat equation with the backward datum $v(T,x) = u_0(x)$. So, if (3.80) admits a solution, the same goes for (3.81). But according to the regularizing property of the heat equation, $u_0(x)$ should necessarily be infinitely differentiable. If not, we deduce that (3.80) does not have a classical (and a weak) solution.

The same conclusion goes for (3.78) in the 1-D case. More precisely, Kichenassamy [191] proved the following result:

Theorem 3.3.6 (Kichenassamy [191]) *Let us suppose that:*

(i) *There exists a constant $K > 0$ such that $b(s) > 0$ for $s < K^2$ and $b(s) < 0$ for $s > K^2$.*

(ii) *Both $c(s)$ and $b(s)$ tend to zero as $s \to +\infty$.*

(iii) *(3.78) has a solution $u(t,x)$ satisfying $K_1 \le u_x(t,x) \le K_2$ for all $x \in [A,B]$ and all $t \in [0,T]$, for some A, B and $K_1 > K$.*

Then $u(t,x)$ is infinitely differentiable at $t = 0$ and for all $x \in]A,B[$. Therefore, if the initial image is not infinitely differentiable, there is no weak solution.

In fact, it follows from [191] that a "solution" must consist of regions in which it has a gradient less than K in absolute value, separated by points of discontinuity where the gradient is infinite. Thus, the notion of solution must be understood in the measure sense.

REGULARIZATION OF THE PERONA AND MALIK MODEL: CATTÉ ET AL.

If we continue to study (3.78) in a backward regime, we have to reconsider our notion of solution. One way to tackle an ill-posed problem like (3.78) is to introduce a regularization that makes the problem well posed. Then, by reducing the amount of regularization and observing the behavior of the solution of the regularized problem, one can obtain precious information about the initial one. This method was followed by Catt et al. [88]. The idea is to substitute in the diffusion coefficient $c(|\nabla u|^2)$ the gradient of the image

∇u by a smooth version of it $G_\sigma * \nabla u$, where G_σ is a smoothing kernel,[5] for example, the Gaussian one (3.46). Since $G_\sigma * \nabla u = \nabla(G_\sigma * u) = \nabla G_\sigma * u$, Catt et al. proposed the regularized model

$$\begin{cases} \dfrac{\partial u}{\partial t}(t,x) = \operatorname{div}(c(|(\nabla G_\sigma * u)(t,x)|^2)\,\nabla u(t,x)), \\ u(0,x) = u_0(x). \end{cases} \qquad (3.82)$$

This model has at least two advantages over the Perona and Malik model:

- If the initial data is very noisy (introducing large oscillations in the gradient of u), then the Perona and Malik model cannot distinguish between "true" edges and "false" edges created by the noise. The proposed model (3.82) avoids this drawback, since now the equation diffuses only if the gradient is estimated to be small. In fact, the model makes the filter insensitive to noise at time $t\sigma$, since $(\nabla G_\sigma * u)\,(t,x)$ is exactly the gradient of the solution at time σ of the solution of the heat equation with initial datum $u(t,x)$.

- As we prove next, equation (3.82) is now well posed.

Let us establish that (3.82) is well posed. Let us note $\Omega =]0,1[\times]0,1[$ and $g(s) = c(s^2)$.

Theorem 3.3.7 [88] *Let $g : R^+ \to R_*^+$ be smooth, decreasing with $g(0) = 1$, $\lim\limits_{s \to +\infty} g(s) = 0$ and $s \to g(\sqrt{s})$ smooth. If $u_0 \in L^2(\Omega)$, then there exists a unique function $u(t,x) \in C([0,T];L^2(\Omega)) \cap L^2((0,T);W^{1,2}(\Omega))$ satisfying in the distributional sense*

$$\begin{cases} \dfrac{\partial u}{\partial t}(t,x) - \operatorname{div}(g(|(\nabla G_\sigma * u)(t,x)|)\,\nabla u(t,x)) = 0 \quad on \ \ ]0,T[\times\Omega, \\ \dfrac{\partial u}{\partial N}(t,x) = 0 \quad on \ \ ]0,T[\times\partial\Omega, \\ u(0,x) = u_0(x). \end{cases}$$

$$(3.83)$$

Moreover, $|u|_{L^\infty((0,T);L^2(\Omega))} \le |u_0|_{L^2(\Omega)}$ and $u \in C^\infty(]0,T[\times\overline{\Omega})$.

Proof We follow [88].
Step 1: Uniqueness of the solution.
Let u_1 and u_2 be two solutions of (3.83). For almost every t in $[0,T]$ and $i = 1,2$, we have

$$\frac{d}{dt}u_i(t) - \operatorname{div}(\alpha_i(t)\nabla u_i(t)) = 0, \quad \frac{\partial u_i}{\partial N} = 0, \quad u_i(0) = u_0,$$

---

[5]Since we need some convolution, u is in fact extended in R^2 as in Section 3.3.1 (by symmetry and periodicity).

where $\alpha_i(t) = g(|\nabla G_\sigma * u_i|)$. Thus

$$\frac{d}{dt}(u_1 - u_2)(t) - \mathrm{div}(\alpha_1(t)(\nabla u_1 - \nabla u_2)(t)) = \mathrm{div}((\alpha_1 - \alpha_2)(t)\nabla u_2(t)).$$

Then, multiplying the above inequality by $(u_1 - u_2)$, integrating over Ω, and using the Neumann boundary condition, we get a.e. t,

$$\frac{1}{2}\frac{d}{dt}\int_\Omega |u_1(t) - u_2(t)|^2 \, dx + \int_\Omega \alpha_1 |\nabla u_1(t) - \nabla u_2(t)|^2 \, dx \qquad (3.84)$$

$$= -\int_\Omega (\alpha_1 - \alpha_2)\nabla u_2(t) \cdot (\nabla u_1(t) - \nabla u_2(t)) \, dx.$$

But since u_1 belongs to $L^\infty((0,T); L^2(\Omega))$, then $|\nabla G_\sigma * u_1|$ belongs to $L^\infty((0,T); C^\infty(\Omega))$ and there exists a constant $M = M\left(G_\sigma, |u_0|_{L^2(\Omega)}\right)$ such that $|\nabla G_\sigma * u_1| \leq M$ a.e. t, $\forall x \in \Omega$. Since g is decreasing and positive, it follows that a.e. in $]0, T[\times\Omega$,

$$\alpha_1(t) = g(|\nabla G_\sigma * u_1|) \geq g(M) = \nu > 0,$$

which implies from (3.84),

$$\frac{1}{2}\frac{d}{dt}\left(|(u_1 - u_2)(t)|^2_{L^2(\Omega)}\right) + \nu |\nabla(u_1 - u_2)(t)|^2_{L^2(\Omega)} \qquad (3.85)$$

$$\leq |\alpha_1 - \alpha_2|_{L^\infty(\Omega)} |\nabla u_2(t)|_{L^2(\Omega)} |\nabla(u_1 - u_2)(t)|_{L^2(\Omega)}.$$

Moreover, since g and G_σ are smooth, we have

$$|\alpha_1(t) - \alpha_2(t)|_{L^\infty(\Omega)} \leq C |u_1(t) - u_2(t)|_{L^2(\Omega)}, \qquad (3.86)$$

where C is a constant that depends only on g and G_σ. From (3.86) and by using Young's inequality, we obtain

$$\frac{1}{2}\frac{d}{dt}\left(|(u_1 - u_2)(t)|^2_{L^2(\Omega)}\right) + \nu |\nabla(u_1 - u_2)(t)|^2_{L^2(\Omega)}$$

$$\leq \frac{2}{\nu}C^2 |(u_1 - u_2)(t)|^2_{L^2(\Omega)} |\nabla u_2(t)|^2_{L^2(\Omega)} + \frac{\nu}{2}|\nabla(u_1 - u_2)(t)|^2_{L^2(\Omega)},$$

from which we deduce

$$\frac{d}{dt}\left(|(u_1 - u_2)(t)|^2_{L^2(\Omega)}\right) \leq \frac{4}{\nu}C^2 |\nabla u_2(t)|^2_{L^2(\Omega)} |(u_1 - u_2)(t)|^2_{L^2(\Omega)}.$$

To conclude, we need Gronwall's inequality (see Section 2.5.1), which we recall here: If $y(t) \geq 0$ satisfies

$$\frac{dy}{dt}(t) \leq c_1(t) y(t) + c_2(t),$$

then

$$y(t) \leq \left(y(0) + \int_0^t c_2(s)\, ds \right) \exp\left(\int_0^t c_1(s)\, ds \right).$$

Applying this inequality to $y(t) = |(u_1 - u_2)(t)|_{L^2(\Omega)}^2$, we get, since $u_1(0) = u_2(0) = u_0$,

$$|(u_1 - u_2)(t)|_{L^2(\Omega)}^2 \leq 0, \text{ i.e., } u_1 = u_2.$$

Step **2**: Existence of a solution.

The proof is based on a classical fixed-point argument. Let us define the space

$$W(0,T) = \left\{ w \in L^2((0,T); W^{1,2}(\Omega));\ \frac{dw}{dt} \in L^2((0,T); W^{1,2}(\Omega)') \right\},$$

where $W^{1,2}(\Omega)'$ is the dual of $W^{1,2}(\Omega)$; $W(0,T)$ is a Hilbert space for the norm

$$|w|_W = |w|_{L^2((0,T); W^{1,2}(\Omega))} + \left| \frac{dw}{dt} \right|_{L^2((0,T); W^{1,2}(\Omega)')}.$$

Let $w \in W(0,T) \cap L^\infty((0,T); L^2(\Omega))$ be such that $|w|_{L^\infty((0,T); L^2(\Omega))} \leq |u_0|_{L^2(\Omega)}$, and let us introduce the variational problem (P_w)

$$\left\langle \frac{du(t)}{dt}, v \right\rangle_{W^{1,2}(\Omega)' \times W^{1,2}(\Omega)} + \int_\Omega g(|(\nabla G_\sigma * \boxed{w})(t)|)\, \nabla u(t)\, \nabla v\, dx = 0$$

for all $v \in W^{1,2}(\Omega)$, a.e. t in $[0,T]$, which is now linear in u. As seen in Step 1, there exists a constant $\nu > 0$ such that $g(|\nabla G_\sigma * w|) \geq \nu$ a.e. in $]0,T[\times\Omega$. Therefore, by applying classical results on parabolic equations (see [148], page 356), we prove that the problem (P_w) has a unique solution u_w in $W(0,T)$ satisfying the estimates

$$|u_w|_{L^2((0,T); W^{1,2}(\Omega))} \leq c_1,$$

$$|u_w|_{L^\infty((0,T); L^2(\Omega))} \leq |u_0|_{L^2(\Omega)}, \tag{3.87}$$

$$\left| \frac{du_w}{dt} \right|_{L^2((0,T); W^{1,2}(\Omega)')} \leq c_2,$$

where c_1 and c_2 are constants depending only on g, G_σ, and u_0. From these estimates we introduce the subspace W_0 of $W(0,T)$ defined by

$$W_0 = \left\{ \begin{array}{l} w \in W(0,T),\ \ w(0) = u_0, \\ |w|_{L^2((0,T); W^{1,2}(\Omega))} \leq c_1, \\ |w|_{L^\infty((0,T); L^2(\Omega))} \leq |u_0|_{L^2(\Omega)}, \\ \left| \dfrac{dw}{dt} \right|_{L^2((0,T); W^{1,2}(\Omega)')} \leq c_2. \end{array} \right\}$$

By construction, $w \to S(w) \equiv u_w$ is a mapping from W_0 into W_0. Moreover, one can prove that W_0 is not empty, convex, and weakly compact in $W(0,T)$. Thus, we can apply Schauder's fixed-point theorem:

Theorem 3.3.8 (Schauder's fixed-point theorem) *If E is a convex, compact subset of a Banach space and if $S : E \to E$ is continuous, then there exists $x \in E$ such that $S(x) = x$.*

So, let us prove that the mapping $S : w \to u_w$ is weakly continuous ($W_0 \to W_0$). Let w_j be a sequence that converges weakly to some w in W_0 and let $u_j = u_{w_j}$. We have to prove that $S(w_j) = u_j$ converges weakly to $S(w) = u_w$. From (3.87) and classical results of compact inclusion in Sobolev spaces [2], we can extract from w_j, respectively from u_j, a subsequence (labeled w_j, respectively u_j) such that for some u, we have

$$\frac{du_j}{dt} \xrightarrow[L^2((0,T);W^{1,2}(\Omega)')]{} \frac{du}{dt},$$

$$u_j \xrightarrow[L^2((0,T);L^2(\Omega))]{} u,$$

$$\frac{\partial u_j}{\partial x_k} \xrightarrow[L^2((0,T);L^2(\Omega))]{} \frac{\partial u}{\partial x_k},$$

$$w_j \xrightarrow[L^2((0,T);L^2(\Omega))]{} w,$$

$$\frac{\partial G_\sigma}{\partial x_k} * w_j \xrightarrow[L^2((0,T);L^2(\Omega))]{} \frac{\partial G_\sigma}{\partial x_k} * w \quad \text{and a.e. on }]0,T[\times\Omega,$$

$$g(|\nabla G_\sigma * w_j|) \xrightarrow[L^2((0,T);L^2(\Omega))]{} g(|\nabla G_\sigma * w|),$$

$$u_j(0) \xrightarrow[W^{1,2}(\Omega)']{} u(0).$$

The above convergences allow us to pass to the limit in (P_w) and obtain $u = u_w = S(w)$. Moreover, since the solution is unique, the whole sequence $u_j = S(w_j)$ converges weakly in W_0 to $u = S(w)$; i.e., S is weakly continuous. Consequently, thanks to Schauder's fixed-point theorem, there exists $w \in W_0$ such that $w = S(w) = u_w$. The function u_w solves (3.83). The regularity follows from the general theory of parabolic equations. ∎

Remark Theorem 3.3.7 provides a natural algorithm for the numerical approximation of the solution. Let $u_0 \in L^2(\Omega)$. We construct a sequence u^n by solving the iterative scheme

$$\begin{cases} \dfrac{\partial u^n}{\partial t}(t,x) - \text{div}(g(|(\nabla G_\sigma * u^n)(t,x)|)\nabla u^{n+1}(t,x)) = 0 \text{ a.e. on }]0,T[\times\Omega, \\ \dfrac{\partial u^{n+1}}{\partial N}(t,x) = 0 \text{ a.e. on }]0,T[\times\partial\Omega, \\ u^{n+1}(0,x) = u_0(x). \end{cases}$$

It is proven in [88] that u^n converges in $C([0, T]; L^2(\Omega))$ to the unique solution of (3.81). ■

Let us mention the existence of other models for regularizing the Perona and Malik equation. For example, Nitzberg–Shiota [256] proposed the coupled system (in 1-D)

$$
\begin{cases}
\dfrac{\partial u}{\partial t} = (c(v)\, u_x)_x, \\[2mm]
\dfrac{\partial v}{\partial t} = \dfrac{1}{\tau}\,(\,|\,u_x|^2 - v\,), \\[2mm]
u(0, x) = u_0(x) \ \text{ and } \ v(0, x) \text{ is a smoothed version of } |(u_0)_x(x)|^2.
\end{cases}
$$

The function v plays the role of time-delay regularization, where the parameter $\tau > 0$ determines the delay. For other models, see Barenblatt et al. [37], Chipot et al. [111], Alvarez et al. [7].

✳ *In spite of the lack of a rigorous mathematical theory concerning the Perona and Malik equation, it has been successfully used in many numerical experiments. This phenomenon is still unexplained. It is likely that the behavior of the associated discrete problem does not reflect the ill-posedness of the continuous version, but this has to be investigated further.*

A first attempt to justify the Perona and Malik model was made by Kichenassamy, who defined in [191] a notion of generalized solution. This direction is promising and should be investigated further.

Once this regularized model is well-defined, a natural question arises: Does equation (3.82) approach equation (3.78) as σ tends to zero? This is a difficult question, and no mathematical response is available today. Perhaps, a clue for tackling this question would be to find a suitable functional framework for which u_σ, the solution of (3.82), and its gradient would be uniformly bounded with respect to σ. Then accumulation points could be considered as good candidates.

Another empirical question, is the choice of the parameter σ. Here again, there is no satisfying answer. In general, this choice is fixed by the user and is related to other parameters, for example those defining the function $c(s)$.

3.3.3 Enhancing PDEs

THE OSHER AND RUDIN SHOCK FILTERS

In this section, we examine edge enhancement via PDEs. In fact, in a way, enhancing and smoothing are opposite processes. In the former case, we want to create discontinuities at places where they have to appear, while in the latter case we want to remove superfluous features and false discontinuities. A typical example of enhancing is deblurring. In this section we show how some nonlinear hyperbolic PDEs (called shock filters) can be used for edge enhancement and deblurring. Let us start with the one-dimensional case. Ideally, an edge can be modeled by the step function

$$u(x) = \begin{cases} 1 & \text{if } x > 0, \\ -1 & \text{if } x < 0. \end{cases}$$

Let us imagine that some process (a convolution, for example) has blurred this edge, so that we have in hand a smooth version $u_0(x)$ of $u(x)$ (see Figure 3.18). The problem is to restore $u(x)$, starting from $u_0(x)$.

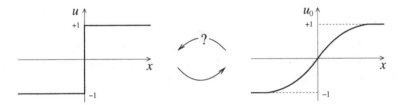

Figure 3.18. Illustration of the one-dimensional case.

To illustrate the reasoning and cover the different possibilities, we consider in this section the following initial condition:

$$u_0(x) = \cos(x).$$

In this case, as depicted in Figure 3.19, we would like to define a family of evolving curves $\{u(t,x)\}_{t>0}$ in order to sharpen the edges.

☞ *As we can observe, the direction of the motion of $u(t,x)$ is a function of x and depends of the sign of the product $u_x(t,x)\, u_{xx}(t,x)$.*

The four cases are indicated in Figure 3.19. Notice that at points x where $u_{xx}(t,x) = 0$ or $u_x(t,x) = 0$, it is desirable that no motion occur. Following this idea, Osher and Rudin [261] proposed to solve

[C++]

$$\begin{cases} u_t(t,x) = -\,|u_x(t,x)|\,\operatorname{sign}\,(u_{xx}(t,x)), \\ u(0,x) = u_0(x), \end{cases} \tag{3.88}$$

where $\operatorname{sign}(u) = 1$ if $u > 0$, $\operatorname{sign}(u) = -1$ if $u < 0$, $\operatorname{sign}(0) = 0$. For example, at points where $u_x(t,x) > 0$ and $u_{xx}(t,x) > 0$, we can verify that (3.88) behaves like $u_t(t,x) + u_x(t,x) = 0$, that is, a transport equation with speed $+1$, which is the desired motion. The same goes for the other cases. Before trying to justify this equation, we are going to consider a simplified version.

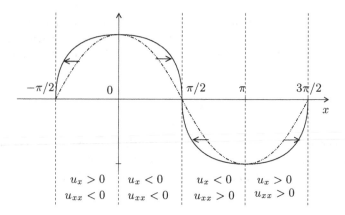

Figure 3.19. Illustration of the deblurring procedure in the 1-D case, for the initial condition $u_0(x) = \cos(x)$, represented on $\left[-\dfrac{\pi}{2}, \dfrac{3\pi}{2}\right]$. The dashed line represents the initial condition, and with the solid line we display the function after some time. Arrows show the direction of displacement.

A CASE STUDY: CONSTRUCTION OF A SOLUTION BY THE METHOD OF CHARACTERISTICS

Let us examine more precisely the following simpler case:

$$\begin{cases} u_t(t,x) = - \mid u_x(t,x)\mid \ \text{sign}(\ (u_0)_{xx}\ (t,x)), \\ u(0,x) = u_0(x), \end{cases} \qquad (3.89)$$

with $u_0(x) = \cos(x)$. We are going to search for an explicit solution. To do this, we use the *method of characteristics*, for which we recall the general formalism.

👁 To know more about the method of characteristics: [148].

Let U be an open subset of R^N and $\Gamma \subset \partial U$, a part of the boundary of U. Let us consider the nonlinear first-order PDE

$$\begin{cases} F(x,u,Du) = 0 \text{ in } U, \\ u = g \text{ on } \Gamma, \end{cases} \qquad (3.90)$$

where F and g are supposed smooth. The idea is to convert the PDE into a system of ordinary differential equations (ODEs). Let us fix any point $x \in U$ and let us suppose that $u \in C^2$ is a solution of (3.90). We would like to calculate $u(x)$ by finding some curve lying within U connecting x with a point $x^0 \in \Gamma$ and along which we can compute u. Since $u = g$ on Γ, we know the value of u at x^0, and we wish to calculate u all along the curve, and in particular at x. Let us suppose that the curve is described parametrically by $x(s) = (x_1(s), \ldots, x_N(s))$. We define

$$z(s) = u(x(s)), \tag{3.91}$$
$$p(s) = Du(x(s)). \tag{3.92}$$

Now, the question is to choose a good curve $x(s)$ in such a way that we can compute $z(s)$ and $p(s)$. In practice, we have to find the equations that are satisfied by $x(s)$, $z(s)$, and $p(s)$, and to solve them.

We first differentiate (3.92) with respect to s:

$$\dot{p}_i(s) = \sum_{j=1}^{N} u_{x_i x_j}(x(s))\,\dot{x}_j(s), \quad i = 1, \ldots, N \quad \left(\dot{p}_i = \frac{dp_i}{ds} \right). \tag{3.93}$$

Then we differentiate (3.90) with respect to x_i:

$$\sum_{j=1}^{N} \frac{\partial F}{\partial p_j}(x, u, Du)\,u_{x_j x_i} + \frac{\partial F}{\partial z}(x, u, Du)\,u_{x_i} + \frac{\partial F}{\partial x_i}(x, u, Du) = 0. \tag{3.94}$$

In order to get rid of second-derivative terms, let us define $x(s)$ as the solution of the ODE system

$$\dot{x}_j(s) = \frac{\partial F}{\partial p_j}(x(s), z(s), p(s)). \tag{3.95}$$

Assuming that $x(s)$ exists and thanks to (3.91)–(3.92), equation (3.94) evaluated at $x = x(s)$ can be written as

$$\sum_{j=1}^{N} \frac{\partial F}{\partial p_j}(x(s), z(s), p(s))\,u_{x_j x_i}(x(s)) + \frac{\partial F}{\partial z}(x(s), z(s), p(s))\,p_i(s)$$

$$+ \frac{\partial F}{\partial x_i}(x(s), z(s), p(s)) = 0.$$

If we replace in this expression $\frac{\partial F}{\partial p_j}(x(s), z(s), p(s))$ by $\dot{x}_j(s)$ (cf. (3.95)), we get from (3.93) and (3.94)

$$\dot{p}_i(s) = -\frac{\partial F}{\partial z}(x(s), z(s), p(s))\,p_i(s) - \frac{\partial F}{\partial x_i}(x(s), z(s), p(s)) = 0.$$

Finally, differentiating (3.91) with respect to s, we obtain

$$\dot{z}(s) = \sum_{j=1}^{N} \frac{\partial u}{\partial x_j}(x(s))\,\dot{x}_j(s) = \sum_{j=1}^{N} p_j(s)\,\frac{\partial F}{\partial p_j}(x(s), z(s), p(s)).$$

In summary, in vector notation, $x(s)$, $z(s)$, and $p(s)$ satisfy the following system of $(2N + 1)$ first-order ODEs (called characteristic equations):

$$\begin{cases} \dot{x}(s) = D_p F(x(s), z(s), p(s)), \\ \dot{z}(s) = D_p F(x(s), z(s), p(s)) \cdot p(s), \\ \dot{p}(s) = -D_x F(x(s), z(s), p(s)) - D_z(x(s), z(s), p(s))\,p(s). \end{cases} \tag{3.96}$$

Naturally, we need in addition to specify the initial conditions. To make things clearer let us return to our initial problem (3.89). Unfortunately, our calculus will be quite formal, since in the case of equation (3.89), the function F is not differentiable! In order to adopt the same notation as previously, we make the change of variables $x_2 = t$ and $x_1 = x$. Therefore, (3.89) can be written as

$$\begin{cases} u_{x_2} + \mathrm{sign}(u_0)_{x_1 x_1} \, |u_{x_1}| = 0 \text{ in } U, \\ u(x_1, 0) = u_0(x_1) \text{ on } \Gamma, \end{cases} \tag{3.97}$$

i.e., $F(x_1, x_2, z, p) = p_2 + \mathrm{sign}(u_0)_{x_1 x_1} \, |p_1|$. We choose $u_0(x_1) = \cos(x_1)$; thus $(u_0(x_1))_{x_1 x_1} = -\cos(x_1)$.

Let $U =]-\frac{\pi}{2}, \frac{3\pi}{2}[\times R^+$ and $\Gamma = \left\{ (x_1, x_2), \, x_1 \in]-\frac{\pi}{2}, \frac{3\pi}{2}[, \, x_2 = 0 \right\}$. To eliminate the sign function, we split the study into two cases.

First case: The equation is studied on $]-\frac{\pi}{2}, \frac{\pi}{2}[\times R^+$.
In this case, $\mathrm{sign}(\cos(x_1))_{x_1 x_1} = -1$. Therefore, we formally have

$$D_{p_1} F = -\frac{p_1}{|p_1|}, \quad D_{p_2} F = 1, \quad D_z F = D_{x_1} F = D_{x_2} F = 0,$$

and (3.96) becomes

$$\begin{cases} \dot{x}_1(s) = -\dfrac{p_1(s)}{|p_1(s)|}, \quad \dot{x}_2(s) = 1, \\ \dot{p}_1(s) = \dot{p}_2(s) = 0, \\ \dot{z}(s) = p_2(s) - |p_1(s)|. \end{cases} \tag{3.98}$$

For $s = 0$, we suppose that $x_1(0) = a \in]-\frac{\pi}{2}, \frac{\pi}{2}[$ and $x_2(0) = 0$. The integration of (3.98) is immediate. We get for some constants p_1^0 and p_2^0,

$$\begin{cases} x_1(s) = -\dfrac{p_1^0}{|p_1^0|} s + a, \quad x_2(s) = s, \\ p(s) = (p_1^0, p_2^0), \\ z(s) = (-|p_1^0| + p_2^0) s + \cos(a). \end{cases}$$

It remains to determine (p_1^0, p_2^0). Since $u(x_1, x_2) = \cos(x_1)$ on Γ, we have $p_1^0 = u_{x_1}(a, 0) = -\sin(a)$ and $p_2^0 = u_{x_2}(a, 0)$. But from equation (3.97), we deduce $p_2^0 = u_{x_2}(a, 0) = |u_{x_1}(a, 0)| = |\sin(a)|$. Therefore, the characteristic curve $x(s)$ is given by

$$\begin{cases} x_1(s) = \dfrac{\sin(a)}{|\sin(a)|} s + a, \\ x_2(s) = s, \end{cases}$$

and $u(x_1(s), x_2(s)) = z(s) = (-|\sin(a)| + |\sin(a)|) + \cos(a) = \cos(a)$, that is, u is constant along characteristics. We have two cases (see also Figure 3.20):

(i) If $\sin(a) > 0$, i.e., $a \in [0, \frac{\pi}{2}[$, then $x_1(s) = s + a$, $x_2(s) = s$, and the characteristics are straight lines. Thus, in this case, the solution of (3.97) is $u(x_1, x_2) = \cos(x_1 - x_2)$ with $x_2 < x_1 < \pi/2$.

(ii) If $\sin(a) < 0$, i.e., $a \in]-\frac{\pi}{2}, 0[$ then $x_1(s) = -s + a$, $x_2(s) = s$ and $u(x_1, x_2) = \cos(x_1 + x_2)$ with $-\pi/2 < x_1 < -x_2$.

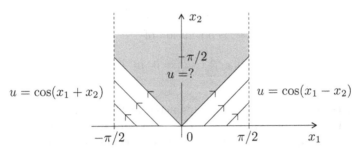

Figure 3.20. Characteristic lines in $]-\frac{\pi}{2}, \frac{\pi}{2}[\times R^+$. Notice that no characteristics go into the gray region.

Second case: The equation is studied on $]\frac{\pi}{2}, \frac{3\pi}{2}[\times R^+$.
In this case the equation becomes $u_{x_2} + |u_{x_1}| = 0$, and a similar study leads to the solution (see also Figure 3.21)

$$u(x_1, x_2) = \begin{cases} \cos(x_1 + x_2) & \text{if } \frac{\pi}{2} < x_1 < \pi - x_2, \\ \cos(x_1 - x_2) & \text{if } x_2 + \pi < x_1 < \frac{3\pi}{2}. \end{cases}$$

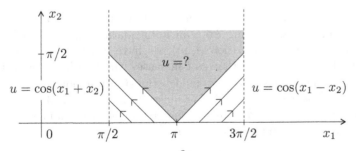

Figure 3.21. Characteristic lines in $]\frac{\pi}{2}, \frac{3\pi}{2}[\times R^+$. Notice that no characteristics go into the gray region.

Thanks to these calculations, we can observe that:

- The function $u(x_1, x_2)$ is discontinuous (a shock) along the line $x_1 = \frac{\pi}{2}$, i.e. at a point where characteristics intersect.

- $u(x_1, x_2)$ is not yet defined for $-x_2 < x_1 < x_2$ and $\pi - x_1 < x_2 < x_1 - \pi$.

- If we do not want to create other discontinuities than those described above, we must set

$$u(x_1, x_2) = \begin{cases} 1 & \text{if } -x_2 < x_1 < x_2, \\ -1 & \text{if } \pi - x_1 < x_2 < x_1 - \pi. \end{cases}$$

In conclusion, we propose as a solution of

$$\begin{cases} u_{x_2} + \text{sign}(-\cos(x_1)) \, |u_{x_1}| = 0 & \text{in }]-\frac{\pi}{2}, \frac{3\pi}{2}[\times R^+, \\ u(x_1, 0) = \cos(x_1), \end{cases} \tag{3.99}$$

the piecewise regular function u depicted in Figure 3.22. It is easy to see, by symmetry, that if $U = R \times R^+$, then we can construct a solution of (3.97) (with $u_0(x_1) = \cos(x_1)$) whose discontinuities develop only at $x_1 = (2k+1)\pi/2$, $k = 0, \pm 1, \pm 2, \ldots$.

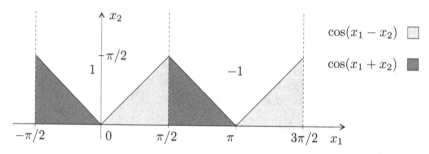

Figure 3.22. Solution proposed for equation (3.99).

This example shows very well why equations like (3.97) acts as edge-enhancement filters. Starting from an initial datum $u_0(x) = \cos(x)$ (we return to our previous notation (t, x)), we have constructed a family of functions $\{u(t, x)\}_{t>0}$ such that as t increases, the limiting process tends to an ideal one-dimensional edge model: the step function $u(x) = (-1)^k$ for $(2k-1)\frac{\pi}{2} < x < (2k+1)\frac{\pi}{2}$. We illustrate in Figure 3.23 the solution at different times. We can observe that edge formation and sharpening processes occur at the places where $(u_0(x))_{xx} = 0$.

COMMENTS ON THE SHOCK-FILTER EQUATION

Now let us return to more general one-dimensional models. As described in [261], let us consider the equation

$$\begin{cases} u_t = -|u_x| \, F(u_{xx}), & x \in R, \, t > 0, \\ u(0, x) = u_0(x). \end{cases} \tag{3.100}$$

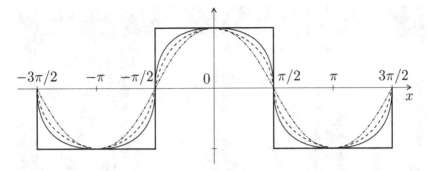

Figure 3.23. Illustration of the function u solution of (3.89) at different times.

Here, F is a Lipschitz continuous function satisfying

$$\begin{cases} F(0) = 0, \\ \text{sign}(s)F(s) > 0, \quad s \neq 0. \end{cases} \qquad (3.101)$$

A typical example of (3.100) is

$$\begin{cases} u_t = -|u_x|\,u_{xx}, \quad x \in R, \ t > 0, \\ u(0, x) = u_0(x), \end{cases} \qquad (3.102)$$

which can be written as

$$\begin{cases} u_t + (u_{xx}\,\text{sign}(u_x))\,u_x = 0, \quad x \in R, \ t > 0, \\ u(0, x) = u_0(x). \end{cases} \qquad (3.103)$$

Equation (3.103) (or (3.102)) can be considered as a transport equation whose speed of propagation is locally given by $c(x) = \text{sign}(u_x)\,u_{xx}$. Moreover, since edges are defined as maximum points of $|u_x|$, then at these points we have necessarily $u_{xx} = 0$, and locally u_{xx} changes of sign. Thus, the speed $c(x)$ plays the role of an edge-detector. From a mathematical point of view this type of equation is severely ill posed.

✳ *As already noticed, up to our knowledge, there is no theoretical justification for this problem. One of the first difficulties is to define a suitable notion of weak solution. We may wonder whether the notion of discontinuous viscosity solutions* [38, 40] *may help in the understanding of this equation.*

Nevertheless, Osher and Rudin [261] have performed very satisfying numerical simulations, and they have conjectured the following result.

Conjecture [261] *The evolution equation* (3.100), *with $u_0(x)$ continuous, has a unique solution that has jumps only at inflection points of $u_0(x)$ and for which the total variation in x of $u(t, x)$ is invariant in time, as well as in the locations and values of local extrema.*

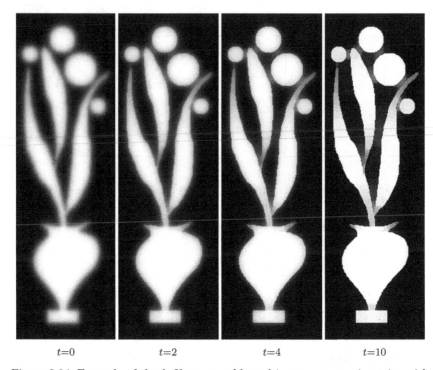

| $t=0$ | $t=2$ | $t=4$ | $t=10$ |

Figure 3.24. Example of shock filters on a blurred image representing a jar with some flowers. The blur has been generated by a convolution with a Gaussian kernel of variance $\sigma = 10$. Some iterations until convergence are then shown. This example shows that degradation due to blur cannot be fully recovered for fine structures (see, for instance, the extremities of the leaves).

original blurred image   result

Figure 3.25. Example of shock filters on the blurred "Borel building" image. The blur has been generated by a convolution with a Gaussian kernel of variance $\sigma = 10$. One can observe the patch effect produced by this algorithm.

The transposition of these ideas to the two-dimensional case is now straightforward. We have to write an equation that is a combination of a propagation term $|\nabla u|$ and an edge-detection term whose desired behavior involves changing sign across edges (or singular features) so that the local flow field is directed toward the edges. A good candidate is

$$\frac{\partial u}{\partial t} = -|\nabla u| \; F(L(u)), \tag{3.104}$$

where F satisfies (3.101) and where L is a nonlinear elliptic operator such that zero crossings define the edges of the processed image. According to Marr's theory [228], the most classical operator is the Laplacian

$$L(u) = \Delta u = u_{xx} + u_{yy}.$$

A better choice would be

$$L(u) = \frac{1}{|\nabla u|^2} \, (u_x^2 u_{xx} + 2u_x u_y u_{xy} + u_y^2 u_{yy}),$$

which corresponds to the second derivative of u in the direction of $\frac{\nabla u}{|\nabla u|}$ (here edges are defined as level curves of u).

The efficiency of this approach is demonstrated in Figures 3.24 and 3.25. We refer the reader to Section A.3.3 of Appendix A for the discretization of (3.104). There are usually two main criticisms of this model. The first is that the results obtained are not realistic from a perceptual point of view. As can be observed in Figure 3.25, the result is a piecewise constant image whose texture and fine details are lost (compare to the clear "Borel building" image in Figure 3.1). However, one cannot expect to recover details not present in the original blurred image. The second criticism is that if we also have some noise present in the image, it will be enhanced. To overcome this difficulty, Alvarez–Mazorra [8] combine shock filters and anisotropic diffusion and add a smoothing kernel for the estimation of the direction of the edges (see also [204]).

3.3.4 Neighborhood Filters, Nonlocal Means Algorithm, and PDEs

We close this chapter, devoted to image restoration by showing how to extend the notion of Gaussian filtering. In [77, 78], Buades et al. revisit the notion of neighborhood filtering. Here the notion of neighborhood must be understood broadly: neighboring pixels, neighboring or similar intensities, or "neighboring neighborhoods." Each of these meanings will correspond to a specific filter. Interestingly, the authors also proved the link between these filters and well-known PDEs such as the heat equation and the Perona–Malik equation.

NEIGHBORHOOD FILTERS

The idea of neighborhood filtering (also called bilateral filtering) is to consider that two pixels are close to each other not only if they occupy nearby spatial locations but also if they have some similarity in the photometric range. The formalization of this idea apparently goes back in the literature to Yaroslavsky [344], then Smith et al. [303] and Tomasi et al. [319]. More recently, Buades et al. [78] have pointed out some connections of these models with PDEs.

A general neighborhood filtering can be described as follows. Let u be an image to be denoised and let $T_h : R^+ \to R^+$ and $w_\rho : R^+ \to R^+$ be two functions whose roles will be to enforce respectively photometric and geometric locality. Parameters h and ρ will measure the amount of filtering for the image u. The filtered image $u_{h,\rho}(x)$ at scale (h, ρ) is given by

$$F_{h,\rho}u(x) = \frac{1}{N(x)} \int_\Omega T_h(|u(y) - u(x)|) \, w_\rho(|x - y|)u(y)dy,$$

where $N(x)$ is a normalization factor

$$N(x) = \int_\Omega T_h(|u(y) - u(x)|) \, w_\rho(|x - y|)dy.$$

For simplicity we suppose that the image has been extended from the domain image Ω (a rectangle) to the whole of R^2, by symmetry and periodicity.

With this formalism we can easily recover the classical spatial linear Gaussian filtering by choosing $T_h \equiv 1$ and $w_\rho(t) = \exp\left(-\frac{t^2}{\rho^2}\right)$. In that case $N(x) = \pi\rho^2$ and the denoised image is defined by

$$G_\rho u(x) = \frac{1}{\pi\rho^2} \int_{R^2} \exp\left(-\frac{|x - y|^2}{\rho^2}\right) u(y)dy.$$

It is interesting to evaluate the filtering residue, i.e., the difference $(u_\rho(x) - u(x))$:

$$u_\rho(x) - u(x) = \frac{1}{\pi\rho^2} \int_{R^2} \exp\left(-\frac{|y|^2}{\rho^2}\right)(u(x - y) - u(x))dy.$$

Rescaling inside the integral we get

$$u_\rho(x) - u(x) = \frac{1}{\pi} \int_{R^2} \exp(-|z|^2)(u(x - \rho z) - u(x))dz.$$

Then using a Taylor expansion of u (and assuming that $u \in C^3(R^2)$), we obtain

$$u_\rho(x) - u(x) = \frac{1}{\pi} \int_{R^2} \exp(-|z|^2)\Big[-\rho Du(x) \cdot z + \frac{\rho^2}{2}D^2u(x)(z, z) - \frac{\rho^3}{6}D^3u(x - \rho\theta z)(z, z, z)\Big]dz$$

where $\theta = \theta(x, \rho, z)$ belongs to $(0, 1)$. But thanks to the moment conditions,

$$\int_{R^2} \exp(-|z|^2)dz = \pi,$$

$$\int_\Omega z_i \exp(-|z|^2)dz = 0,$$

$$\int_{R^2} z_i z_j \exp(-|z|^2)dz = 0 \quad (i, j = 1, 2, i \neq j),$$

$$\int_{R^2} z_i^2 \exp(-|z|^2)dz = \frac{\pi\sqrt{\pi}}{2}.$$

We finally obtain

$$u_\rho(x) - u(x) = \frac{\rho^2}{2}\sqrt{\pi}\Delta u(x) + o(\rho^2). \tag{3.105}$$

From this result, we can deduce that:

- The residue of a Gaussian denoising is zero (up to $o(\rho^2)$) in regions where $\Delta u = 0$.

- The residue will be significant near singular parts of the image, namely at edges or textured regions where the Laplacian is large.

The approximation (3.105) is established for a fixed ρ and a single iteration of the filter. In the coming lemma we examine, in a more general setting, the asymptotic behavior of the solution obtained by iterating infinitely the filtering procedure.

Lemma 3.3.2 *Let $g \in L^1(R^N)$ satisfying the moment conditions*

$$\int_{R^N} g(x) \, dx = 1,$$

$$\int_{R^N} x_i g(x) \, dx = 0 \;\; \forall i,$$

$$\int_{R^N} x_i x_j g(x) \, dx = 0 \;\; \forall i \neq j,$$

$$\int_{R^N} x_i^2 g(x) \, dx = 2 \;\; \forall i,$$

$$\int_{R^N} |x|^3 |g(x)| \, dx < \infty.$$

We define

$$g_\rho(x) = \frac{1}{\rho^{N/2}} g\left(\frac{x}{\sqrt{\rho}}\right),$$

$$T_\rho u = g_\rho \star u.$$

Then we have

$$((T_\rho)^n u_0)(x) \xrightarrow[L^1(C)]{} u(t,x), \quad \text{when } n \to \infty \text{ and } n\rho \to t,$$

where $u(t,x)$ is the solution of the heat equation with initial condition u_0, and $C =]-1,1[^N$.

Proof Let us estimate

$$T_\rho u(t,x) - u(t,x) = g_\rho \star u(t,x) - u(t,x)$$

$$= \int_{R^N} \rho^{-N/2} g\left(\frac{y}{\sqrt{\rho}}\right) (u(t,x-y) - u(t,x))\, dy$$

$$= \int_{R^N} g(z)(u(t,x-\sqrt{\rho}\,z) - u(t,x))\, dz.$$

With a Taylor expansion, one obtains

$$T_\rho u(t,x) - u(t,x) = \int_{R^N} g(z)\left(-\sqrt{\rho}Du(t,x)\cdot z + \frac{\rho}{2}D^2u(t,x)(z,z)\right)dz$$

$$- \frac{1}{6}\rho^{3/2}\int_{R^N} g(z)D^3u(t,x-\sqrt{\rho}\,\theta\, z)(z,z,z)\, dz, \tag{3.106}$$

where $\theta = \theta(x,z,\rho) \in (0,1)$. With moment conditions, and if we assume that u is regular enough (at least C^3), then $|D^3u|(t,x)$ can be uniformly bounded on any compact $[t_1,t_2] \times C$, $t_1 > 0$, and (3.106) becomes

$$T_\rho u(t,x) - u(t,x) = \rho \triangle u(t,x) + O(\rho^{3/2}). \tag{3.107}$$

But if u satisfies the heat equation, we have

$$u(t+\rho,x) - u(t,x) = \rho\frac{\partial u}{\partial t}(t,x) + O(\rho^2),$$

$$= \rho\triangle u(t,x) + O(\rho^2) \quad \text{on } [t_1,t_2] \times C. \tag{3.108}$$

By subtracting equalities (3.107) and (3.108), we have

$$T_\rho u(t,x) - u(t+\rho,x) = O(\rho^{3/2}). \tag{3.109}$$

If we apply the operator T_h to (3.109), since $T_\rho(O(\rho^{3/2})) = O(\rho^{3/2})$ (u is bounded), we get

$$(T_\rho)^2u(t,x) - T_\rho u(t+\rho,x) = O(\rho^{3/2}),$$

and with (3.109) we have

$$(T_\rho)^2u(t,x) - u(t+2\rho,x) = 2O(\rho^{3/2}).$$

In fact, this computation can be iterated so that

$$(T_\rho)^n u(t,x) - u(t+n\rho,x) = nO(\rho^{3/2}) \quad \text{with } t_1 \le t+n\rho \le t_2.$$

As n tends to infinity, if we define $\rho = \tau/n$, we have

$$(T_\rho)^n u(t,x) - u(t+\tau,x) = O\left(\frac{\tau^{3/2}}{n^{1/2}}\right) \quad \text{with} \quad 0 < t_1 \le t+\tau \le t_2. \quad (3.110)$$

To complete the proof, it remains to show that (3.110) is also true when $t_1 = t = 0$. If $u_0 \in L^1(C), \forall \varepsilon > 0, \forall t_1$ small enough,

$$|u(t_1) - u_0|_{L^1(C)} = \int_C |u(t_1,x) - u_0(x)| \, dx < \varepsilon. \quad (3.111)$$

We also have

$$|g_\rho \star u|_{L^1(C)} \le |g_\rho|_{L^1(C)} |u|_{L^1(C)}. \quad (3.112)$$

But since $\int_{R^N} g_\rho = 1$, we deduce from (3.111) and (3.112), and by iterating,

$$|(T_\rho)^n u(t_1,.) - (T_\rho)^n u_0|_{L^1(C)} \le \varepsilon. \quad (3.113)$$

Integrating (3.110) on C with $t = t_1$ we have

$$|(T_\rho)^n u(t_1,x) - u(t_1 + \tau, x)|_{L^1(C)} = O\left(\frac{\tau^{3/2}}{n^{1/2}}\right) < \varepsilon, \quad (3.114)$$

for n high enough. Then (3.113) and (3.114) give

$$|(T_\rho)^n u_0(x) - u(t_1 + \tau, x)|_{L^1(C)} \le 2\varepsilon. \quad (3.115)$$

Having t_1 tend to zero in (3.115) achieves the proof. ∎

Now let us consider bilateral filters. As mentioned before, the idea is to take an average of the values of pixels that are both close in gray level value and spatial distance. Of course many choices are possible for the kernels T_h and w_ρ. Classical choices are

$$T_h(t) = \exp\left(-\frac{t^2}{h^2}\right)$$

and

$$w_\rho(t) = \exp\left(-\frac{t^2}{\rho^2}\right) \quad \text{or} \quad w_\rho(t) = \chi_{B(x,\rho)}(t),$$

where $\chi_{B(x,\rho)}$ denotes the characteristic function of the ball of center x and radius ρ. With the former choice of w_ρ, we get the SUSAN filter [303] or the bilateral filter [319]:

$$S_{\rho,h}u(x) = \frac{1}{N(x)} \int_{R^2} \exp\left(-\frac{|u(y) - u(x)|^2}{h^2}\right) \exp\left(-\frac{|y - x|^2}{\rho^2}\right) u(y) dy.$$

With the latter choice of w_ρ, we recover the Yaroslavsky filter

$$Y_{\rho,h}u(x) = \frac{1}{N(x)} \int_{B(x,\rho)} \exp\left(-\frac{|u(y) - u(x)|^2}{h^2}\right) u(y) dy. \quad (3.116)$$

The SUSAN and Yaroslavsky filters have similar behaviors. Inside a homogeneous region, the gray level values slightly fluctuate because of the noise. Nearby sharp boundaries, between a dark and a bright region, both filters compute averages of pixels belonging to the same region as the reference pixel: edges are not blurred.

Interestingly, the estimation of the residue $u_{\rho,h}(x) - u(x)$ gives some analogies with well-known PDEs.

Theorem 3.3.9 *Suppose $u \in C^2(\Omega)$ and let ρ, h, and $\alpha > 0$ such that $\rho, h \to 0$ and $h = O(\rho^\alpha)$. Let us consider the continuous function*

$$g(t) = \frac{1}{3} \frac{t \exp(-t^2)}{E(t)} \; \text{ for } t \neq 0, g(0) = \frac{1}{6} \; \text{ where } E(t) = \int_0^t \exp(-s^2) ds.$$

Let f be the continuous function

$$f(t) = 3g(t) + 3\frac{g(t)}{t^2} - \frac{1}{2t^2}, \; \text{ for } t \neq 0 \text{ and } f(0) = \frac{1}{6}.$$

Then for $x \in \Omega$,

- *if $\alpha < 1$, $Y_{\rho,h}u(x) - u(x) \approx \frac{\Delta u(x)}{6} \rho^2$,*
- *if $\alpha = 1$, $Y_{\rho,h}u(x) - u(x) \approx \left[g(\frac{\rho}{h}|Du(x)|)u_{TT}(x) + f(\frac{\rho}{h}|Du(x)|)u_{NN}(x) \right] \rho^2$,*
- *if $1 < \alpha < \frac{3}{2}$, $Y_{\rho,h}u(x) - u(x) \approx g(\rho^{1-\alpha}|Du(x)|) \left[u_{TT}(x) + 3u_{NN}(x) \right] \rho^2$,*

where $u_{TT} = D^2 u \left(\frac{Du^\perp}{|Du|}, \frac{Du^\perp}{|Du|} \right)$ and $u_{NN} = D^2 u \left(\frac{Du}{|Du|}, \frac{Du}{|Du|} \right)$.

We refer to [78] for the proof of the theorem. It is not difficult, somewhat technical, and relies on a Taylor expansion of $u(y)$ and the exponential function. More interesting is the interpretation of this theorem. For α ranging from 1 to $\frac{3}{2}$ an iterated procedure of the Yaroslavsky filter behaves asymptotically as an evolution PDE involving two terms respectively proportional to the direction $T = \frac{Du^\perp(x)}{|Du(x)|}$, which is tangent to the level passing through x and to the direction $N = \frac{Du(x)}{|Du(x)|}$, which is orthogonal to the level passing through x. In fact, we may write

$$\frac{Y_{\rho,h}u(x) - u(x)}{\rho^2} = c_1 u_{TT} + c_2 u_{NN}.$$

The filtering or enhancing properties of the model depend on the sign of c_1 and c_2. Following Theorem 3.3.9, we have:

- If $\alpha < 1$, then $\frac{Y_{\rho,h}u(x) - u(x)}{\rho^2} \approx \frac{\Delta u(x)}{6}$, which corresponds to a Gaussian filtering.

- If $\alpha = 1$, the neighborhood filter acts as a filtering/enhancing algorithm. Since the function g is positive (and decreasing) there is always a diffusion in the tangent direction, but since the function f can take positive or negative values, we may have filtering/enhancing effects depending of the values of $|Du(x)|$. For example, if $|Du(x)| > a\frac{h}{\rho}$, where a is such that $f(a) = 0$, then we get an enhancing effect. Let us remark since $g(t) \to 0$ as $t \to \infty$, points with large gradient are preserved.

- If $1 < \alpha < \frac{3}{2}$, then $\frac{\rho}{h}$ tends to infinity and $g(\frac{\rho}{h}|Du|)$ tends to zero and consequently the original image is hardly deteriorated.

Finally, let us observe that when $\alpha = 1$, the Yaroslavsky filter behaves asymptotically like the Perona–Malik equation (see Section 3.3.2)

$$\frac{\partial u}{\partial t} = c(Du^2)u_{TT} + b(Du^2)u_{NN}.$$

By choosing $c(s) = g(\sqrt{s})$ in (3.78) we get

$$\frac{\partial u}{\partial t} = g(Du^2)u_{TT} + h(Du^2)u_{NN},$$

with $h(s) = g(s) + sg'(s)$. We have $h(s) \neq f(s)$ but the coefficients in the tangent direction for the Perona–Malik equation and the Yaroslavsky filter are equal, and the functions h and f have the same behavior. Therefore both models share the same qualitative properties, which can be observed in Figure 3.26. In particular, the staircase effect appears.

How to Suppress the Staircase Effect?

When $\alpha = 1$, the Yaroslavsky filter (3.116) can create unexpected features such as artificial contours inside flat zones, also called the staircase effect (Figure 3.26). The main reason is the difficulty to choose an appropriate threshold for the gradient. The origin of the staircase effect can be explained with a 1-D convex increasing signal (respectively a 1-D increasing concave signal) (Figure 3.27).

For each x, the number of points y such that $u(x) - h < u(y) \leq u(x)$ is larger (respectively smaller) than the number of points satisfying $u(x) \leq u(y) \leq u(x)+h$. Thus, the average value $Y_{\rho,h}$ is smaller (respectively larger) than $u(x)$. Since edges correspond to inflection points (i.e., points where $u'' = 0$), the signal is enhanced at inflection points; the discontinuities become more marked. To overcome this difficulty, Buades et al. [78] introduced an intermediate regression correction in order to better approximate the signal locally. For every x in a 2-D image, one searches for a triplet (a, b, c) minimizing

$$\int_{B(x,\rho)} w(x,y)(u(y) - ay_1 - by_2 - c)^2 dy, \qquad (3.117)$$

where $w(x,y) = \exp \frac{-|u(y)-u(x)|^2}{h^2}$, and then replacing $u(x)$ by $(ax_1 - bx_2 - c)$. Let us denote this improved version of the original Yaroslavsky filter by $Ly_{\rho,h}$.

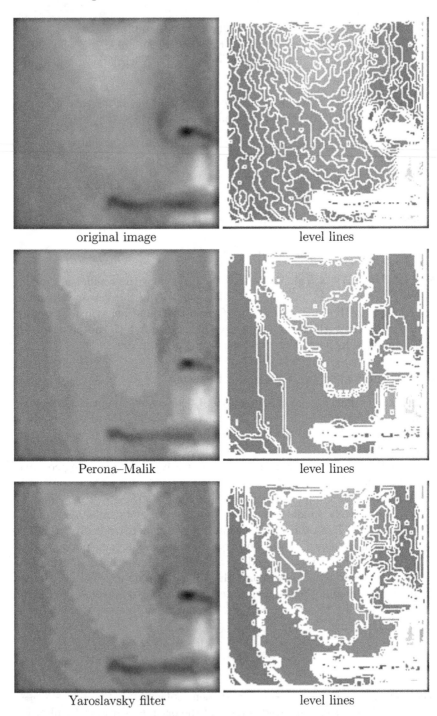

original image   level lines

Perona–Malik   level lines

Yaroslavsky filter   level lines

Figure 3.26. The Perona–Malik model and the Yaroslavsky filter have the same behavior and create a staircase effect.

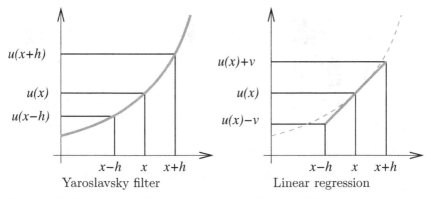

Yaroslavsky filter   Linear regression

Figure 3.27. Yaroslavsky filter creates stepwise functions: The reason is that for each x, the number of points y such that $u(x) - h < u(y) \leq u(x)$, is larger than the number of points satisfying $u(x) \leq u(y) \leq u(x) + h$. This can be avoided with a local linear approximation.

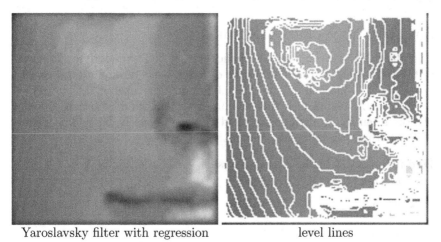

Yaroslavsky filter with regression   level lines

Figure 3.28. The staircase effect can be eliminated with regression (see Figure 3.26 for the original image and results without regression).

Theorem 3.3.10 *Suppose $u \in C^2(\Omega)$, and let $\rho, h > 0$ be such that $\rho, h \to 0$ and $O(\rho) = O(h)$. Let g be the continuous function defined by $g(0) = \frac{1}{6}$ and $g(t) = \frac{8t^2 e^{-t^2} - 8te^{-t^2} E(t) + 2E(t)^2}{t^2(4E(t)^2 - 8te^{-t^2}E(t))}$, for $t \neq 0$, where $E(t) = \int_0^t e^{-s^2}\, ds$. Then*

$$Ly_{\rho,h}u(x) - u(x) \approx \left[\frac{1}{6} u_{\xi\xi} + g\left(\frac{\rho}{h}|Du| \right) u_{\eta\eta} \right] \rho^2. \qquad (3.118)$$

According to Theorem 3.3.10, the enhancing effect has disappeared; the coefficient in the normal direction is now always positive and decreasing.

original image   neighborhood filters   NL-means

Figure 3.29. NL-means performs better than neighborhood filters on periodic textures.

When the gradient is large, the weighting function in the normal direction tends to 0 and the image is filtered only in the tangent direction. Figure 3.28 shows how regression can improve the results.

Non Local Means Filter (NL-Means)

This algorithm was proposed by Buades et al. [77]. The underlying idea was to take advantage of the high degree of redundancy of any natural image. Following [77] one defines the "neighborhood of a pixel x" as any set of pixels y in the image such that the intensity in a window around y looks like the intensity in a window around x. More precisely, let us first define a neighborhood system.

Definition 3.3.3 *A neighborhood system on Ω is a family $\mathcal{N} = \{\mathcal{N}\}_{x \in \Omega}$ such that for all $x \in \Omega$,*

- $x \in \mathcal{N}_x$,

- $y \in \mathcal{N}_x \Rightarrow x \in \mathcal{N}_y$.

For an image u, we set $U(\mathcal{N}_x) = \{u(y), y \in \mathcal{N}_x\}$. The similarity between two pixels x and y will be based on the similarity of the intensity gray level between $\mathcal{N}_x$ and $\mathcal{N}_y$. This similarity measure is computed with a Gaussian distance

$$|u(\mathcal{N}_x) - u(\mathcal{N}_y)|^2 = \int_{R^2} G_a(t)(u(x+t) - u(y+t))^2 dt,$$

where G_a is a Gaussian kernel of variance a, i.e.,

$$|u(\mathcal{N}_x) - u(\mathcal{N}_y)|^2 = (G_a \star (u(x+.) - u(y+.))^2)(0).$$

Then, if u is a noisy image, the nonlocal means filtering of u is defined as

$$NL_h u(x) = \frac{1}{C(x)} \int_{\Omega} \exp(-(G_a \star (u(x+.) - u(y+.))^2)(0))u(y)dy, \quad (3.119)$$

where $C(x)$ is a normalization factor:

$$C(x) = \int_\Omega \exp(-(G_a \star (u(x + .) - u(y + .))^2)(0))dy.$$

The filtering algorithm is nonlocal since all pixels in the image are used for the estimation at a pixel x. In practice, the size of the neighborhood is controlled by the parameter a and the amount of averaging (i.e. the magnitude of the weights) is controlled by the parameter h. In the limiting case, if the window reduces to one pixel ($a \to 0$), we recover the Yaroslavsky filter.

In general, windows of size 7×7 or 9×9 give good results. Of course, due to the nature of the algorithm, most favorable cases for the NL-means are the periodic and the textured patterns (Figure 3.29). Note that introducing regression correction can also be applied to the NL-means algorithm [78].

Remark This idea of spanning other neighborhoods in order to find the suitable intensity can be found in some texture synthesis approaches. We refer the reader to Section 5.1.4, where we show some recent work about inpainting by Criminisi et al. [120] is discussed. ∎

4
The Segmentation Problem

How to Read This Chapter

This chapter is concerned with image segmentation, which plays a very important role in many applications. The aim is to find a partition of an image into its constituent parts. As we will see, the main difficulty is that one needs to manipulate objects of different kinds: functions, domains in R^2, and curves.

- We first try in Section 4.1 to define more precisely what image segmentation is, and we briefly survey some classical ideas in image segmentation. In fact, the notion of segmentation depends on the kind of image we have to process and what we want to do. In the last decade two main approaches have been developed: the Mumford and Shah approach and the geodesic active contours method.

- Section 4.2 is concerned with the Mumford and Shah functional. Here the idea is to find a close image of the initial one compounded of several regions with nearly constant intensity. The difficulty in studying the Mumford and Shah functional is that it involves two unknowns: the intensity function and the set K of edges. This difficulty is tackled in Sections 4.2.2 and 4.2.3, which are concerned with the mathematical study of this problem (definition of a suitable mathematical framework, optimality conditions, regularity of the edge set). Section 4.2.4 is a survey of some approaches for approximating the Mumford and Shah functional. Most of them are based on Γ-convergence theory. Finally, we present in Section 4.2.5 some experimental results.

• Section 4.3 deals with the geodesic active contours and the level sets method. Here, the objective is to find the boundaries of objects in an image. The idea is to model those contours as curves that should match the highest gradients. We start in Section 4.3.1 by recalling the Kass, Witkin, and Terzopoulos snakes model, which is one of the first efforts in this direction. This model, which has some drawbacks, has been revisited by Caselles, Kimmel, and Sapiro, who proposed a geodesic active contour strategy (Section 4.3.2). We clearly establish the connection between these two formulations. One of the main interests in the latter model is that it can be rewritten using a level sets formulation. This is detailed in Section 4.3.3, where we prove the well-posedness of this model in the viscosity sense. This section is rather technical, but it shows a complete and classical proof using viscosity solutions. The numerical implementation of these methods requires to use a reinitialization equation periodically which will preserve some regularity of the level sets during their evolution. This is performed by an Hamilton-Jacobi type equation, with a discontinuous Hamiltonian, which is theoretically studied in Section 4.3.4. We illustrate this kind of segmentation approcah in Section 4.3.5, and we refer to the Appendix for the details regarding the discretization. We finally present in Section 4.3.6 some extensions of this framework.

4.1 Definition and Objectives

As a first definition, we could say that segmenting an image means dividing it into its contituent parts. However, this definition is rather unsatisfactory and ambiguous. Let us have a look at the images presented in Figure 4.1. In the left-hand image, every piece of contour (edge) information is important, and it would be interesting to have an identification of all the contours separating two regions of different intensities. Equivalently, one would like to have a simplified version of the original image, compounded of homogeneous regions separated by sharp edges. The right-hand image illustrates the notion of objects. Some contours (the boundaries an the object) may have more importance (depending on the application), and it would be interesting to find an approach to detecting them. These two examples show that the notion of segmentation is not unique and may depend on the kind of image we have to deal with.

Still, in each case, the important features are edges. Edge detection has been an early concern in computer vision. Classical approaches are based on local differential properties of an edge, for instance on the first and second derivatives of the image (see Figure 4.2).

"Borel building"   "objects"

Figure 4.1. Two examples of images suggesting different notions of segmentation.

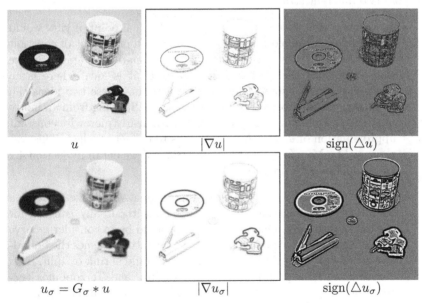

Figure 4.2. Edges and derivatives. First- and second-order derivatives of the image "objects" and a smoothed version of it are displayed. The gradient is represented using inverted colors for better visualization. The Laplacian is negative in the black areas, positive in the white ones, zero otherwise. Edges can be seen as the locations where the gradient is locally maximum, or where the Laplacian changes sign. Notice how the smoothing allows one to obtain a cleaner description of the edges.

With first-order derivatives, the earliest methods were based on the application of some convolution masks, thus enhancing edges [278, 277, 304]. Then, Canny [82] proposed an edge detector that is still widely used. The starting point was to define criteria that an edge detector should satisfy: reliability of the detection, accuracy of the localization, and the requirement of one response per edge. This leads to an optimal filter that is in fact a very close approximation to the first derivative of a Gaussian. This has been further developed by Deriche [132] and Shen and Castan [301], who proposed sharper filters implemented recursively.

With second-order derivatives, the important starting point was the method proposed by Marr and Hildreth [229] based on zero-crossing detection of a Laplacian of a Gaussian, abbreviated LoG (see also [180, 257, 342]). This kind of approach produces closed contours; the corners are rounded and the connectivity at the junctions is poor.

These approaches are local and combine derivatives at different scales. The goal is to identify the edges that are characterized by sharp variations of the intensity. If we consider the examples from Figure 4.1, we may propose two different strategies:

- To segment the "Borel building" image, the dual point of view would be to find a simplified image as a combination of regions of constant intensities. By constructing such an approximation of an image, we would also have the segmentation. Also, as it is local, there is no concern about the smoothness of the contours. These two ideas can be incorporated into a variational framework: Starting from an image u_0, we look for a pair (u, K) such that u is a nearly piecewise constant approximation of u_0 and K corresponds to the set of edges. This was proposed by Mumford and Shah in 1989, and we detail in Section 4.2 this model and its properties.

- Now if we consider the "objects" image, one would like to have a technique for separating the five objects, without any concern for the internal texture. One intuitive idea should be to consider a curve enclosing all the objects and make it evolve until it reaches the boundaries of the objects. Eventually, this curve could shrink or split. This idea was initially proposed by Kass, Witkin, and Terzopoulos (active contours), and it is based on an energy minimization depending on the curve. This is presented in Section 4.3 together with further developments such as the level sets formulation.

We will work afterwards with the "objects" image, which has interesting properties with regard to the previous discussion. First of all, it is compounded of five different objects of different sizes, shapes, or textures (see Figure 4.3), so that both aspects of segmentation can be tested (in terms of edge detection or object segmentation): The mug is covered with several images, one part of the stapler is quite elongated and fine, several similar keys are superimposed, the disk has a hole, and the coin is small and presents small contrasts. It will be interesting to observe what kind of results we obtain with the different approaches.

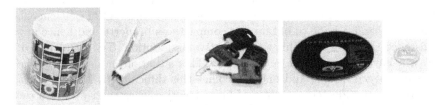

Figure 4.3. Different parts of the "objects" image.

4.2   The Mumford and Shah Functional

4.2.1   A Minimization Problem

Let us present the model introduced by Mumford and Shah in 1989 [247]. In this section Ω is a bounded open set of R^N, $N = 2, 3$, and $u_0(x)$ is the initial image. Without loss of generality we can always assume that $0 \leq u_0(x) \leq 1$ a.e. $x \in \Omega$. We search for a pair (u, K), where $K \subset \Omega$ is the set of discontinuities, minimizing

$$F(u, K) = \int_{\Omega - K} (u - u_0)^2 dx + \alpha \int_{\Omega - K} |\nabla u|^2 \, dx + \beta \int_K d\sigma, \qquad \text{(MS)}$$

where α and β are nonnegative constants and $\int_K d\sigma$ is the length of K. In their seminal paper [247], the authors made the following conjecture:

Conjecture 4.2.1 *There exists a minimizer of F such that the edges (the discontinuity set K) are the union of a finite set of $C^{1,1}$ embedded curves. Moreover, each curve may end either as a crack tip (a free extremity, i.e., K looks like a half-line) or in triple junction, that is, three curves meeting at their endpoints with $2\pi/3$ angle between each pair.*

The purpose of this section is to analyze this conjecture and the recent advances in this subject. We first concentrate on the existence problem (Section 4.2.2) and then study the geometric properties of the set K (Section 4.2.3). We conclude by giving some details on the approximation of this functional (Section 4.2.4).

4.2.2   The Mathematical Framework for the Existence of a Solution

Before studying this problem, we need to correctly define the functional $F(u, K)$ and in particular decide which class of K to consider. It is clear that we cannot a priori impose that K is made of a finite set of $C^{1,1}$-curves, since one cannot hope to obtain any compactness property and hence any existence theorem with this too-restrictive assumption. The regularity of K will have to be proven a posteriori. This situation is classical in the calculus of variations, and one overcomes this difficulty by looking for a solution in a wider class of sets of finite length rather than just in a set of $C^{1,1}$-curves. This is achieved by defining the length of K as its $(N-1)$-dimensional Hausdorff measure $\mathcal{H}^{N-1}(K)$, which is the most natural way of extending the notion of length to nonsmooth sets (see Definition 2.2.8, Section 2.2). Therefore, we rewrite the Mumford and Shah functional as

$$F(u, K) = \int_{\Omega - K} (u - u_0)^2 \, dx + \alpha \int_{\Omega - K} |\nabla u|^2 \, dx + \beta \, \mathcal{H}^{N-1}(K),$$

where for the moment $K \subset \Omega$ is a closed set and u belongs to the Sobolev space $W^{1,2}(\Omega - K)$. We can observe that $F(u, K)$ is minimal in the sense that removing one of the three terms would imply that $\inf F(u, K) = 0$ and we could obtain trivial solutions. For example, if we drop the first integral in F, then $u = 0$ and $K = \emptyset$ are solutions, or if we drop the second term, then $u = u_0$ and $K = \emptyset$ are solutions. Nevertheless, if we reject trivial solutions, it is interesting to study the latter case which is the reduced Mumford and Shah functional

$$E(u, K) = \int_{\Omega - K} (u - u_0)^2 \, dx + \beta \, \mathcal{H}^{N-1}(K).$$

It is easy to see that if the set of discontinuity K is fixed ($K \neq \emptyset$), then a solution u is piecewise constant (u is equal to the mean of u_0 on the connected components of $\Omega - K$), and so E becomes a function only of K. Much work has been devoted to this particular case. Let us mention contributions from Mumford and Shah [247], Morel and Solimini [245, 243, 244], Massari and Tamanini [231].

☛ *The difficulty in studying F is that it involves two unknowns u and K of different natures: u is a function defined on an N-dimensional space, while K is an $(N-1)$-dimensional set.*

In order to apply the direct method of the calculus of variations, it is necessary to find a topology that ensures at the same time lower semicontinuity of F and compactness of the minimizing sequences. The difficulty comes from $\mathcal{H}^{N-1}(K)$. Indeed, let E be a Borel set of R^N with topological boundary ∂E. It is easy to convince oneself that

☞ *the map $E \rightarrow \mathcal{H}^{N-1}(\partial E)$ is not lower semicontinuous with respect to any compact topology.*

Let us consider the following example. Let $\{x_i\}$ be the sequence of all rational points in R^N and let

$$B_i = \{x \in R^N \; ; \; |x - x_i| \leq 2^{-i}\}$$

$$E_k = \bigcup_{i=0}^{k} B_i, \quad E = \bigcup_{i=0}^{\infty} B_i.$$

Denoting by $|E|$ the N-dimensional Lebesgue measure of E and by w_N the Lebesgue measure of the unit ball in R^N, we get

$$|E| \leq \sum_{i=0}^{\infty} |B_i| = w_N \sum_{i=0}^{\infty} 2^{-iN} = \frac{w_N}{1 - 2^{-N}} < \infty.$$

Since rational points are dense in R^N, we have $\overline{E} = R^N$, and thus $\partial E = \overline{E} - E = R^N - E$ has infinite Lebesgue measure, which implies

$$\mathcal{H}^{N-1}(\partial E) = +\infty.$$

On the other hand,

$$\mathcal{H}^{N-1}(\partial E_k) \leq \mathcal{H}^{N-1} \left(\bigcup_{i=0}^{k} \partial B_i \right) = N w_{N-1} \sum_{i=0}^{k} 2^{-i(N-1)}$$

$$\leq N \frac{w_{N-1}}{1 - 2^{-(N-1)}} < +\infty.$$

Therefore, the sequence $\{\mathcal{H}^{N-1}(\partial E_k)\}$ is bounded, $E_k \rightarrow E$ $(k \rightarrow +\infty)$ in the sense of measures, or equivalently, the sequence of characteristic functions χ_{E_k} converges in L^1 to χ_E; however, we do not have

$$\mathcal{H}^{N-1}(\partial E) \leq \varliminf_{k \rightarrow +\infty} \mathcal{H}^{N-1}(\partial E_k).$$

This shows the necessity of finding another formulation of $F(u, K)$. The new formulation involves the space $BV(\Omega)$ of functions of bounded variation in Ω (Section 2.2). The idea is to identify the set of edges K with the jump set S_u of u, which allows us to eliminate the unknown K. So the idea is to consider the functional

$$G(u) = \int_{\Omega} (u - u_0)^2 \; dx + \alpha \int_{\Omega} |\nabla u|^2 \; dx + \beta \, \mathcal{H}^{N-1}(S_u). \tag{4.1}$$

If we do not have a lower semicontinuity property with sets (see the above remark), we are going to show that it can be obtained with functions. Now, it is tempting to minimize G on the space $BV(\Omega)$. Unfortunately, the space

$BV(\Omega)$ may contain pathological nonconstant functions that are continuous and have approximate differential equal to zero almost everywhere (a well-known example is the Cantor–Vitali function [12]). For such a function v we have

$$G(v) = \int_{\Omega} (v - u_0)^2 \, dx \geq \inf_{u \in BV(\Omega)} G(u) \geq 0,$$

and since these pathological functions are dense in $L^2(\Omega)$, we get

$$\inf_{u \in BV(\Omega)} G(u) = 0,$$

which implies that the infimum of G cannot be achieved in $BV(\Omega)$ in general. To avoid this phenomenon we must eliminate these pathological functions, which have the peculiarity that their distributional derivatives are measures concentrated on Cantor sets. Let us recall that the distributional derivative Du of a $BV(\Omega)$ function can be split into three mutually singular measures:

$$Du = \nabla u \, dx + (u^+ - u^-) \, n_u \, \mathcal{H}^{N-1}_{|S_u} + C_u,$$

where $J(u) = (u^+ - u^-) \, n_u \, \mathcal{H}^{N-1}_{|S_u}$ is the jump part and C_u the Cantor part. Following Di Giorgi [127, 126] we define $SBV(\Omega)$ as the space of special functions of bounded variation, which is the space of $BV(\Omega)$ functions such that $C_u = 0$. Note that the Cantor–Vitali functions mentioned above do not belong to $SBV(\Omega)$, since their supports are mainly based on Cantor sets. Consequently, the suitable functional space to minimize (4.1) seems to be $SBV(\Omega)$. The natural question is now to establish the relation between the following two problems:

$$\inf_{u,K} \left\{ \begin{array}{c} F(u, K), \ u \in W^{1,2}(\Omega - K) \cap L^\infty(\Omega), \\ K \subset \Omega, \ K \ \text{closed} \ , \ \mathcal{H}^{N-1}(K) < \infty \end{array} \right\}, \qquad (P_1)$$

$$\inf_{u} \left\{ G(u), \ u \in SBV(\Omega) \cap L^\infty(\Omega) \right\}. \qquad (P_2)$$

The answer can be found in Ambrosio [11] and is the consequence of the following theorem:

Theorem 4.2.1 [11] *Let $K \subset \Omega$ be a closed set such that $\mathcal{H}^{N-1}(K) < \infty$ and let $u \in W^{1,2}(\Omega - K) \cap L^\infty(\Omega)$. Then $u \in SBV(\Omega)$ and $S_u \subset K \cup L$ with $\mathcal{H}^{N-1}(L) = 0$.*

From Theorem 4.2.1 it follows that inf $P_2 \leq$ inf P_1. By using compactness and lower semicontinuity theorems (see below) it can be shown that (P_2) has a solution u. For such a minimizer De Giorgi–Carriero–Leaci [128] proved that

$$\mathcal{H}^{N-1}(\Omega \cap (\overline{S_u} - S_u)) = 0.$$

So by setting $K = \Omega \cap \overline{S_u}$ we get a solution of (P_1) and min (P_1) = min (P_2). It remains to show that (P_2) has a solution. This is a direct consequence of the following theorem:

Theorem 4.2.2 [13] *Let $u_n \in SBV(\Omega)$ be a sequence of functions such that there exists a constant C with $|u_n(x)| \leq C < \infty$ a.e. $x \in \Omega$ and $\int_\Omega |\nabla u_n|^2 \, dx + \mathcal{H}^{N-1}(S_{u_n}) \leq C$. Then there exists a subsequence u_{n_k} converging a.e. x to a function $u \in SBV(\Omega)$. Moreover, ∇u_{n_k} converges weakly in $L^2(\Omega)^N$ to ∇u, and $\underline{\lim} \, \mathcal{H}^{N-1}(S_{u_{n_k}}) \geq \mathcal{H}^{N-1}(S_u)$.*

We obtain a solution for (P_2) by applying Theorem 4.2.2 to any minimizing sequence of (P_2) and by observing beforehand that we can restrict our attention to minimizing sequences satisfying $|u_n|_{L^\infty(\Omega)} \leq |u_0|_{L^\infty(\Omega)}$ (using a truncation argument).

☛ *The SBV cluster points of sequences as defined in Theorem 4.2.2 are solutions of (P_2). Observe that no uniqueness result is available. This will be illustrated in the coming section.*

Having established the existence of a minimizer, we would like now to compute it. The natural way to do so is to search for optimality conditions. Curiously, it is easier to establish them with $F(u, K)$ than with $G(u)$. So, let us suppose that there exists a pair (u, K), a solution of (P_1), i.e.,

$$F(u, K) \leq F(v, K'), \tag{4.2}$$

for all $v \in W^{1,2}(\Omega - K') \cap L^\infty(\Omega)$, $K' \subset \Omega$, K' closed, $\mathcal{H}^{N-1}(K') < \infty$. Moreover, let us suppose that (u, K) satisfies the Mumford and Shah conjecture:

(C_1) K consists of a finite number of $C^{1,1}$-curves γ_i, meeting $\partial \Omega$ and meeting each other only at their endpoints.

(C_2) u is C^1 on each connected component of $\Omega - K$.

Theorem 4.2.3 *Let (u, K) be a solution of (P_1) satisfying (C_1) and (C_2). Then*

$$\alpha \, \Delta u = u - u_0 \text{ on } \Omega - K, \tag{4.3}$$

$$\frac{\partial u}{\partial N} = 0 \text{ on } \partial \Omega \text{ and on the two sides } \gamma_i^\pm \text{ of each } \gamma_i, \tag{4.4}$$

$$e(u^+) - e(u^-) + \beta \text{ curv } \gamma_i = 0 \text{ on } \gamma_i, \tag{4.5}$$

where $e(u) = (u - u_0)^2 + \alpha |\nabla u|^2$, u^+ and u^- are the traces of u on each side of K (each side of γ_i), curv γ_i is the curvature of γ_i.

Proof The proof of (4.3) and (4.4) is standard. We first look at the variations of F with respect to u. In (4.2) we choose $K' = K$ and $v = u + \theta \varphi$ with $\theta \in R$, and φ is a test function with compact support. Then

$$0 \le F(u + \theta\varphi, K) - F(u, K) = \theta^2 \int_{\Omega - K} (\varphi^2 + \alpha\, |\nabla\varphi|^2)\, dx \qquad (4.6)$$

$$+ 2\theta \int_{\Omega - K} (\varphi\,(u - u_0) + \alpha\, \nabla\varphi \cdot \nabla u)\, dx.$$

Dividing (4.6) by $\theta > 0$ (respectively by $\theta < 0$) and letting $\theta \to 0^+$ (respectively $\theta \to 0^-$) we get

$$0 = \int_{\Omega - K} \varphi\,(u - u_0)\, dx + \alpha \int_{\Omega - K} \nabla\varphi \cdot \nabla u\, dx, \quad \forall \varphi. \qquad (4.7)$$

Choosing φ with compact support in $\Omega - K$ and integrating the second integral by parts in (4.7) we obtain

$$0 = \int_{\Omega - K} \varphi\,(u - u_0 - \alpha\, \Delta u)\, dx \quad \forall \varphi,$$

i.e.,

$$u - u_0 - \alpha\, \Delta u = 0 \quad \text{on}\ \ \Omega - K.$$

Now, multiplying (4.3) by a function $\varphi \in C^1(\Omega)$, we easily obtain (4.4).

To prove (4.5), the idea is to look at the variation of F with respect to K. We propose to give a slightly different proof than that given by Mumford and Shah. The arguments we are going to use will be useful later for active contours. For the sake of clarity, we also look at a simpler version of the Mumford and Shah problem. We suppose that there is only one object in the scene, and that K is a closed $C^{1,1}$-curve.

Let Ω_{int} be the open set enclosed by K and $\Omega_{\text{ext}} = \Omega - \Omega_{\text{int}} - K$. Our aim is to consider variations of K according to the flow $\frac{dx}{dt} = v(t, x)$, where v is an arbitrary velocity. We denote by $K(t)$ such a variation, $t \ge 0$, with $K(0) = K$. Since u varies as K moves, we denote by $u(t, x)$ the unique solution of $\inf\limits_u F(u, K(t))$ and $u_{\text{int}}(t, x) = u(t, x)|_{\Omega_{\text{int}}}(t)$, $u_{\text{ext}}(t, x) = u(t, x)|_{\Omega_{\text{ext}}(t)}$. So let

$$f(t) = \int_{\Omega - K(t)} [(u(t, x) - u_0(x))^2 + \alpha\, |\nabla u(t, x)|^2]\, dx + \beta \int_{K(t)} d\sigma.$$

Writing $\Omega = \Omega_{\text{int}}(t) \cup \Omega_{\text{ext}}(t) \cup K(t)$, we have

$$f(t) = \int_{\Omega_{\text{int}} - K(t)} [(u_{\text{int}}(t,x) - u_0(x))^2 + \alpha \, |\nabla u_{\text{int}}(t,x)|^2] \, dx$$

$$+ \int_{\Omega_{\text{ext}} - K(t)} [(u_{\text{ext}}(t,x) - u_0(x))^2 + \alpha \, |\nabla u_{\text{ext}}(t,x)|^2] \, dx + \beta \int_{K(t)} d\sigma.$$

We remark that in the expression of $f(t)$ both the domain of integration and the integrands depend on t. As we are interested in the first variation of F, we need to estimate $f'(t)$.

- For the first two integrals we need to use a classical result on the *derivative of a domain integral*: If $l(t,x)$ is a regular function defined on a bounded regular domain $w(t)$ of R^N and if we set

$$g(t) = \int_{w(t)} l(t,x) \, dx, \qquad (4.8)$$

  then

$$g'(t) = \int_{w(t)} \frac{\partial l}{\partial t}(t,x) \, dx + \int_{\partial w(t)} l(t,x) \, v \cdot N \, d\sigma, \qquad (4.9)$$

  where $\partial w(t)$ is the boundary of $w(t)$, N is the unit outward normal to $\partial w(t)$, and v is the velocity of $\partial w(t)$.

- Regarding the last term, we also need to know how to estimate the derivative of the length. We can show that

$$\frac{d}{dt}\left(\int_{K(t)} d\sigma \right) = \int_{K(t)} \text{curv} K(t) \, v \cdot N d\sigma.$$

The proof can be found in Section 4.3.2 (see (4.22)–(4.23)).

By applying the above results we get

$$f'(t) = 2 \int_{\Omega_{int}(t)} (u_{int} - u_0) \frac{\partial u_{int}}{\partial t} \, dx + \int_{K(t)} (u_{int} - u_0)^2 v \cdot N \, d\sigma \qquad (4.10)$$

$$+ 2\alpha \int_{\Omega_{int}(t)} \nabla u_{int} \cdot \nabla \left(\frac{\partial u_{int}}{\partial t} \right) dx + \alpha \int_{K(t)} |\nabla u_{int}|^2 \, v \cdot N \, d\sigma$$

$$+ 2 \int_{\Omega_{ext}(t)} (u_{ext} - u_0)^2 \frac{\partial u_{ext}}{\partial t} \, dx - \int_{K(t)} (u_{ext} - u_0)^2 v \cdot N \, d\sigma$$

$$+ 2\alpha \int_{\Omega_{ext}(t)} \nabla u_{ext} \cdot \nabla \left(\frac{\partial u_{ext}}{\partial t} \right) dx - \alpha \int_{K(t)} |\nabla u_{ext}|^2 \, v \cdot N \, d\sigma$$

$$+ \beta \int_{K(t)} \operatorname{curv} K(t) \, v \cdot N \, d\sigma.$$

Thanks to Green's formula we have

$$\int_{\Omega_{int}(t)} \nabla u_{int} \cdot \nabla \left(\frac{\partial u_{int}}{\partial t} \right) dx = - \int_{\Omega_{int}} \Delta u_{int} \frac{\partial u_{int}}{\partial t} \, dx + \int_{K(t)} \frac{\partial u_{int}}{\partial t} \frac{\partial u_{int}}{\partial N} \, d\sigma,$$

but $u_{int}(t, x)$ is the solution of

$$\begin{cases} \alpha \, \Delta u_{int}(t, x) = u_{int}(t, x) - u_0(x) \text{ in } \Omega_{int}(t), \\ \dfrac{\partial u_{int}}{\partial N} = 0 \text{ on } K(t). \end{cases}$$

Thus

$$\alpha \int_{\Omega_{int}(t)} \nabla u_{int} \cdot \nabla \left(\frac{\partial u_{int}}{\partial t} \right) dx = - \int_{\Omega_{int}(t)} (u_{int}(t, x) - u_0(x)) \frac{\partial u_{int}}{\partial t}(t, x) \, dx.$$

The same goes for

$$\alpha \int_{\Omega_{ext}(t)} \nabla u_{ext} \cdot \nabla \left(\frac{\partial u_{ext}}{\partial t} \right) dx = - \int_{\Omega_{ext}(t)} (u_{ext}(t, x) - u_0(x)) \frac{\partial u_{ext}}{\partial t}(t, x) \, dx.$$

Therefore, by replacing these last expressions in (4.10) we get

$$f'(t) = \int_{K(t)} ((u_{int} - u_0)^2 + \alpha \, |\nabla u_{int}|^2) \, v \cdot N \, d\sigma$$

$$- \int_{K(t)} ((u_{ext} - u_0)^2 + \alpha \, |\nabla u_{ext}|^2) \, v \cdot N \, d\sigma + \beta \int_{K(t)} \operatorname{curv} K(t) \, v \cdot N \, d\sigma,$$

or, with the notation of Theorem 4.2.3,

$$f'(t) = \int\limits_{K(t)} (e(u_{\text{int}}) - e(u_{\text{ext}}) + \beta \, \text{curv} K(t)) \, v \cdot N \, d\sigma.$$

Until now we have not specified the variations of $K(t)$. We choose that K moves along its outward normal according to the following differential equation:

$$\frac{\partial x}{\partial t} = v(t, x) = V(x(t)) \, N,$$
$$x(0) = K,$$

where V is an arbitrary velocity. Since $|N|^2 = 1$, $f'(t)$ can be written as

$$f'(t) = \int\limits_{K(t)} (e(u_{\text{int}}(t, x)) - e(u_{\text{ext}}(t, x)) + \beta \, \text{curv} K(t)) \, V(x(t)) \, d\sigma.$$

If (u, K) is a minimizer of the Mumford and Shah functional, we have necessarily $f'(0) = 0$, i.e.,

$$0 = \int\limits_{K} (e(u_{\text{int}}(x)) - e(u_{\text{ext}}(x)) + \beta \, \text{curv} K) \, V(x) d\sigma,$$

and since V is arbitrary, we obtain,

$$e(u_{\text{int}}) - e(u_{\text{ext}}) + \beta \, \text{curv} K = 0 \text{ on } K.$$

This proves (4.5) in this simpler case. We let the reader convince himself that the above proof also runs in the general case. ∎

Remark Before analyzing the edge set K it would be appropriate to say some words about the regularity of the function u. If K is supposed $C^{1,1}$ and if u_0 is continuous, then the standard theory of elliptic operators, [148, 161, 173] implies that:

- u is C^1 on the open set $\Omega - K$ and at all simple boundary points of K and $\partial \Omega$.

- u extends locally to a C^1-function on the region plus its boundary.

Connected component of Ω−K

However, problems can arise at corners: If P is a corner with an angle α such that $\pi < \alpha \leq 2\pi$ (including the exterior of a crack, i.e., when P is the endpoint of a $C^{1,1}$-curve that is not continued by any other arc), then u has the form (in polar coordinates centered at P)

$$u(r, \theta) = c \, r^{\pi/\alpha} \sin\left(\frac{\pi}{\alpha}(\theta - \theta_0)\right) + \hat{v}(r, \theta), \qquad (4.11)$$

where $\hat{v}(r, \theta)$ is C^1 and c, θ_0 are suitable constants. ∎

4.2.3 Regularity of the Edge Set

In this section we look at qualitative properties satisfied by a minimizer of the Mumford and Shah functional, and we examine in particular the structure of the set K. The first result was given by Mumford and Shah in 1989 [247]. It was a first step toward a solution of their own conjecture (see Section 4.2.1).

Theorem 4.2.4 [247] *Let $N = 2$. If (u, K) is a minimizer of $F(u, K)$ such that K is a union of simple $C^{1,1}$-curves γ_i meeting $\partial\Omega$ and each other only at their endpoints, then the only vertices of K are:*

(i) *Points P on $\partial\Omega$ where one γ_i meets $\partial\Omega$ perpendicularly.*

(ii) *Triple points P where three γ_i meet with angles $2\pi/3$.*

(iii) *Crack-tip where a γ_i ends and meets nothing.*

Proof *(Sketch of the proof).* As we can imagine, proving (i)–(iii) is not simple, and we do not reproduce all the details. We refer the interested reader to [247] for instructive and illuminating constructions. Five steps can be distinguished:

(A) Since we are concerned with the regularity of K, we deal with a local phenomenon and we have only to consider the energy inside a ball $B(P, \varepsilon)$. In fact, we say that (u, K) minimizes F if no change altering (u, K) inside a ball B, and leaving it unchanged outside, can decrease F.

(B) One proves that K has no kinks, i.e., points P where two edges γ_i and γ_j meet at an angle other than π.

(C) One shows that γ_i meets $\partial\Omega$ perpendicularly.

(D) At triple points, that is, points P where three curves γ_i ,γ_j, γ_k meet with angles $\theta_{i,j}$, $\theta_{j,k}$,$\theta_{k,i}$, then we necessarily have $\theta_{i,j} = \theta_{j,k} = \theta_{k,i} = 2\pi/3$.

(E) Finally, one proves that there is no point where four or more γ_i meet at positive angles and no cuspidal corners, i.e., corners where two arcs are tangent.

The way to prove (B), (C), (D), or (E) is always the same: by contradiction if (B), (C), (D), or (E) were not satisfied, then we could locally construct from (u, K) another pair (u', K') that strictly decreases $F(u, K)$, thus contradicting that (u, K) is a minimizer in the sense of (A).

To illustrate the above discussion let us show how (B) can be proved. Let P be a kink point and let $B_\varepsilon = B(P, \varepsilon)$ be the ball of center P and radius ε. Let us suppose that $\gamma_i \cup \gamma_j$ divides B_ε into sectors B_ε^+ with angle α^+,

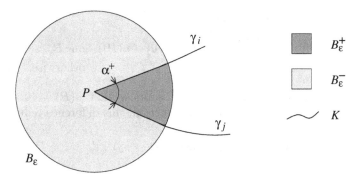

Figure 4.4. Definition of B_ε^+ and B_ε^+.

$0 < \alpha^+ < \pi$ and B_ε^- with angle α^-, $\pi < \alpha^- < 2\pi$ (see Figure 4.4).
Let us define the C^∞-function $\phi(x,y)$, $0 \leq \phi \leq 1$, by

$$\phi(x,y) = \begin{cases} 1 & \text{if } x^2 + y^2 \geq 1, \\ 0 & \text{if } x^2 + y^2 \leq \frac{1}{2}, \end{cases}$$

and let

$$\phi_\varepsilon(x,y) = \phi\left(\frac{x - x(P)}{\varepsilon}, \frac{y - y(P)}{\varepsilon}\right).$$

Now we are going to construct from (u, K) another admissible pair (u', K').
What we are doing is cutting the corner at P at a distance of $\varepsilon/2$, shrinking
B_ε^+ and expanding B_ε^-. More precisely, the only change we are making in
K is to remove the curvilinear triangle PMN from B_ε^+ and to add it to the
set B_ε^-, leaving unchanged the rest of K (see Figure 4.5). We call this new
set of edges K', and $B_\varepsilon'^+$ denotes the new B_ε^+ (respectively $B_\varepsilon'^-$ the new
B_ε^-).

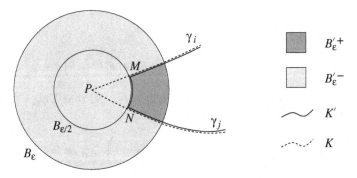

Figure 4.5. Construction of the other solution (u', K').

Then we define $u'(x, y)$ as follows:

$$u'(x, y) = \begin{cases} u(P) + \phi_\varepsilon(x, y)(u(x, y) - u(P)) & \text{on } B'^-_\varepsilon, \\ u(x, y) & \text{otherwise,} \end{cases}$$

and without loss of generality we can suppose that $u(P) = 0$. Let us evaluate $F(u', K') - F(u, K)$. By construction, this difference reduces to

$$F(u', K') - F(u, K) = \int\limits_{B'^-_\varepsilon} [(\phi_\varepsilon u - u_0)^2 - (u - u_0)^2] \, dx$$

$$+ \alpha \int\limits_{B'^-_\varepsilon} [|\nabla(\phi_\varepsilon u)|^2 - |\nabla u|^2] \, dx + \beta \int\limits_{K'} d\sigma - \beta \int\limits_{K} d\sigma.$$

We examine the first two terms separately (we will denote by c a universal constant):

$$A_{1\varepsilon} = \int\limits_{B'^-_\varepsilon} [(\phi_\varepsilon u - u_0)^2 - (u - u_0)^2] \, dx$$

$$= \int\limits_{B'^-_\varepsilon} \left[\frac{1 - \phi_\varepsilon}{1 + \phi_\varepsilon} u_0^2 - (1 - \phi_\varepsilon)\left(\sqrt{1 + \phi_\varepsilon}\, u - \frac{u_0}{\sqrt{1 + \phi_\varepsilon}} \right)^2 \right] dx$$

$$\leq \int\limits_{B'^-_\varepsilon} u_0^2 \, dx \leq c\varepsilon^2,$$

$$A_{2\varepsilon} = \alpha \int\limits_{B'^-_\varepsilon} \left[|\nabla(\phi_\varepsilon u)|^2 - |\nabla u|^2 \right] dx$$

$$= \alpha \int\limits_{B'^-_\varepsilon} \left[(\phi_\varepsilon^2 - 1)|\nabla u|^2 + u^2 |\nabla \phi_\varepsilon|^2 + 2u\phi_\varepsilon \nabla \phi_\varepsilon \cdot \nabla u \right] dx.$$

But thanks to (4.11) we have

$$u = O\left(r^{\pi/\alpha^-} \right) \quad \text{and} \quad |\nabla u| = O\left(r^{\pi/\alpha^- - 1} \right),$$

and taking into account that $\phi_\varepsilon^2 \leq 1$, $|\nabla \phi_\varepsilon| \leq c/\varepsilon$, and that $\nabla \phi_\varepsilon = 0$ in the ball $B(P, \varepsilon/2)$, we obtain

$$A_{2\varepsilon} \leq c\varepsilon^2 \left(\frac{\varepsilon^{2\pi/\alpha^-}}{\varepsilon^2} + \varepsilon^{\pi/\alpha^-} \frac{1}{\varepsilon} \varepsilon^{\pi/\alpha^- - 1} \right) = c\varepsilon^{2\pi/\alpha^-}.$$

Finally, in the third term, $\int_{K'} d\sigma - \int_K d\sigma$, we are replacing, asymptotically as $\varepsilon \to 0$, the equal sides of an isosceles triangle with angle α^+ by the third side, so

$$A_{3\varepsilon} = \int_{K'} d\sigma - \int_{K} d\sigma \le \varepsilon\left(\sin\frac{\alpha^+}{2} - 1\right) \le 0,$$

and we obtain

$$F(u', K') - F(u, K) = A_{1\varepsilon} + A_{2\varepsilon} + A_{3\varepsilon} \le c\left(\varepsilon^2 + \varepsilon^{\frac{2\pi}{\alpha^-}} + \varepsilon\left(\sin\frac{\alpha^+}{2} - 1\right)\right).$$
(4.12)

Since $0 < \alpha^+ < \pi$ and $\pi < \alpha^- < 2\pi$, (4.12) shows that the energy decreases by order ε if ε is sufficiently small, which contradicts that (u, K) is a minimizer. We note that the data term $A_{1\varepsilon} = \int_{B'_{\varepsilon^-}} [(\phi_\varepsilon u - u_0)^2 - (u - u_0)^2]\, dx$ is of order ε^2, and thus negligible with respect to the two other terms independently of the data u_0. ∎

Remark These regularity conditions are very interesting, and they constrain the segmentation to satisfy some properties, at the cost of fidelity to the image. For instance, if we consider the simple case, depicted in Figure 4.6, Theorem 4.2.4 shows that we cannot get the exact segmentation with lines crossing at $\pi/2$. Qualitatively, we may obtain one of the two configurations described in Figure 4.6. This is a simplified illustration of why uniqueness may not hold.

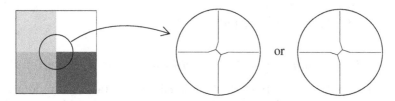

Figure 4.6. Illustration of "equivalent" segmentations for a given image. This is to show why regularity conditions on the edge set may have some influence on the uniqueness of the solution. ∎.

As said before, Theorem 4.2.4 is a first step in the proof of the Mumford and Shah conjecture. To go further we have to remove the assumption that K is made up of a finite union of simple $C^{1,1}$-curves meeting $\partial\Omega$ and each other only at their endpoints. A. Bonnet made important progress in this direction. We state only his results, and we refer the reader to [62, 63, 64] for the proofs.

Theorem 4.2.5 [62, 64] *If (u, K) is a minimizer of F such that K is connected, then (u, K) is one of the following:*

(i) *K is empty and u is constant.*

(ii) *K is a straight line defining two half-planes and u is constant on each half-plane.*

(iii) *K is the union of three half lines with $2\pi/3$ angles and u is constant on each sector.*

(iv) *In a polar set of coordinates (r, θ), $u(r, \theta) = \sqrt{2/\pi}\sqrt{r}\cos(\theta/2)$ for $\theta \in [0, 2\pi[$ and K is the half-axis $\theta = 0$ (a crack-tip).*

We note that the conclusions of Theorem 4.2.5 and Theorem 4.2.4 are very similar. The major difference lies in the assumptions. In Theorem 4.2.4, K is supposed to consist of a finite union of $C^{1,1}$-arcs, while in Theorem 4.2.5, K is assumed to be connected. Bonnet also proved the following result:

Theorem 4.2.6 [63, 64] *Every isolated connected component of K is the union of a finite set of C^1-arcs. These arcs are $C^{1,1}$ away from crack-tips and can merge through triple junctions with $2\pi/3$ angles.*

We notice that Theorem 4.2.6 does not allow a minimizer to have an infinite number of arbitrarily small pieces connected to each other. The proof of Theorem 4.2.6 relies on a characterization of the minimizers, which is very similar to the one given in Theorem 4.2.4.

✳ *We conclude this section by saying that, to the best of our knowledge, the Mumford and Shah conjecture remains an open question. However there has been a substantial progress in its understanding. For example, we refer the reader to [124] for more details.*

4.2.4 Approximations of the Mumford and Shah Functional

The lack of differentiability of the functional for a suitable norm does not allow us to use, as is classical, Euler–Lagrange equations. Moreover, the discretization of the unknown discontinuity set is a very complex task. A commonly used method is to approximate $F(u, K)$ (or $G(u)$) by a sequence F_ε of regular functionals defined on Sobolev spaces, the convergence of F_ε to F as $\varepsilon \to 0$ being understood in the Γ-convergence framework (see Section 2.1.4). Of course, if we want to get an efficient approximation, the set K must not appear in F_ε. Four classes of approaches may be distinguished:

(A) *Approximation by elliptic functionals* [17].
  In this approach the set S_u (or K) is replaced by an auxiliary variable v (a function) that approximates the characteristic function $(1 - \chi_{S_u})$, i.e., $v(x) \approx 0$ if $x \in S_u$ and $v(x) \approx 1$ otherwise. Ambrosio and Tortorelli [17] proposed the following sequence of functionals:

$$F_\varepsilon(u, v) = \int_\Omega (u - u_0)^2 \, dx + \int_\Omega v^2 \, |\nabla u|^2 \, dx + \int_\Omega \left(\varepsilon \, |\nabla v|^2 + \frac{1}{4\varepsilon}(v - 1)^2 \right) dx \, ,$$

which is discussed afterwards.

(B) *Approximations by introducing second order singular perturbations*
    [71] (see also [45, 84])
    For instance

$$F_\varepsilon(u) = \begin{cases} \displaystyle\int_\Omega (u - u_0)^2 \, dx + \frac{1}{\varepsilon}\int_\Omega f(\varepsilon |\nabla u|^2) \, dx + \varepsilon^3 \int_\Omega |\nabla^2 u|^2 \, dx \\ \qquad\qquad\qquad\qquad\qquad\qquad\qquad\qquad\text{if } u \in W^{2,2}(\Omega), \\ +\infty \text{ if } u \in L^1(\Omega) - W^{2,2}(\Omega), \end{cases}$$

where $W^{2,2}(\Omega)$ is the Sobolev space of L^2-functions whose distributional derivatives up to second order belong to $L^2(\Omega)$ and where f is a lower semicontinuous increasing function from $[0, +\infty[$ to $[0, +\infty[$ such that there exist $\alpha, \beta \in R$ such that

$$\alpha = \lim_{s \to 0^+} \frac{f(s)}{s}, \qquad \beta = \lim_{s \to +\infty} f(s),$$

and $\nabla^2 u$ denotes the Hessian matrix of u equipped with the norm $|A| = \max(\langle A\xi, \xi\rangle, |\xi| = 1)$. In fact, for this kind of approximation one can prove [71] that F_ε Γ-converges in the L^1-topology to a variant of the Mumford and Shah functional:

$$\widetilde{G}(u) = \begin{cases} \displaystyle\int_\Omega (u - u_0)^2 \, dx + \alpha \int_\Omega |\nabla u|^2 \, dx + m(\beta) \int_{S_u} \sqrt{u^+ - u^-} d\mathcal{H}^{N-1} \\ \qquad\qquad\qquad\qquad\qquad\qquad\qquad\quad \text{if } u \in GSBV(\Omega), \\ +\infty \text{ otherwise,} \end{cases}$$

where $m(\beta) = \beta^{3/4}\left(2\sqrt{3/2} + \sqrt{2/3}\right)$ and where $GSBV(\Omega)$ is the space of L^1-functions u for which the truncated function $u_T = -T \vee u \wedge T$ belongs to $SBV(\Omega)$ for all $T > 0$ ($\vee$, respectively $\wedge$, denotes the sup, respectively inf, operator).

(C) *Approximation by introducing nonlocal terms* [72]
    A typical example is

$$F_\varepsilon(u) = \int_\Omega (u - u_0)^2 dx + \frac{1}{\varepsilon}\int_\Omega f\left(\varepsilon \fint_{B(x,\varepsilon)} |\nabla u(y)|^2 dy\right) dx,$$

where f is a suitable nondecreasing continuous function and $\fint_{B(x,\varepsilon)} h(y)\,dy$ denotes the mean value of h on the ball $B(x, \varepsilon)$.
Perhaps the motivation of introducing nonlocal approximation comes from the impossibility, as pointed out in [72] (see also [122]), of obtaining a variational approximation by means of local integral functionals

of the form

$$E_\varepsilon(u) = \int_\Omega (u - u_0)^2 dx + \int_\Omega f_\varepsilon(\nabla u(x))\, dx.$$

Indeed, if such an approximation existed, the Mumford and Shah functional would be also the Γ-limit of the relaxed sequence

$$RE_\varepsilon(u) = \int_\Omega (u - u_0)^2 dx + \int_\Omega f_\varepsilon^{**}(\nabla u(x))\, dx,$$

where f_ε^{**} is the convex envelope of f_ε (see Section 2.1.3). Therefore, the Mumford and Shah functional would also be convex!

(D) *Approximation by finite-difference schemes* [90, 165]

This kind of approximation is perhaps the most natural one from a numerical point of view. The method consists in considering $u(x)$ as a discrete image defined on a mesh of step-size $h > 0$ and F^h as a discrete version of the Mumford and Shah functional. To the best of our knowledge, Chambolle proposed the first theoretical work in that direction [90] following earlier ideas of Blake and Zissermann [59]. In the 1-D case, let

$$g_k^h = \frac{1}{h}\int_{kh}^{(k+1)h} u_0(t)dt,$$

$$u^h = (u_k^h)_{kh \in \Omega} \quad \text{a given discrete signal.}$$

Then Chambolle proposed the following discrete functional:

$$F^h(\,u^h\,) = h \sum_k W_h \left(\frac{u_{k+1}^h - u_k^h}{h}\right) + h \sum_k (u_k^h - g_k^h)^2, \qquad (4.13)$$

where $W_h(t) = \min\left(t^2, 1/h\right)$, and proved that F^h Γ-converges to

$$F(u) = \int_\Omega (u - u_0)^2 dx + \int_{\Omega - S_u} u'^2 dx + \text{card}(S_u) \quad \text{for } u \in SBV(\Omega).$$

This approximation can be adapted in dimension two for

$$F^h(u^h) = h^2 \sum_{k,l} W_h \left(\frac{u_{k+1,l}^h - u_{k,l}^h}{h}\right) + h^2 \sum_{k,l} W_h \left(\frac{u_{k,l+1}^h - u_{k,l}^h}{h}\right)$$

$$+ h^2 \sum_{k,l} (u_{k,l}^h - g_{k,l}^h)^2. \qquad (4.14)$$

A similar Γ-convergence result can be proved, but the 1-D Hausdorff measure is changed into an anisotropic 1-D measure that takes into account the lack of rotational invariance of the natural 2-D extension of (4.14). We refer the reader to [90] for more complete proofs as well as other related works [280, 281, 69, 46, 92].

It is beyond the scope of this book to examine in detail these four ways of approximating the Mumford and Shah functional. That would surely take more than one hundred pages! We choose instead to focus on the approximation (A), proposed by Ambrosio and Tortorelli [17], which is the first that appeared in literature, and which is commonly used in vision. We consider here the case $N = 2$, and we set the parameters α and β to 1:

$$F_\varepsilon(u, v) = \int_\Omega (u - u_0)^2 dx + \int_\Omega v^2 |\nabla u|^2 dx + \int_\Omega \left(\varepsilon |\nabla v|^2 + \frac{1}{4\varepsilon}(v - 1)^2 \right) dx.$$

Before stating rigorous mathematical results, let us show thanks to intuitive arguments how F_ε approaches the Mumford and Shah functional. We reproduce a good explanation given by March [227]. Since the discontinuity set S_u is of zero Lebesgue measure (and so $v(x)$ would be equal to 1 a.e.), our aim is to construct a sequence of functions $(u_\varepsilon, v_\varepsilon)$ converging to $(u, 1)$ such that the sequence $F_\varepsilon(u_\varepsilon, v_\varepsilon)$ converges to $G(u)$ defined in (4.1).

Let us fix some notation. We denote by: $\tau(x)$ the distance of the point x to S_u, and $A_\varepsilon = \{x \, ; \, \tau(x) < \eta(\varepsilon)\}$, $B_\varepsilon = \{x \, ; \, \eta(\varepsilon) < \tau(x) < \gamma(\varepsilon)\}$, with $\lim_{\varepsilon \to 0} \eta(\varepsilon) = \lim_{\varepsilon \to 0} \gamma(\varepsilon) = 0$. We restrict our construction to functions u_ε such that u_ε are smooth on A_ε and $u_\varepsilon = u$ outside A_ε. For the control functions, since ∇u_ε blows up near S_u, the v_ε have to be small in A_ε. We choose $v_\varepsilon = 0$ on A_ε, $v_\varepsilon = 1 - w(\varepsilon)$ outside $A_\varepsilon \cup B_\varepsilon$, with $\lim_{\varepsilon \to 0} w(\varepsilon) = 0$. Finally, we require that v_ε be smooth in the whole domain Ω.

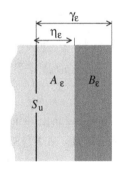

It is easy to verify that the first two integrals in F_ε converge to the first two integrals in G. The treatment of the third term is more delicate. By construction we have

$$\int_\Omega \varepsilon |\nabla v_\varepsilon|^2 \, dx + \int_\Omega \frac{1}{4\varepsilon}(1 - v_\varepsilon)^2 \, dx \qquad (4.15)$$

$$= \frac{1}{4\varepsilon} \left[\int_{A_\varepsilon} dx + w(\varepsilon)^2 \int_{\Omega - A_\varepsilon \cup B_\varepsilon} dx \right] + \int_{B_\varepsilon} \left(\varepsilon |\nabla v_\varepsilon|^2 + \frac{1}{4\varepsilon}(1 - v_\varepsilon)^2 \right) dx.$$

If we choose $\eta(\varepsilon)$ and $w(\varepsilon)$ such that $\eta(\varepsilon)/\varepsilon$ and $w(\varepsilon)^2/(4\varepsilon)$ go to zero as $\varepsilon \to 0$, then the first two integrals on the right-hand side of (4.15) converge to zero. Thus, it remains to study the limit of

$$R_\varepsilon = \int_{B_\varepsilon} \left(\varepsilon |\nabla v_\varepsilon|^2 + \frac{1}{4\varepsilon}(1 - v_\varepsilon)^2 \right) dx.$$

We search for v_ε in B_ε as a function of the form $v_\varepsilon(x) = \sigma_\varepsilon(\tau(x))$ with $\sigma_\varepsilon : R^+ \to R^+$. With this choice, since $|\nabla\tau| = 1$, R_ε reduces to

$$R_\varepsilon = \int_{B_\varepsilon} \left(\varepsilon\, \sigma_\varepsilon'^2(\tau(x)) \right) + \frac{1}{4\varepsilon}\left(1 - \sigma_\varepsilon(\tau(x))^2\right)\, dx.$$

By setting $t = \tau(x)$ and $g(t) = \mathcal{H}^1\{x;\ \tau(x) = t\}$, we obtain

$$R_\varepsilon = \int_{\eta(\varepsilon)}^{\gamma(\varepsilon)} \left(\varepsilon\, \sigma_\varepsilon'^2(t) + \frac{1}{4\varepsilon}(1 - \sigma_\varepsilon(t)^2) \right) g(t)dt.$$

Then let us define σ_ε as the solution of the ordinary differential equation

$$\begin{cases} \sigma_\varepsilon'(t) = \dfrac{1}{2\varepsilon}(1 - \sigma_\varepsilon(t)), \\ \sigma_\varepsilon(\eta(\varepsilon)) = 0. \end{cases}$$

An elementary calculation gives

$$\sigma_\varepsilon(t) = 1 - \exp\left(\frac{\eta(\varepsilon) - t}{2\varepsilon} \right),$$

and R_ε can be rewritten as

$$R_\varepsilon = \frac{1}{2\varepsilon} \int_{\eta(\varepsilon)}^{\gamma(\varepsilon)} \exp\left(\frac{\eta(\varepsilon) - t}{\varepsilon} \right) g(t)\, dt.$$

Thanks to the mean value theorem, there exists $t_0 \in]\eta(\varepsilon), \gamma(\varepsilon)]$ such that

$$R_\varepsilon = \frac{g(t_0)}{2\varepsilon} \int_{\eta(\varepsilon)}^{\gamma(\varepsilon)} \exp\left(\frac{\eta(\varepsilon) - t}{\varepsilon} \right) dt = \frac{g(t_0)}{2} \left(1 - \exp\left(\frac{\eta(\varepsilon) - \gamma(\varepsilon)}{\varepsilon} \right) \right).$$

Choosing $\gamma(\varepsilon)$ such that $\lim_{\varepsilon\to 0} \gamma(\varepsilon)/\varepsilon = +\infty$, and observing that $g(t_0)$ converges to $\mathcal{H}^1(S_u)$, we get

$$\lim_{\varepsilon\to 0} R_\varepsilon = \mathcal{H}^1(S_u).$$

To conclude, we have just constructed a sequence $(u_\varepsilon, v_\varepsilon)$ approaching $(u, 1)$ and such that $\lim_{\varepsilon\to 0} F_\varepsilon(u_\varepsilon, v_\varepsilon) = G(u)$. Naturally, this does not prove the Γ-convergence of F_ε to G, but it gives an idea of the methodology. It is also a way to check how we formally obtain the expected limit.

Now let us return to a more rigorous discussion. We first have to prove that $F_\varepsilon(u, v)$ admits a minimizer and then that $F_\varepsilon(u, v)$ Γ-converges to the Mumford and Shah functional. Let us fix $\varepsilon > 0$. Then $F_\varepsilon(u, v)$ is well-defined on the space $V = \{(u, v) \in W^{1,2}(\Omega)^2\,;\ 0 \leq v \leq 1\}$, and it is weakly (i.e., for the weak topology) lower semicontinuous on this space. To obtain the existence of a minimizer it suffices to bound on V any minimizing sequence $(u_\varepsilon^n, v_\varepsilon^n)$ independently of n. We easily bound in $L^2(\Omega)$ the sequences u_ε^n, v_ε^n, ∇v_ε^n, but a difficulty arises when we want to bound ∇u_ε^n, since we have no control on the term $\int_\Omega (v_\varepsilon^n)^2 |\nabla u_\varepsilon^n|^2 \, dx$. To bypass this difficulty we slightly modify $F_\varepsilon(u,v)$ by adding a perturbation:

$$\widetilde{F}_\varepsilon(u,v) = F_\varepsilon(u,v) + h(\varepsilon) \int_\Omega |\nabla u|^2 \, dx,$$

where $h(\varepsilon) > 0$ is a suitable constant such that $\lim_{\varepsilon \to 0} h(\varepsilon) = 0$. With this modification, it is now clear that $\widetilde{F}_\varepsilon(u,v)$ is coercive on V, and we have proved the following theorem:

Theorem 4.2.7 *Let us suppose that $u_0 \in L^\infty(\Omega)$. Then the problem $\inf_V \widetilde{F}_\varepsilon(u,v)$ admits a solution $(u_\varepsilon, v_\varepsilon)$ with $|u_\varepsilon|_{L^\infty(\Omega)} \le |u_0|_{L^\infty(\Omega)}$.*

When $\varepsilon \to 0$, we have the following Γ-convergence result:

Theorem 4.2.8 [17, 71] *Let $\widetilde{F}_\varepsilon : L^1(\Omega) \times L^1(\Omega) \to [0, +\infty]$ be defined by*

$$\widetilde{F}_\varepsilon(u,v)$$
$$= \begin{cases} \displaystyle\int_\Omega (u - u_0)^2 \, dx + \int_\Omega (v^2 + h(\varepsilon)) |\nabla u|^2 \, dx \\ \qquad + \displaystyle\int_\Omega \left(\varepsilon |\nabla v|^2 + \frac{1}{4\varepsilon}(1 - v)^2 \right) dx & \text{if } (u,v) \in W^{1,2}(\Omega)^2, \ 0 \le v \le 1, \\ +\infty & \text{otherwise,} \end{cases}$$

and let $G : L^1(\Omega) \times L^1(\Omega) \to [0, +\infty]$ be defined by

$$G(u,v) = \begin{cases} \displaystyle\int_\Omega (u - u_0)^2 \, dx + \int_\Omega |\nabla u|^2 \, dx + \mathcal{H}^1(S_u) & \text{if } u \in GSBV(\Omega) \\ & \text{and } v = 1 \text{ a.e.,} \\ +\infty & \text{otherwise.} \end{cases}$$

If $h(\varepsilon) = o(\varepsilon)$, then $\widetilde{F}_\varepsilon(u,v)$ Γ-converges to $G(u,v)$ in the $L^1(\Omega)^2$-strong topology. Moreover, $\widetilde{F}_\varepsilon$ admits a minimizer $(u_\varepsilon, v_\varepsilon)$ such that up to subsequences, u_ε converges in $L^1(\Omega)$ to a minimizer of G, $u \in SBV(\Omega)$, and $\inf \widetilde{F}_\varepsilon \to \inf G(u,v)$ $(\varepsilon \to 0)$.

The proof of Theorem 4.2.8 is long and rather technical. We refer the interested reader to Braides [71].

4.2.5 Experimental Results

A natural method to compute numerically a solution of the Mumford and Shah functional is to consider one of the approximations (A)–(D) described in Section 4.2.4, and then to write the discretized version of it. This can be achieved by using a finite difference scheme [90, 165, 315] (see also the Appendix) or a finite element scheme [46, 69].

We present in Figure 4.7 some experimental results obtained with a modified Mumford and Shah functional, namely the piecewise constant model.

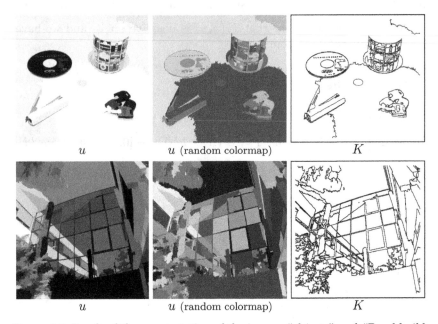

u u (random colormap) K

u u (random colormap) K

Figure 4.7. Result of the segmentation of the images "objects" and "Borel building" using the region-growing approach from *Megawave2*. The solution (u, K) is displayed. Notice that the middle image is displayed with a random colormap just to give a better idea of the different regions. Created in 1993 by Jacques Froment at the CEREMADE, University of Paris 9 Dauphine, *MegaWave2* is now directed by the CMLA laboratory of the Ecole Normale Supieure de Cachan. It can be downloaded from http://www.cmla.ens-cachan.fr/Cmla/Megawave/.

The only change with respect to the Mumford and Shah functional is that we consider only two terms in the energy: the fitting term to the data and the measure of the discontinuity set K:

$$F(u, K) = \int_{\Omega - K} (u - u_0)^2 \, dx + \beta \int_K ds.$$

The strategy to minimize F is relatively simple and belongs to the class of region-merging (or region-growing) methods (see [282]). It consists in:

- Observing that, given a segmentation $\Omega^0 = \bigcup_{i=1}^{M} (\Omega_i^0 \cup K^0)$, then the corresponding minimizer u of $F(u, K^0)$ is a piecewise constant function where the constants are the averages of u_0 over each Ω_i^0.

- Merging recursively all pairs of regions whose merging decreases the energy F, the coefficient $\beta > 0$ playing the role of a scale parameter.

4.3 Geodesic Active Contours and the Level-Set Method

In this section we examine the snake and the geodesic active contours models. Unlike that for the Mumford and Shah functional, the aim is no longer to find a partition of the image but to automatically detect contours of objects. This raises two questions: how a contour may be represented and that criteria would permit one to select the true contours. In many methods of image detection one supposes a sharp variation of the image intensity $I(x)$ between the background of the scene and the objects. Therefore, the magnitude of the gradient of I is high across the boundaries of objects, and we may choose $|\nabla I(x)|$ (or a function of it) as a detector of contours. There exists an extensive literature on snakes and geodesic active contours, and the method is by now widely used in image analysis. Though the theory may be applicable in both two and three dimensions, we develop only the 2-D case for the sake of simplicity.

☞ To know more about active contours, snakes and level sets: [58, 300, 260, 293].

4.3.1 The Kass–Witkin–Terzopoulos model

We begin by describing the Kass–Witkin–Terzopoulos model [190], which is to the best of our knowledge one of the first efforts in this direction. We let Γ denote the set of the image edges (the boundaries of objects). We suppose that $\Gamma = \bigcup_{j \in J} C_j$, J finite or countable, where each C_j is a piecewise $\mathcal{C}^1$ closed curve in R^2. Concerning the intensity $I: \Omega \subset R^2 \to R$ (Ω bounded) we assume that the function $x = (x_1, x_2) \to |\nabla I(x_1, x_2)|$ belongs to $W^{1,\infty}(\Omega)$. In order to characterize edges by zero values rather than by infinite values we define a function $g : [0, +\infty[\to]0, +\infty[$ satisfying

(i) g is regular monotonic decreasing.

(ii) $g(0) = 1$, $\lim_{s \to +\infty} g(s) = 0$.

The function $x \to g(|\nabla I(x)|)$ is called an edge-detector function. A typical choice of g is $g(s) = 1/(1 + s^2)$ (see Figure 4.8).

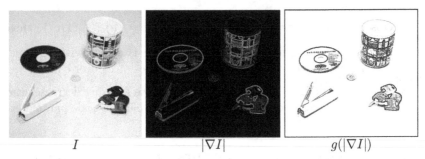

$$I \qquad\qquad |\nabla I| \qquad\qquad g(|\nabla I|)$$

Figure 4.8. Example of edge-detector function. $g(s) = 1/(1+s^2)$ has been chosen for the right-hand image. Noise is reduced, and contours are enhanced. Also note the inversion of colors due to the function g.

In Kass et al. [190], boundary detection consists in matching a deformable model to an image by means of energy minimization. Because of the way the contours move while the energy is minimized they called them *snakes*. Let C be the set of curves of R^2 defined by

$$C = \{c : [a,b] \to \Omega,\ c \text{ piecewise } C^1,\ c(a) = c(b)\}.$$

Then for $c \in C$ let $J(c)$ be the following energy:

$$J(c) = \underbrace{\int_a^b |c'(q)|^2\, dq + \beta \int_a^b |c''(q)|^2\, dq}_{\text{internal energy}} + \underbrace{\lambda \int_a^b g^2(|\nabla I(c(q))|)\, dq}_{\text{external energy}}, \quad (4.16)$$

where

$$c(q) = (c_1(q), c_2(q)),\ c'(q) = \left(\frac{dc_1}{dq}, \frac{dc_2}{dq}\right),\ |c'(q)| = \sqrt{\left(\frac{dc_1}{dq}\right)^2 + \left(\frac{dc_2}{dq}\right)^2},$$

and with the same notation for c''. The first two terms, called spline or internal energy, are used to impose a smoothness constraint. The first-order term makes the curve act like a membrane, and the second-order term makes it act like a thin plate. Setting $\beta = 0$ allows second-order discontinuities as corners. The third term, the external energy, attracts the curve toward the edges of the objects. Since Ω is bounded, it is easy to show that the energy $J(c)$ admits at least a global minimum in the Sobolev space $(W^{2,2}(a,b))^2$. The Euler–Lagrange equations associated with $J(c)$ are a fourth-order system:

$$\begin{cases} -c'' + \beta\, c^{(iv)} + \lambda\, \nabla F|_c(c) = 0, \\ c(a) = c(b), \end{cases} \quad (4.17)$$

where $c^{(iv)}$ is the fourth-order derivative and $F(c_1, c_2) = g^2(|\nabla I(c_1, c_2)|)$. Other boundary conditions can be added.

Unfortunately, since $J(c)$ is nonconvex (it is lower semicontinuous on $(W^{2,2}(a,b))^2)$, no uniqueness result is available, and by solving (4.17) (as done in Kass et al. [190]) we can hope to reach only a local minimum.

☞ *In this approach, the main idea was to formulate the problem as a minimization one.*

However, this approach has significant drawbacks:

- The functional $J(c)$ is not intrinsic, since it depends on the parametrization of c. We could obtain different solutions by changing the parametrization while preserving the same initial curve.

- Because of the regularity constraint, the model does not handle changes of topology. In fact, it is impossible to detect more than one object. Moreover, this object has to be convex.

- In practice, to solve the problem numerically we embed (4.17) into a dynamical scheme by making the curve depend on an artificial parameter (the time) $t \geq 0$; that is, we solve

$$\begin{cases} \dfrac{\partial c}{\partial t}(t,q) = -c''(t,q) + \beta\, c^{(iv)}(t,q) + \lambda\, \nabla F|_c(c(t,q)), \\ c(0,q) = c_0(q), \\ c(t,a) = c(t,b), \end{cases} \tag{4.18}$$

where $c_0(q)$ is an initial curve surrounding the object to be detected. Numerical problems arise in solving (4.18). Since we can reach only a local minimum, we have to choose $c_0(q)$ close enough to the object to be detected. Another difficult task is the choice of a set of marker points for discretizing the parametrized evolving curve. The positions of the marker points have to be updated in time according to the approximations in the equations of the motion. For large and complex motions several problems occur. For example, concentration, or to the contrary, void regions can be created, causing numerical instabilities and false detections. A good explanation of such phenomena is given in [298].

We are now going to show how the above difficulties can be overcome with the *geodesic active contour model*.

4.3.2 The Geodesic Active Contours Model

In the Kass et al. model (4.16), the term $\beta \int_a^b |c''(q)|^2\, dq$ (the elasticity term) is a second-order smoothness component that minimizes the squared curvature. As we will see later, the model with $\beta = 0$ also decreases the curvature, making this term redundant (see (4.24)). It is then natural to introduce the functional J_1 defined by

$$J_1(c) = \int_a^b |c'(q)|^2 \, dq + \lambda \int_a^b g^2(|\nabla I(c(q))|) dq \qquad (4.19)$$

on the set $C = \{c : [a, b] \to \Omega, \ c \text{ piecewise } C^1, \ c(a) = c(b)\}$. Still, the functional J_1 is not yet satisfactory, because it is not intrinsic; that is, it depends on the parametrization of c. So the idea is to introduce the functional J_2 defined by [87, 86, 193, 192]

$$J_2(c) = 2\sqrt{\lambda} \int_a^b g(|\nabla I(c(q))|) \, |c'(q)| \, dq. \qquad (4.20)$$

It is easy to see that J_2 is now intrinsic: If we define a new parametrization of the curve via $q = \phi(r), \phi : [a', b'] \to [a, b], \phi' > 0$, we obtain

$$J_2(c) = 2\sqrt{\lambda} \int_c^d g(|\nabla I(\bar{c}(r))|) |\bar{c}\,'(r)| \, dr$$

with $\bar{c}(r) = c(\phi(r))$, i.e., there is no change in the energy. Therefore, if we compare J_2 to the classical length definition of a curve,[1] we observe that J_2 can be seen as a new length by weighting the Euclidean length. The weight is $g(|\nabla I(c(q))|)$, which contains information regarding the objects' boundaries. In other words, we have defined a new metric (a Riemannian metric) for which we search for geodesics. Beyond this geometric argument, we will see that this formulation also enables us to apply very efficient numerical schemes.

☞ *Starting from J_1, we have introduced a functional J_2 that is intrinsic and that can be interpreted as a weighted Euclidean length. Now the question is to understand the link between the two minimization problems.*

In [87, 86], Caselles et al. have shown, by using concepts of *Hamiltonian theory*, that minimizing J_1 is "equivalent" to minimizing J_2. This idea has been widely reused in the sequel to justify this choice. However, it is unsatisfactory in two ways. The first is that the notion of equivalence is not even clear. It would be natural to say that two minimization problems are equivalent if they have the same solutions or possibly if they have the same extremals. In our case we cannot apply these notions, since it is not clear whether the problem $\inf_c J_2(c)$ has a solution or not. The edge function g

---

[1] The length of a curve is defined by $L = \int_a^b |c'(q)| \, dq$.

being degenerate in any neighborhood of an edge, it would be difficult to bound minimizing sequences in any reasonable space. The second is that we may wonder why it is necessary to use concepts from Hamiltonian theory and whether it could be possible instead to use classical techniques of the calculus of variations.

In this direction, Aubert and Blanc-Féraud [22] defined a precise notion of equivalence and proved it in this context. We do not reproduce the whole discussion of [22], but we will state only what can be, in our opinion, a correct definition of equivalence. Before formulating that definition we need to study the variations of the energies in a neighborhood of a given curve $c(q)$. Calculations developed below are presented in detail, since they will be useful afterwards.

Let $c(q) \in C$ and let $c(t, q)$ be a family of curves, where $t \geq 0$ is an exterior parameter (the time) such that $c(0, q) = c(q)$. Let us set $J_i(t) = J_i(c(t, q))$, $i = 1, 2$. The first step consists in computing $J_i'(t)$, $i = 1, 2$.

- Calculus of $J_1'(t)$. We have

$$J_1(c) = \int_a^b \left| \frac{\partial c}{\partial q}(t, q) \right|^2 dq + \lambda \int_a^b g^2(|\nabla I(c(t, q))|) dq.$$

In order to simplify the notations we write c instead of $c(t, q)$, g for $g(|\nabla I(c(t, q))|)$, and we suppose that $\lambda = 1$. We will also write $u \cdot v = \langle u, v \rangle$. Thus

$$\frac{1}{2} J_1'(t) = \int_a^b \left\langle \frac{\partial c}{\partial q}, \frac{\partial^2 c}{\partial t \partial q} \right\rangle dq + \int_a^b \left\langle \frac{\partial c}{\partial t}, g \nabla g \right\rangle dq.$$

By integrating the first integral by parts with respect to q (we assume that $c(t, a) = c(t, b)$ and $\frac{\partial c}{\partial q}(a, t) = \frac{\partial c}{\partial q}(t, b)$ for all $t > 0$) we get

$$\frac{1}{2} J_1'(t) = \int_a^b \left\langle \frac{\partial c}{\partial t}, -\frac{\partial^2 c}{\partial q^2} + g \nabla g \right\rangle dq.$$

Denoting the arc length by $s = s(q) = \int_a^q \left| \frac{\partial c}{\partial q}(\tau) \right| d\tau$, we always have for any curve c,

$$\frac{\partial^2 c}{\partial q^2} = \left| \frac{\partial c}{\partial q} \right|^2 \frac{\partial^2 c}{\partial s^2} + \left\langle T, \frac{\partial^2 c}{\partial q^2} \right\rangle T,$$

where T denotes the unit tangent vector: $T = \frac{\partial c}{\partial q} / \left| \frac{\partial c}{\partial q} \right|$. Let N be the unit normal vector to the curve and let κ be the curvature. Then we have (see section 2.4)

$$\frac{\partial^2 c}{\partial s^2} = \kappa N.$$

Thus $J_1'(t)$ becomes

$$\frac{1}{2}J_1'(t) = \int_a^b \left\langle \frac{\partial c}{\partial t}, -\kappa \left|\frac{\partial c}{\partial q}\right|^2 N - \left\langle T, \frac{\partial^2 c}{\partial q^2}\right\rangle T + g\nabla g \right\rangle dq.$$

If we decompose the vector ∇g in the tangential and normal directions

$$\nabla g = \langle \nabla g, N\rangle N + \langle \nabla g, T\rangle T,$$

we obtain

$$\frac{1}{2}J_1'(t) = \int_a^b \Bigg\langle \frac{\partial c}{\partial t}, \left[-\kappa \left|\frac{\partial c}{\partial q}\right|^2 + \langle g\nabla g, N\rangle\right]N$$

$$+ \left[\langle g\nabla g, T\rangle - \left\langle T, \frac{\partial^2 c}{\partial q^2}\right\rangle\right]T \Bigg\rangle dq.$$

Therefore, thanks to the Cauchy–Schwarz inequality the flow for which $J_1(t)$ decreases most rapidly is given by

$$\frac{\partial c}{\partial t} = \left[\kappa \left|\frac{\partial c}{\partial q}\right|^2 - \langle g\nabla g, N\rangle\right]N - \left[\langle g\nabla g, T\rangle - \left\langle T, \frac{\partial^2 c}{\partial q^2}\right\rangle\right]T. \quad (4.21)$$

• Calculus of $J_2'(t)$. We have

$$\frac{1}{2}J_2(t) = \int_a^b g(|\nabla I(c(t,q))|) \left|\frac{\partial c}{\partial q}(t,q)\right| dq. \quad (4.22)$$

Then

$$\frac{1}{2}J_2'(t) = \int_a^b g \left\langle \frac{\frac{\partial c}{\partial q}}{\left|\frac{\partial c}{\partial q}\right|}, \frac{\partial^2 c}{\partial t\partial q}\right\rangle dq + \int_a^b \left|\frac{\partial c}{\partial q}\right| \left\langle \nabla g, \frac{\partial c}{\partial t}\right\rangle dq.$$

The first integral on the right-hand side is integrated by parts with respect to q:

$$\frac{1}{2}J_2'(t) = -\int_a^b \left[\left\langle g\frac{\partial}{\partial q}\left(\frac{\frac{\partial c}{\partial q}}{\left|\frac{\partial c}{\partial q}\right|}\right) + \frac{\frac{\partial c}{\partial q}}{\left|\frac{\partial c}{\partial q}\right|}\left\langle \nabla g, \frac{\partial c}{\partial q}\right\rangle, \frac{\partial c}{\partial t}\right\rangle\right] dq$$

$$+ \int_a^b \left|\frac{\partial c}{\partial q}\right| \left\langle \nabla g, \frac{\partial c}{\partial t}\right\rangle dq.$$

This equation can be rewritten as

$$\frac{1}{2}J_2'(t) = \int_a^b \left|\frac{\partial c}{\partial q}\right| \left\langle \frac{\partial c}{\partial t}, \nabla g - \frac{1}{\left|\frac{\partial c}{\partial q}\right|}\frac{\partial}{\partial q}\left(\frac{\frac{\partial c}{\partial q}}{\left|\frac{\partial c}{\partial q}\right|}\right) g \right.$$
$$\left. - \frac{\frac{\partial c}{\partial q}}{\left|\frac{\partial c}{\partial q}\right|}\left\langle \nabla g, \frac{\frac{\partial c}{\partial q}}{\left|\frac{\partial c}{\partial q}\right|}\right\rangle \right\rangle dq,$$

and recalling the definitions of T, N, and κ, we get

$$\frac{1}{2}J_2'(t) = \int_a^b \left|\frac{\partial c}{\partial q}\right| \left\langle \frac{\partial c}{\partial t}, \nabla g - \kappa g N - \langle T, \nabla g\rangle T \right\rangle dq.$$

Again decomposing ∇g on the basis (N, T), we finally obtain

$$\frac{1}{2}J_2'(t) = \int_a^b \left|\frac{\partial c}{\partial q}\right| \left\langle \frac{\partial c}{\partial t}, \langle \nabla g, N\rangle N - \kappa g N \right\rangle dq, \qquad (4.23)$$

so the direction for which $J_2(t)$ decreases most rapidly is given by

$$\frac{\partial c}{\partial t} = (\kappa g - \langle \nabla g, N\rangle)N. \qquad (4.24)$$

Remark Note that if $g \equiv 1$, then the flow (4.24) reduces to

$$\frac{\partial c}{\partial t} = \kappa N, \qquad (4.25)$$

which is the well-known mean curvature motion (or shortening flow). This `C++`
flow decreases the total curvature as well as the number of zero crossings
and the value of maxima/minima curvature. Therefore, it has the properties of shortening (an initial curve shrinks under (4.25) to a point in finite
time with asymptotically circular shape) as well as smoothing (points with
high curvature evolve faster and disappear asymptotically). An example is
shown in Figure 4.9 (see also Section A.3.4 in the Appendix). For more
geometric details about (4.25) we refer to [147].   ∎

We are now in position to state what we mean by saying that the two
minimization problems $\inf_c J_1(c)$ and $\inf_c J_2(c)$ are equivalent.

Definition 4.3.1 (equivalence between $\inf_c J_1(c)$ and $\inf_c J_2(c)$)
*The problems $\inf_c J_1(c)$ and $\inf_c J_2(c)$ are equivalent if for any curve $c \in C$
there exists a neighborhood $V(c)$ of c such that in $V(c)$ the flow that most
decreases J_1 is also a decreasing flow for J_2 and vice versa.*

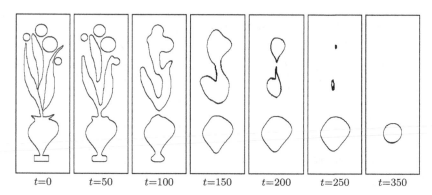

$$t=0 \quad t=50 \quad t=100 \quad t=150 \quad t=200 \quad t=250 \quad t=350$$

Figure 4.9. Example of the mean curvature motion.

In order to apply this criterion, we need a more precise definition of an edge.

Definition 4.3.2 (edge) *Let $c \in C$. We say that c is an edge of the image I if there exists ε_0 such that $\forall \varepsilon < \varepsilon_0$, $\exists\, \alpha_\varepsilon$, $\lim_{\varepsilon \to 0} \alpha_\varepsilon = 0$, such that $|\nabla I(x)| \geq 1/\varepsilon$ if $x \in V_\varepsilon = \{x : \; d((x); c) \leq \alpha_\varepsilon\}$, where d is the distance function.*

Theorem 4.3.1 *Let us assume that the edge-detection function g satisfies the following: For any edge $c \in C$ there exist l, $l' \in Q$, $l < l'$, such that $\forall \varepsilon < \varepsilon_0, \forall x \in V_\varepsilon$:*

(i) $g(|\nabla I(x)|) = O(\varepsilon^l)$.

(ii) $|\nabla g(|\nabla I(x)|)| = O(\varepsilon^{l'})$.

Then the two minimization problems $\inf_c J_1(c)$ and $\inf_c J_2(c)$ are equivalent in the sense of Definition 4.3.1.

Proof We have to prove that the flow (4.21) that makes J_1 decrease most rapidly also makes J_2 decrease, and conversely, the flow (4.24) that makes J_2 decrease most rapidly also makes J_1 decrease. In order to do that we replace $\partial c/\partial t$ given by (4.21) in the expression of $J_2'(t)$ and vice versa. When $\partial c/\partial t$ is given by (4.21), $J_2'(t)$ becomes

$$J_2'(t) = \int_a^b -g(\langle \nabla g, N\rangle - \kappa g)\left(\langle \nabla g, N\rangle - \frac{\kappa}{g}\left|\frac{\partial c}{\partial q}\right|^2\right)\left|\frac{\partial c}{\partial q}\right| dq. \quad (4.26)$$

In the same way, if we replace $\partial c/\partial t$ by its expression in (4.24), then $J_1'(t)$ becomes

$$J_1'(t) = \int_a^b -g(\langle \nabla g, N\rangle - \kappa g)\left(\langle \nabla g, N\rangle - \frac{\kappa}{g}\left|\frac{\partial c}{\partial q}\right|^2\right)\left|\frac{\partial c}{\partial q}\right| dq. \quad (4.27)$$

In order to determine the sign of $J_2'(t)$ in (4.26) or the sign of $J_1'(t)$ in (4.27) it suffices to study a.e. q the sign of the integrand

$$z(t,q) = -g(\langle \nabla g, N \rangle - \kappa g)\left(\langle \nabla g, N \rangle - \frac{\kappa}{g}\left|\frac{\partial c}{\partial q}\right|^2\right).$$

By developing this expression we obtain:

$$z(t,q) = -\kappa^2\left|\frac{\partial c}{\partial q}\right|^2 g + \kappa\left(\left|\frac{\partial c}{\partial q}\right|^2 + g^2\right)\langle \nabla g, N \rangle - g(\langle \nabla g, N \rangle)^2.$$

Let us recall now that g stands for $g(|\nabla I(c(t,q))|)$ and let us assume that $c(t,q)$ is in a neighborhood V_ε of an edge $c(0,q)$ of the image I. Then we have, thanks to the assumptions (i) and (ii) on g

$$z(t,q) \approx -a\varepsilon^l \pm b\varepsilon^{l'} - c\varepsilon^{2l'}$$

with $a, b, c \geq 0$. When ε is small enough the sign of $z(t,q)$ is given by the sign of $-a\varepsilon^l$, and so it is negative (we suppose that κ is bounded; otherwise, as soon as $|\kappa| \gg 1$ we have $\kappa^2 \geq |\kappa|$, and z would remain nonpositive). ∎

The Caselles et al. model [87, 86] may be improved by adding to the right-hand side of (4.24) a supplementary term

$$\frac{\partial c}{\partial t} = (\kappa g - \langle \nabla g, N \rangle + \alpha g)N. \tag{4.28}$$

The main interest in adding αg to the velocity is that it makes the detection of nonconvex objects easier and it increases the speed of convergence. In fact, $\alpha \geq 0$ must be chosen large enough so that the coefficient $(\kappa + \alpha)$ remains of constant sign. Consequently, the curvature κ can have a nonconstant sign, and nonconvex shapes can then be detected.

Remark   It is worth noting that equation (4.28) does not come from any energy unless the function g is a constant (and this not the case, since g is a detector function). As a matter of fact, the flow $\partial c/\partial t = \alpha N$ with $\alpha \geq 0$ (a constant) is the flow deduced from the area energy

$$A(t) = -\frac{1}{2}\int_a^b \left\langle c(t,q), \begin{pmatrix} -\partial c_2/\partial q \\ \partial c_1/\partial q \end{pmatrix} \right\rangle dq.$$

For each $t > 0$, $A(t)$ is the area enclosed by $c(t,q) = (c_1(t,q), c_2(t,q))$, and it is easy to verify that

$$A'(t) = -\int_a^b \left\langle \frac{\partial c}{\partial t}, N \right\rangle \left|\frac{\partial c}{\partial q}\right| dq.$$

Thus the direction in which $A(t)$ is decreasing most rapidly is $\partial c/\partial t = N$. We refer to [302] for a complete study of area minimizing flows. ∎

To summarize the situation, starting from the initial formulation of Kass et al. (4.16), we introduced the energy J_1 and an intrinsic (geometric) functional J_2. We clearly defined the link between these two optimization problems. From a numerical point of view, we can wonder which formulation is the best to choose. As we are going to see in the next section, the Euler–Lagrange equations associated with J_2 can be written in an Eulerian formulation by using a *level set approach* (which is not the case for J_1). The level set approach is based on the description of the curve as the zero crossing of a higher-dimensional function and allows major simplifications.

4.3.3   The Level-Set Method

The aim of this section is to find an efficient algorithm to solve (4.28). Naturally, one could parametrize the curve c and discretize the equation, but this direct approach faces difficulties that we will emphasize later. More generally, we are interested in flows governed by equations of the form

$$\begin{cases} \dfrac{\partial c}{\partial t} = F\,N, \\ c(0,q) = c_0(q). \end{cases} \tag{4.29}$$

Equation (4.29) says that the curve $c(t,q)$ moves along its normal with a speed F, which may depend on t, c, c', c''. The *level set formulation* is based on the following observation due to [134, 262]:

☛ *A curve can be seen as the zero-level of a function in higher dimension.*

For example, a curve in R^2 can be represented as the zero-level line of a function $R^2 \to R$ (see Figure 4.10). More precisely, let us suppose that there exists a function $u : R^+ \times R^2 \to R$ such that

$$u(t, c(t,q)) = 0 \quad \forall q,\ \forall t \geq 0. \tag{4.30}$$

Then if u is sufficiently regular, by differentiating (4.30) with respect to t we obtain

$$\frac{\partial u}{\partial t} + \left\langle \nabla u\,, \frac{\partial c}{\partial t} \right\rangle = 0,$$

and by replacing the expression for the speed given in (4.29) we get

$$\frac{\partial u}{\partial t} + \langle \nabla u\,, F\,N \rangle = 0. \tag{4.31}$$

But recalling that the unit inward normal to the front defined by (4.30) is given by $N = -\frac{\nabla u}{|\nabla u|}$ (we suppose that u is negative inside the curve and positive outside), then (4.31) can be rewritten as

$$\frac{\partial u}{\partial t}(t, c(t,q)) = F\,|\nabla u(t, c(t,q))|. \tag{4.32}$$

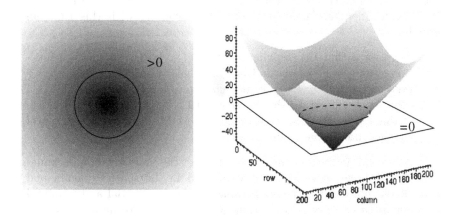

Figure 4.10. Basis of the level set approach: A closed curve can be seen as the zero-level of a function in higher dimension. For instance, the function can be the signed distance to the curve, negative inside and positive outside.

According to the way we have established (4.32), this equation is a priori valid only for the zero-level set of u. But one of the advantages of the method is that u may be regarded as defined on the whole domain $R^+ \times \Omega$. So, we can solve the PDE

$$\frac{\partial u}{\partial t}(t,x) = F \, |\nabla u(t,x)|$$

for $t \geq 0$ and $x \in \Omega$ as soon as F is well-defined off the front, i.e., on the whole space. Then once u is calculated on $R^+ \times \Omega$ we just need to extract the zero-level set of u to get the curve. We will return to this question later. Of course, we have to add the following conditions.

(i) A boundary condition: One generally chooses that the normal derivative vanishes on $\partial\Omega$ i.e., $\frac{\partial u}{\partial N} = 0$ on $\partial\Omega$.

(ii) An initial condition at $t = 0$. A good candidate is the signed distance function to an initial given curve $c_0(q)$ surrounding the objects:

$$u(0,x) = \bar{d}(x,c_0) = \begin{cases} +d(x,c_0) & \text{if } x \text{ is outside } c_0, \\ -d(x,c_0) & \text{if } x \text{ is inside } c_0, \end{cases}$$

where $d(x,c_0)$ is the Euclidean distance to c_0.

Therefore, the final model is

$$\begin{cases} \dfrac{\partial u}{\partial t}(t,x) = F \, |\nabla u(t,x)| \text{ for } (t,x) \in \,]0,+\infty[\, \times \, \Omega, \\ u(0,x) = \bar{d}(x,c_0), \\ \dfrac{\partial u}{\partial N} = 0 \text{ for } (t,x) \in \,]0,+\infty[\, \times \, \partial\Omega. \end{cases} \qquad (4.33)$$

The equation (4.33) is called a Hamilton–Jacobi equation (see also Section 2.3.2). There are many advantages to working with this Eulerian formulation:

- The first is that the evolving function $u(t, x)$ always remains a function as long as F is smooth. But if we consider only the level set $u = 0$ (and so the front $c(t, q)$), it may change topology, break, merge as u evolves. We illustrate this in Figure 4.11. This is a main advantage of this representation, since we do not need to take these topological changes into account numerically.

- A second important interest concerns the numerical approximation: We can use a fixed discrete grid in the spatial domain and choose finite-difference approximations for the spatial and temporal derivatives. We refer to Sections 4.3.5 and A.3.4 for more details.

- Another advantage is that intrinsic geometric elements of the front such as the normal vector and the curvature can be easily expressed with respect to u. Notice that this is a necessary condition for any representation to be useful.

- Finally, this above level set formulation can be extended and applied in any dimension. For instance, a surface can be represented implicitly by the zero-level set of a function defined in a volume.

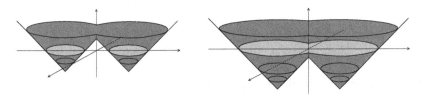

Figure 4.11. Illustration of the change of topology.

☛ *This representation is useful as soon as the evolution of a hypersurface is involved and when the motion can be expressed as a velocity along the normal.*

This approach has been extensively used in computer vision as well as in other domains.

👁 To know more about the level set method: [300, 260, 293].

In image analysis, [85, 222, 192, 223, 86, 193, 87] are some of the papers that first appeared in this direction. All these papers rely on the same ideas. The major difference between them is that [87, 193, 192, 86] start from an energy concept (the minimization of the weighted length), while [85, 223, 222] formulate directly their problem in terms of level sets.

To give an example, let us return to the segmentation problem. We can show that the level sets expression of (4.28) is

$$\frac{\partial u}{\partial t} = \left((\kappa + \alpha)g + \left\langle \nabla g, \frac{\nabla u}{|\nabla u|} \right\rangle \right) |\nabla u|,$$

or, remembering that the curvature κ is given by $\kappa = \text{div}\left(\frac{\nabla u}{|\nabla u|}\right)$,

$$\frac{\partial u}{\partial t} = g(|\nabla I|)\left(\text{div}\left(\frac{\nabla u}{|\nabla u|}\right) + \alpha\right)|\nabla u| + \langle \nabla g, \nabla u\rangle, \qquad (4.34)$$

with the boundary and initial conditions of (4.33). In the first term, the coefficient $g(|\nabla I|)$ permits one to stop the evolving curve when it arrives at the object boundaries.[2] The action of the second term, $\langle \nabla g, \nabla u\rangle$, is less obvious. To better understand its contribution, let us consider the following one-dimensional example. Let $I(x)$ be the Heaviside function, $I(x) = 1$ if $x \geq 0$ and $I(x) = 0$ otherwise, and let I_ε be a regularization of I by a cubic function:

$$I_\varepsilon(x) = \begin{cases} 1 & \text{if } x \geq \varepsilon, \\ -\dfrac{x^3}{4\varepsilon^3} + \dfrac{3x}{4\varepsilon} + \dfrac{1}{2} & \text{if } -\varepsilon \leq x \leq \varepsilon, \\ 0 & \text{if } x \leq -\varepsilon. \end{cases}$$

Then, if we set $g_\varepsilon(x) = \frac{1}{1+|I'_\varepsilon(x)|^2}$, it is easy to verify that in a neighborhood of $x = 0$ we have $g_\varepsilon(x) \approx \varepsilon^2$ and $g'_\varepsilon(x) \approx \frac{9}{4\varepsilon^4}x$ (see Figure 4.12). Therefore, the leading term in (4.34) is (in this one-dimensional case) the

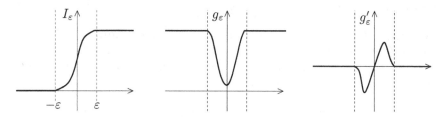

Figure 4.12. The initial signal I_ε and the functions $g_\varepsilon(x)$, $g'_\varepsilon(x)$.

transport term $g'_\varepsilon(x)\,u'(x)$. Thus, the front evolves from right to left for $x > 0$ and from left to right for $x < 0$. The point $x = 0$ (the discontinuity front) can be seen as an attractive point. This effect is the same for images.

---

[2]Notice that in practice, because of the presence of additional noise in the image, we use a smoothed version of I.

☛ *the term $\langle \nabla g, \nabla u \rangle$ increases the attraction of the deforming contour toward the boundary of objects.*

Before proceeding with the approximation we now present results regarding the existence and uniqueness of a solution for (4.34), using the theory of viscosity solutions (see Section 2.3).
According to the identity

$$\operatorname{div}\left(\frac{\nabla u}{|\nabla u|}\right) = \frac{1}{|\nabla u|^3}\left[(u_{x_1})^2 u_{x_2 x_2} + (u_{x_2})^2 u_{x_1 x_1} - 2u_{x_1} u_{x_2} u_{x_1 x_2}\right],$$

where $u_{x_i} = \frac{\partial u}{\partial x_i}$, the equation (4.34) can be rewritten as

$$\frac{\partial u}{\partial t} = g(x) \sum_{i,j=1}^{2} a_{i,j}(\nabla u)\, u_{x_i x_j} + H(x, \nabla u), \qquad (4.35)$$

where

$$a_{i,j}(p) = \delta_{i,j} - \frac{p_i p_j}{|p|^2} \quad \text{if } p \neq 0 \quad (\delta_{i,j} \text{ is the Kronecker symbol}), \quad (4.36)$$

$$H(x,p) = \alpha\, g(x)\, |p| + \sum_{i=1}^{2} \frac{\partial g}{\partial x_i}(x)\, p_i \quad \text{with } g(x) = g(|\nabla I(x)|). \qquad (4.37)$$

To avoid some tedious technicalities at corners, instead of the Neumann boundary condition $\frac{\partial u}{\partial N} = 0$ on $]0, +\infty[\times \partial\Omega$, we will work with periodic solutions (see Section 3.3.1 to extend the function u defined on Ω to a periodic function defined on R^2). Of course, we also suppose that the initial condition $u(0, x) = u_0(x)$ is periodic.

Let us recall the definition of viscosity solutions, adapting Definition 2.3.1 for the parabolic equation (4.35):

Definition 4.3.3 (viscosity subsolution, supersolution, solution) *Let u be in $C([0, T] \times R^2)$, $0 < T < \infty$. Then u is a viscosity subsolution of (4.35) if for $\varphi \in C^2([0, T] \times R^2)$ the following condition holds: At any point (t_0, x_0) in $]0, T[\times R^2$ that is a local maximum of $(u - \varphi)$,*

$$\frac{\partial \varphi}{\partial t}(t_0, x_0) - g(x_0) \sum_{i,j=1}^{2} a_{i,j}(\nabla\varphi(t_0, x_0))\, \varphi_{x_i x_j}(t_0, x_0) - H(x_0, \nabla\varphi(t_0, x_0)) \leq 0$$

$$\text{if } |\nabla\varphi(t_0, x_0)| \neq 0,$$

$$\frac{\partial \varphi}{\partial t}(t_0, x_0) - g(x_0) \overline{\lim_{p \to 0}} \sum_{i,j=1}^{2} a_{i,j}(p)\, \varphi_{x_i x_j}(t_0, x_0) \leq 0 \quad \text{if } |\nabla\varphi(t_0, x_0)| = 0.$$

Similarly, we define the notion of a viscosity supersolution by changing local maximum to local minimum, ≤ 0 to ≥ 0, and $\overline{\lim}$ to $\underline{\lim}$. A viscosity solution is a continuous function that is both a sub- and a supersolution.

If V is a Banach space, we recall that $L^\infty((0,T);V)$ is the space defined by

$$L^\infty((0,T);V) = \left\{ \begin{array}{l} f : (0,T) \to V \text{ such that} \\ |f|_{L^\infty((0,T);V)} = \inf\{c; |f|_V \le c, \text{ a.e. } t\} < \infty \end{array} \right\}.$$

We may now state the main result.

Theorem 4.3.2 *Let us assume that $g \ge 0$, with g and $\sqrt{g}$ Lipschitz continuous.*

(i) *Let $u_0(x)$ be the initial condition, Lipschitz continuous. Then (4.35) has a unique viscosity solution in $C([0,\infty[\times R^2) \cap L^\infty((0,T); W^{1,\infty}(R^2))$ for any $T < \infty$. Moreover,*

$$\inf_{R^2} u_0(x) \le u(t,x) \le \sup_{R^2} u_0(x).$$

(ii) *Let v be a viscosity solution of (4.35) with u_0 replaced by v_0. Then for all T in $[0,+\infty[$ we have,*

$$\sup_{0 \le t \le T} |u(t,.) - v(t,.)|_{L^\infty(R^2)} \le |u_0 - v_0|_{L^\infty(R^2)}. \tag{4.38}$$

Proof The proof follows [85]. It is rather technical and long, and we divide it into three steps.

Step 1: Stability and uniqueness
We begin with the stability estimate (4.38), from which we will deduce the uniqueness of a solution. Let u and v be two (viscosity) solutions associated respectively with u_0 and v_0. We are interested in the maximum of $|u(t,x) - v(t,x)|$. If u and v are smooth, we can proceed as described in Section 2.3.4. Otherwise, we need a trick, the duplication of variables. Let us define

$$l(t,x,y) = u(t,x) - v(t,y) - \frac{1}{4\varepsilon}|x-y|^4 - \lambda t, \ t \in [0,T], \ x,y \in R^2, \tag{4.39}$$

where $\varepsilon, \lambda \in]0,+\infty[$ will be determined later. Now we are interested in the maximum of l. Let (t_0, x_0, y_0) be a maximum point of l on $[0,T] \times R^2 \times R^2$ (ε fixed).

We claim that $t_0 = 0$. Otherwise, the maximum would be attained at some point (t_0, x_0, y_0) with $t_0 > 0$. In that case thanks to the Crandall–Ishii lemma (Lemma 2.3.2), we can find for any $\mu > 0$ two real numbers a, b and two symmetric (2×2) matrices X and Y such that

$$a - b = \lambda, \quad \begin{pmatrix} X & 0 \\ 0 & -Y \end{pmatrix} \le \begin{pmatrix} B + \mu B^2 & -B - \mu B^2 \\ -B - \mu B^2 & B + \mu B^2 \end{pmatrix} \tag{4.40}$$

with

$$B_{i,j} = \varepsilon^{-1}|x_0 - y_0|^2 \delta_{i,j} + 2\varepsilon^{-1}(x_0 - y_0)_i(x_0 - y_0)_j.$$

Moreover, if $x_0 \neq y_0$, one has

$$a - g(x_0)a_{i,j}(\varepsilon^{-1}|x_0 - y_0|^2(x_0 - y_0))X_{i,j} \tag{4.41}$$

$$- H(x_0, \varepsilon^{-1}|x_0 - y_0|^2(x_0 - y_0)) \leq 0,$$

$$b - g(y_0)a_{i,j}(\varepsilon^{-1}|x_0 - y_0|^2(x_0 - y_0))Y_{i,j} \tag{4.42}$$

$$- H(y_0, \varepsilon^{-1}|x_0 - y_0|^2(x_0 - y_0)) \geq 0.$$

In fact, if $x_0 = y_0$, the two last inequalities have to be interpreted in terms of suitable limits. In that case $B = 0$, $X \leq 0$, $Y \geq 0$ and (4.41), (4.42) can be rewritten as

$$a - g(x_0)\varlimsup_{p \to 0} a_{i,j}(p)\, X_{i,j} \leq 0,$$

$$b - g(x_0)\varliminf_{p \to 0} a_{i,j}(p)\, Y_{i,j} \geq 0.$$

Hence in particular, $a \leq 0$, $b \geq 0$, which contradicts $a - b = \lambda > 0$. So we have $x_0 \neq y_0$.

Next, we choose $\mu = \varepsilon |x_0 - y_0|^{-2}$ (which is now possible) and deduce from (4.40)

$$\begin{pmatrix} X & 0 \\ 0 & -Y \end{pmatrix} \leq 2\varepsilon^{-1}\begin{pmatrix} C & -C \\ -C & C \end{pmatrix}, \tag{4.43}$$

where $c_{i,j} = |x_0 - y_0|^2 \delta_{i,j} + 5(x_0 - y_0)_i(x_0 - y_0)_j$. Then we set

$$G = \begin{pmatrix} g(x_0)A & \sqrt{g(x_0)g(y_0)}A \\ \sqrt{g(x_0)g(y_0)}A & g(y_0)A \end{pmatrix},$$

where $A = a_{i,j}(\varepsilon^{-1}|x_0 - y_0|^2(x_0 - y_0))$. G is a nonnegative symmetric matrix, so that multiplying (4.43) on the left by G and taking the trace we get

$$g(x_0)\sum_{i,j} a_{i,j}X_{i,j} - g(y_0)\sum_{i,j} a_{i,j}Y_{i,j} \leq 2\varepsilon^{-1}(\sqrt{g(x_0)} - \sqrt{g(y_0)})^2\text{trace}(AC).$$

$$\tag{4.44}$$

Next, combining (4.40), (4.41), (4.42), (4.44) and carrying out some manipulations we obtain

$$\lambda \leq \frac{c_1}{\varepsilon}|x_0 - y_0|^4, \tag{4.45}$$

where c_1 is a constant depending only on $a_{i,j}(p)$ and g. We now estimate $|x_0 - y_0|$. According to the definition of (t_0, x_0, y_0) we have

$$u(t_0, x_0) - v(t_0, y_0) - \frac{1}{4\varepsilon}|x_0 - y_0|^4 - \lambda t_0 \geq u(t_0, y_0) - v(t_0, y_0) - \lambda t_0$$

and thus

$$\frac{1}{4\varepsilon}|x_0 - y_0|^4 \leq c_2 |x_0 - y_0|, \tag{4.46}$$

where c_2 is the Lipschitz constant of $u(t_0, .)$ on $[0,T] \times R^2$. Therefore, from (4.45)

$$\lambda \leq c_3 \varepsilon^{1/3} \text{ with } c_3 = c_1 4^{4/3} c_2^{4/3}.$$

Now, recall that λ and ε are arbitrary. Without loss of generality we may suppose that $\sup_{[0,T] \times R^2} |u - v| \neq 0$ (otherwise, we are done), and we choose

$$\varepsilon^{1/3} = \delta \sup_{[0,T] \times R^2} |u - v| \quad (\delta > 0), \tag{4.47}$$

$$\lambda = 2\delta c_3 \sup_{[0,T] \times R^2} |u - v|. \tag{4.48}$$

This choice contradicts (4.46), and so $t_0 = 0$.

Next let us estimate $\sup_{[0,T] \times R^2} |u - v|$. We fix λ and ε as before. Since $t_0 = 0$, we have for all (t, x, y),

$$u(t,x) - v(t,y) - \frac{1}{4\varepsilon}|x - y|^4 - \lambda t \leq u_0(x_0) - v_0(y_0) - \frac{1}{4\varepsilon}|x_0 - y_0|^4, \tag{4.49}$$

but observing that

$$u_0(x_0) - v_0(y_0) = u_0(y_0) - v_0(y_0) + u_0(x_0) - u_0(y_0)$$
$$\leq |u_0(y_0) - v_0(y_0)| + c_2|x_0 - y_0|$$

and letting $x = y$ in (4.49), we get

$$\sup_{[0,T] \times R^2} (u(t,x) - v(t,x)) - \lambda t \leq \sup_{y \in R^2} |u_0(y) - v_0(y)| + \sup_{r > 0} \left(c_2 r - \frac{1}{4\varepsilon}r^4 \right),$$

i.e.,

$$\sup_{[0,T] \times R^2} (u(t,x) - v(t,x)) - \lambda t \leq \sup_{y \in R^2} |u_0(y) - v_0(y)| + \frac{3}{4}c_2^{4/3}\varepsilon^{1/3}.$$

By (4.47) and (4.48) we obtain

$$\sup_{[0,T] \times R^2} (u-v) \leq \sup_{y} |u_0(y) - v_0(y)| + \frac{3}{4}c_2^{4/3}\delta \sup_{[0,T] \times R^2} |u - v| + 2\delta c_3 \sup_{[0,T] \times R^2} |u - v| T.$$

Exchanging the role of u and v and letting $\delta \to 0$, we obtain

$$\sup_{[0,T] \times R^2} |u - v| \leq \sup_{y} |u_0(y) - v_0(y)|.$$

This proves part (ii) of Theorem 4.3.2 and the uniqueness in part (i). Observe that by using the definition of viscosity solutions we may also obtain the following bound: $\inf u_0(x) \leq u(t,x) \leq \sup u_0(x)$ for all (t,x).

We now prove the existence part of the theorem. We begin by setting an a priori estimate on ∇u.

Step 2: Estimate of ∇u on $L^\infty(R^2)$

In this step we suppose that all the coefficients of (4.35) have been smoothed as much as necessary (notation is unchanged) and that (4.35) admits a regular solution (we will return to this question in step 3). Thus let u be a regular solution of

$$\frac{\partial u}{\partial t} = g(x) \sum_{i,j} a_{i,j}(\nabla u) u_{x_i x_j} + H(x, \nabla u).$$

We are going to apply the classical *Bernstein method* and derive a parabolic inequality for $w = |\nabla u|^2$. To this end we differentiate (4.35) with respect to x_l:

$$\frac{\partial u_{x_l}}{\partial t} - \frac{\partial g}{\partial x_l} \sum_{i,j} a_{i,j}(\nabla u) u_{x_i x_j} - g \left[\sum_{i,j} \sum_k \frac{\partial a_{i,j}(\nabla u)}{\partial p_k} u_{x_k x_l} u_{x_i x_j} \right.$$

$$\left. + a_{i,j}(\nabla u) u_{x_l x_i x_j} \right] - \frac{\partial H}{\partial x_l} - \sum_k \frac{\partial H}{\partial p_k} u_{x_k x_l} = 0.$$

We multiply this equation by $2u_{x_l}$ and sum over l to obtain

$$\frac{\partial w}{\partial t} - g \left[\sum_{i,j} a_{i,j}(\nabla u) w_{x_i x_j} + \sum_{i,j} \sum_k \frac{\partial a_{i,j}(\nabla u)}{\partial p_k} u_{x_i x_j} w_{x_k} \right] - \sum_k \frac{\partial H}{\partial p_k} w_{x_k}$$

$$\tag{4.50}$$

$$= 2 \sum_l \frac{\partial g}{\partial x_l} u_{x_l} \sum_{i,j} a_{i,j}(\nabla u) u_{x_i x_j} - 2 \sum_l \frac{\partial H}{\partial x_l} u_{x_l} - 2g \sum_l \sum_{i,j} a_{i,j}(\nabla u) u_{x_l x_i} u_{x_l x_j},$$

where $w = |\nabla u|^2$. We denote by Lw the left-hand side of (4.50). On the other hand, for the right-hand side of (4.50), since g is smooth, we have the following:

(i) $\sum_l u_{x_l} \frac{\partial H}{\partial x_l} = \sum_l u_{x_l} \left[\alpha \frac{\partial g}{\partial x_l} |\nabla u| + \sum_i \frac{\partial^2 g}{\partial x_i \partial x_l} u_{x_i} \right] \le k_1 |\nabla u|^2 = k_1 w,$

where k_1 is a constant depending only on g.

(ii) Since the matrix $a_{i,j}$ is symmetric semidefinite, we can reduce it to its principal axes, and we easily obtain

$$\left(\sum_{i,j} a_{i,j} u_{x_i} u_{x_j} \right)^2 \le k_2 \sum_k \sum_{i,j} a_{i,j} u_{x_i x_k} u_{x_j x_k} \quad \text{for some constant } k_2.$$

Using the Cauchy–Schwarz inequality, the obvious inequality $2ab \le a^2 + b^2$, and again that $a_{i,j}$ is symmetric semidefinite, we obtain

$$2 \sum_k \frac{\partial g}{\partial x_k} u_{x_k} \sum_{i,j} a_{i,j} u_{x_i} u_{x_j} - 2g \sum_k \sum_{i,j} a_{i,j} u_{x_i x_k} u_{x_j x_k} \le k_3 w.$$

Inserting these bounds in (4.50), we have just proved the existence of a constant $k_3 > 0$ such that $Lw \le k_3 w$. Unfortunately, since $k_3 > 0$, we cannot directly apply the maximum principle to get a result of the kind

$$|w(t,.)|_{L^\infty(R^2)} \leq |w_0(.)|_{L^\infty(R^2)} \quad \text{a.e.}$$

However, we can still show that (see the remark at the end this proof)

$$|w(t,.)|_{L^\infty(R^2)} \leq (1 + ct\, e^{ct})\, |w_0(.)|_{L^\infty(R^2)} \quad \text{a.e.,}$$

where c is a constant depending only on k_3. For u this means

$$|\nabla u(t,.)|_{L^\infty(R^2)} \leq k\,|\nabla u_0|_{L^\infty(R^2)}, \tag{4.51}$$

where the constant k depends only on $|g|_{L^\infty(R^2)}$, $|\nabla g|_{L^\infty(R^2)}$, and T. In particular, k does not depend on the way the coefficients have been regularized. This inequality is proved in the remark following the proof. Notice that this result is interesting only because we consider t bounded.

Step 3: Approximation and existence of a viscosity solution
In order to conclude, we are going to approximate (4.35) by a similar equation for which we are able to prove the existence of a smooth solution satisfying (4.51). To this end we consider a periodic C^∞ function u_0^ε such that $u_0^\varepsilon \to u_0$ uniformly satisfying

$$|\nabla u_0^\varepsilon|_{L^\infty(R^2)} \leq |\nabla u_0|_{L^\infty(R^2)}, \quad |u_0^\varepsilon|_{L^\infty(R^2)} \leq |u_0|_{L^\infty(R^2)}.$$

We also replace $a_{i,j}$, g, and H respectively by

$$a_{i,j}^\varepsilon(p) = \varepsilon\delta_{i,j} + \delta_{i,j} - \frac{p_i p_j}{|p|^2 + \varepsilon},$$

$$g_\varepsilon = g + \varepsilon,$$

$$H_\varepsilon(x,p) = \alpha g_\varepsilon(x)\sqrt{|p|^2 + \varepsilon} + \sum_i \frac{\partial g_\varepsilon}{\partial x_i}(x)p_i.$$

According to the general theory of quasilinear parabolic equations [208], we know there exists a smooth solution u^ε of

$$\begin{cases} \dfrac{\partial u}{\partial t} = g_\varepsilon(x)\sum_{i,j} a_{i,j}^\varepsilon(\nabla u)u_{x_i x_j} + H_\varepsilon(x, \nabla u), \\ u(0,x) = u_0^\varepsilon(x). \end{cases} \tag{4.52}$$

Thanks to (4.51) we have

$$|\nabla u^\varepsilon(t,.)|_{L^\infty(R^2)} \leq k|\nabla u_0|_{L^\infty(R^2)} \equiv c_T.$$

This means that

$$|u^\varepsilon(t,x) - u^\varepsilon(t,y)| \leq c_T\,|x - y| \quad \text{for all } t \in [0,T] \text{ and } x, y \in R^2. \tag{4.53}$$

We can also show from (4.52) and (4.53) that

$$|u^\varepsilon(t,x) - u^\varepsilon(s,x)| \leq c_T\,|t - s|^{\frac{1}{2}} \quad \text{for all } t, s \in [0,T] \text{ and } x \in R^2. \tag{4.54}$$

Inequalities (4.53) and (4.54) together with $|u|_{L^\infty((0,T)\times R^2)} \leq cte$ allow us to conclude by means of the Arzelà–Ascoli theorem (see Section 2.5.4) that there exists a subsequence of u^ε converging uniformly on $[0,T] \times R^2$ to a function $u \in C[(0,T) \times R^2] \cap L^\infty[(0,T); W^{1,\infty}(R^2)]$ for $T < \infty$. Then by applying a stability result for viscosity solutions (see Lemma 2.3.1) we conclude that u is a viscosity of (4.35).   ∎

Remark (About the maximum principle in the parabolic case)
Most uniqueness results for linear parabolic (or elliptic) PDEs follow from maximum or comparison principles. Roughly speaking, they state that if u is a solution of a parabolic PDE on $U_T = (0,T) \times U$, then the maximum and the minimum of u are attained on the parabolic boundary of U_T defined by $\Gamma_T = (0,T) \times \partial U \cup \{t = 0\} \times U$. More precisely, let us consider the operator

$$Lu = -\sum_{i,j} a_{i,j} u_{x_i x_j} + \sum_i b_i u_{x_i}$$

with $\sum_{i,j} a_{i,j} \xi_i \xi_j \geq \theta|\xi|^2 \ \forall(t,x) \in U_T$, $a_{i,j}$, b_i continuous and U bounded. If u is a regular solution of $u_t + Lu \leq 0$, then we have [148]

$$\max_{\overline{U}_T} u = \max_{\Gamma_T} u.$$

Likewise, if $u_t + Lu \geq 0$, then we have [148]

$$\min_{\overline{U}_T} u = \min_{\Gamma_T} u.$$

So it is clear that if two solutions coincide on Γ_T, they coincide on $\overline{U}_T$. From the above result we can deduce the following corollaries:

- Let $\tilde{u}$ be such that

$$\begin{cases} \tilde{u}_t + L\tilde{u} = g(t) \geq 0, \\ \tilde{u}(0,x) = u_0(x). \end{cases}$$

Then $\tilde{u} = v + \int_0^t g(s)ds$, where v is the solution of

$$\begin{cases} v_t + Lv = 0, \\ v(0,x) = u_0(x). \end{cases}$$

So we have $\min_x u_0(x) \leq v(t,x) \leq \max_x u_0(x)$.

- Let us consider

$$\begin{cases} u_t + Lu \leq g(t) = \tilde{u}_t + L\tilde{u}, \\ \tilde{u}(0,x) = u_0(x). \end{cases}$$

This can be rewritten as

$$\begin{cases} (u - \tilde{u})_t + L(u - \tilde{u}) \leq 0, \\ (u - \tilde{u})(0,x) = 0. \end{cases}$$

Thanks to the maximum principle, we have $\max\limits_{\overline{U}_T} (u - \tilde{u}) = \max\limits_{\Gamma_T} (u - \tilde{u}) = 0$. That is, $u \le \tilde{u}$. But

$$\tilde{u} = v + \int_0^t g(s)ds \le |u_0|_{L^\infty(U_T)} + \int_0^t g(s)ds,$$

so we have

$$u \le |u_0|_{L^\infty(U_T)} + \int_0^t g(s) \ ds.$$

- Finally, let us consider

$$\begin{cases} w_t + Lw \le cw \ \ \text{with} \ \ c \ge 0, \\ w(0,x) = u_0(x). \end{cases}$$

The classical maximum principle can no longer be applied (because $c \ge 0$). However, we have $w_t + Lw \le c \max\limits_x w(t,x) = g(t)$, so

$$w(t,x) \le |u_0|_{L^\infty(U)} + c \int_0^t \max_x w(x,s)ds,$$

and then

$$|w|_{L^\infty(U)}(t) \le |u_0|_{L^\infty(U)} + c \int_0^t |w|_{L^\infty(U)}(s)ds.$$

By applying the Gronwall inequality (see Section 2.5), we obtain

$$|w|_{L^\infty(U)}(t) \le |u_0|_{L^\infty(U)}\left(1 + ct \ e^{ct}\right) \le k \, |u_0|_{L^\infty(U)},$$

with k depending on c and T. Observe that on $\partial(0,T) \times U$ we can choose Neumann or periodic boundary conditions. ∎

Finally, we would like to check the correctness of the geometric model, that is, to show that the zero-level set of $u(t,x)$ asymptotically fits the desired contour $\Gamma = \{x \in [0,1]^2; g(x) = 0\}$. Let us recall a result proven in [85]. We assume that Γ is smooth and separates $[0,1]^2$ into two regions: the inside $I(\Gamma)$ and the outside $E(\Gamma)$ (i.e., Γ is a Jordan curve). We recall that $\mathcal{P}(R^N)$, the set of all subsets of R^N, can be equipped with the Hausdorff metric

$$d(A,B) = \max\left(\sup_{x \in A} d(x,B), \ \sup_{x \in B} d(x,A)\right),$$

where $d(x,A)$ denotes as usual the distance of the point x to the set A:

$$d(x,A) = \inf_{y \in A} d(x,y).$$

Theorem 4.3.3 [85] *Let* $\Gamma = \{x \in [0,1]^2; \ g(x) = 0\}$ *be a Jordan curve of class* C^2 *and* $\nabla g(x) = 0$ *on* Γ. *Assume that* $u_0(x)$ *is smooth and bounded* *and that the set $\{x \in R^2; \; u_0(x) \le 0\}$ contains Γ and its interior. Let $u(t,x)$ be the viscosity solution of (4.35) and $\Gamma(t) = \{x; \; u(t,x) = 0\}$. Then if α (the constant component of the velocity) is sufficiently large, $\Gamma(t) \to \Gamma$ as $t \to +\infty$ in the Hausdorff metric.*

4.3.4   The Reinitialization Equation

In practice, when we want to use equation (4.34) to segment an image, we choose as an initial condition at $t = 0$ the signed distance function to an initial given curve $c_0(q)$ surrounding the objects. Unfortunately, equation (4.34) governing the evolution does not guarantee that the evolving function $v(x,t)$ will remain a signed distance function. In particular, its gradient can become unbounded, and this is the origin of serious numerical problems.

To overcome this difficulty, when we have the estimation of the level sets $v(x,t_0)$ at time t_0, the idea is to build another distance function having the same zero level as $v(x,t_0)$. To this end, Morel proposed to solve the following PDE:

C++

$$\begin{cases} \dfrac{\partial u}{\partial t} + \text{sign}(v(x,t_0))\,(|Du| - 1) = 0, \\ u(x,0) = v(x,t_0), \end{cases} \tag{4.55}$$

where

$$\text{sign}(w) = \begin{cases} 1 & \text{if } w > 0, \\ 0 & \text{if } w = 0, \\ -1 & \text{if } w < 0, \end{cases} \tag{4.56}$$

and $v(x,t_0)$ is the evolving function at time t_0. Practically, this reinitialization procedure needs to be done regularly. We refer the reader to the abundant literature on this subject [1, 223, 312, 348] for more precise information, and to Section A.3.4 for numerical details.

In this section, we present work by Aujol and Aubert [29] concerning the theoretical justification of this PDE. As we will see, the main result is that a Hopf–Lax formula can be found that is an explicit expression of the solution.

At first glance, the PDE (4.55) is of Hamilton–Jacobi type, i.e., of the form

$$\frac{\partial u}{\partial t} + H(x, Du) = 0, \tag{4.57}$$

and then the theory of viscosity solutions seems to be the right tool to state existence and uniqueness results. This framework is revisited in Section 2.3 and we refer the reader to Section 3.3.1 (see the Alvarez–Guichard–Lions–Morel scale space theory) for an example where it is useful. However, most work is conducted under the assumption that the Hamiltonian H is continuous with respect to all its variables, whereas PDE (4.55) has a singularity on $\Gamma = \{x/v(x,t_0) = 0\}$. As a consequence, the classical theory of viscosity solutions does not allow us to tackle this problem, and one has to investigate the extension of the theory to discontinuous Hamiltonians. The existing literature on on this subject is not very large. The case of discontinuous solutions is discussed in [10], and the case of a discontinuity in the second member of the PDE in [187, 307, 308].

CHARACTERIZATION OF THE DISTANCE FUNCTION

Let Γ be a closed curve in R^2 and $\text{dist}(x,\Gamma)$ the distance of a point x to a curve Γ, i.e.,

$$\text{dist}(x,\Gamma) = \inf_{z \in \Gamma} |x - z|.$$

Let us start with some definitions.

Definition 4.3.4 *Let* $u : R^2 \to R$ *such that* $u(\Gamma) = 0$. *We will say that* u *is a signed distance function to the curve* Γ *if*

$$u(x) = \begin{cases} \varepsilon \, \text{dist}\,(x,\Gamma) & \text{if } x \text{ lies outside } \Gamma, \\ -\varepsilon \, \text{dist}\,(x,\Gamma) & \text{if } x \text{ lies inside } \Gamma, \end{cases}$$

where

(i) $\text{dist}(x,\Gamma) = \inf_{y \in \Gamma} |x - y|$,

(ii) $\varepsilon \in \{+1, -1\}$,

(iii) $|x|$ *is here the Euclidean norm of* x.

In what follows, we will take the convention that $\varepsilon = 1$, and we will call u the signed distance function.

Definition 4.3.5 *The skeleton of* Γ, *denoted by* S, *is the set of points* $x \in R^2$ *such that there exist at least two distinct points* y *and* z *in* Γ *satisfying*

$$|x - y| = |x - z| = \text{dist}(x,\Gamma).$$

We denote by $\bar{S}$ *the closure of* S.

We give below some elementary properties of the signed distance function.

Proposition 4.3.1

(i) *The signed distance function* u *to a closed curve* Γ *in* R^2 *is 1-Lipschitz.*

(ii) *If* u *is a signed distance function to a closed curve* Γ *in* R^2, *and if* $u \in C^1(R^2 \backslash \bar{S} \to R)$ *(where* $\bar{S}$ *stands for the closure of the skeleton of* Γ*), then we have for all* $x \in R^2 \backslash \bar{S}$,

$$|\nabla u(x)| = 1.$$

The next proposition gives a characterization of the signed distance function, proposed in [29], that will be the starting point for discovering an explicit solution of the reinitialization equation (4.55).

Proposition 4.3.2 *Let Γ be a closed curve in R^2. Then:*

(i) *u is the distance function to Γ if and only if*

$$\begin{cases} u(x) \geq 0 & \forall x \in R^2, \\ u(x) = 0 & if\ x \in \Gamma, \\ u(x) = \inf_{|y|=h} (u(x+y) + h) & \forall h \in [0, \mathrm{dist}(x, \Gamma)]. \end{cases}$$

(ii) *u is the opposite of the distance function to Γ if and only if*

$$\begin{cases} u(x) \leq 0 & \forall x \in R^2, \\ u(x) = 0 & if\ x \in \Gamma, \\ u(x) = \sup_{|y|=h} (u(x+y) - h) \\ \quad\ = -\inf_{|y|=h} (-u(x+y) + h) & \forall h \in [0, \mathrm{dist}(x, \Gamma)]. \end{cases}$$

(iii) *u is the signed distance function to Γ if and only if*

$$\begin{cases} u(x) = 0 & if\ x \in \Gamma, \\ u(x) \geq 0 & if\ x\ lies\ in\ the\ exterior\ of\ \Gamma, \\ u(x) \leq 0 & if\ x\ lies\ in\ the\ interior\ of\ \Gamma, \\ u(x) = \varepsilon_x \inf_{|y|=h} (\varepsilon_x u(x+y) + h) & \forall h \in [0, \mathrm{dist}(x, \Gamma)], \end{cases}$$

where $\varepsilon_x = \begin{cases} 1 & if\ x\ lies\ in\ the\ exterior\ of\ \Gamma, \\ -1 & if\ x\ lies\ in\ the\ interior\ of\ \Gamma. \end{cases}$

Proof We show only the first point of the proposition. The demonstration for the two other points is similar.

Step 1: We assume that u is the signed distance function to Γ.
It is therefore obvious that $u(x) = 0 \Leftrightarrow x \in \Gamma$ (since Γ is closed). Let $x \in R^2 \backslash \Gamma$. We want to show that

$$u(x) = \inf_{|y|=h} (u(x+y) + h) , \forall h \in [0, u(x)].$$

Since u is he distance function, this is equivalent to

$$\inf_{z \in \Gamma} |x - z| = \inf_{|y|=h} (\inf_{z \in \Gamma} |x + y - z| + h) , \forall h \in [0, u(x)],$$

i.e.,

$$\inf_{z \in \Gamma} |x - z| = \inf_{z \in \Gamma} (\inf_{|y|=h} |x + y - z| + h) , \forall h \in [0, u(x)]. \qquad (4.58)$$

Let us prove this last equality.

- Thanks to the triangular inequality, we have

$$\underbrace{|x - z| - |y| + h}_{=|x-z| \text{ if } |y|=h} \leq |x + y - z| + h.$$

So

$$|x - z| \leq \inf_{|y|=h} (|x + y - z|) + h,$$

and

$$\inf_{z \in \Gamma} |x - z| \leq \inf_{z \in \Gamma} \left(\inf_{|y|=h} |x + y - z| + h \right), \forall h \in [0, u(x)].$$

Thus we have shown that

$$u(x) \leq \inf_{|y|=h} (u(x + y) + h) , \forall h \in [0, u(x)]. \tag{4.59}$$

- To reach a contradiction, let us assume that there exists $h \in]0, u(x)]$ such that (4.59) is strict, i.e.,

$$u(x) = \inf_{z \in \Gamma} |x - z| < \inf_{z \in \Gamma} \left(\inf_{|y|=h} |x + y - z| + h \right),$$

or

$$u(x) < h + \underbrace{\inf_{z \in \Gamma} \left(\inf_{|y|=h} |x + y - z| \right)}_{\inf_{z \in \Gamma} \left(\inf_{|w-x|=h} |w-z| \right)}.$$

Since Γ is closed, we know that there exists $\bar{z} \in \Gamma$ such that $u(x) = \text{dist}(x, \Gamma) = |x - \bar{z}|$, so

$$u(x) < h + \inf_{|w-x|=h} |w - \bar{z}|. \tag{4.60}$$

Let us choose

$$\bar{w} = \frac{h}{u(x)}\bar{z} + \frac{-h + u(x)}{u(x)}x,$$

that is,

$$\bar{w} - x = \frac{h}{u(x)}(\bar{z} - x),$$

which implies

$$|\bar{w} - x| = h\frac{|\bar{z} - x|}{u(x)} = h.$$

And so $\bar{w}$ is admissible (according to (4.60)). Moreover, we have

$$|\bar{w} - \bar{z}| = \frac{u(x) - h}{u(x)}|x - \bar{z}| = u(x) - h,$$

so, with (4.60), we obtain that $u(x) < u(x)$, which is obviously wrong. This completes the proof of (4.58).

Step 2: Conversely, to show that $u(x)$ is a distance function, by choosing $h = \text{dist}(x, \Gamma)$, we get

$$u(x) = \underbrace{\inf_{|y| = \text{dist}(x, \Gamma)} (u(x + y))}_{=0} + \text{dist}(x, \Gamma).$$

∎

EXISTENCE AND UNIQUENESS OF A UNIFORMLY CONTINUOUS SOLUTION

We are now in position to study the reinitialization equation

$$\begin{cases} \dfrac{\partial u}{\partial t} + \text{sign}(u_0(x)) \left(|Du| - 1 \right) = 0, \\ u(., 0) = u_0(x), \end{cases} \qquad (4.61)$$

where u_0 is the initial condition (see for example (4.55)). We remark that in equation (4.61), the Hamiltonian is discontinuous with respect to the x variable. So we need to extend the classical theory given in Section 2.3.

So, let us consider an equation of the form

$$F(x, u, Du, D^2u) = 0, \qquad (4.62)$$

where Du and D^2u stand respectively for the first and second derivatives in the space variable. We also need the notion of lower (respectively upper) semicontinuous envelope:

Definition 4.3.6 *Let f be a locally bounded function. Furthermore, f_* and f^* will denote respectively the lower semicontinuous envelope (l.s.c.) and the upper semicontinuous envelope (u.s.c.) of f:*

$$f_*(x) = \varliminf_{y \to x} f(y),$$

$$f^*(x) = \varlimsup_{y \to x} f(y).$$

With this definition, the notion of viscosity solutions for discontinuous Hamiltonian can be established.

Definition 4.3.7 *Let $F\colon \Omega \times R \times R^N \times S^N \to R$ be defined everywhere and locally bounded.*

(i) *A function u locally bounded, u.s.c. on Ω, is a viscosity subsolution of (4.62) if and only if for all $\phi \in C^2(\Omega)$ and $x_0 \in \Omega$ a local maximum point of $u - \phi$,*

$$F_*(x_0, u(x_0), D\phi(x_0), D^2\phi(x_0)) \leq 0. \qquad (4.63)$$

(ii) *A function u locally bounded, l.s.c. on Ω, is a viscosity supersolution of (4.62) if and only if for all $\phi \in C^2(\Omega)$ and $x_0 \in \Omega$ a local minimum point of $u - \phi$,*

$$F^*(x_0, u(x_0), D\phi(x_0), D^2\phi(x_0)) \geq 0. \qquad (4.64)$$

(iii) *We will call every continuous function satisfying (4.63) and (4.64) a viscosity solution of (4.62).*

We can give now the main result of this section, the existence and the uniqueness of a solution for the reinitialization equation (4.61). In the sequel, the initial condition u_0 will be supposed uniformly continuous on R^2 and we will define $\Gamma = \{x/u_0(x) = 0\}$ and $\varepsilon_x = \text{sign}(u_0(x))$. The aim is to prove that the function $u : R^2 \times R_+ \to R^2$ defined by

$$u(x,t) = \begin{cases} \varepsilon_x \underset{|y| \leq t}{\inf} (\varepsilon_x u_0(x+y) + t) & \text{if } t \leq t_x, \\ \varepsilon_x \text{dist}(x, \Gamma) & \text{if } t > t_x, \end{cases} \qquad (4.65)$$

where

$$t_x = \inf \left\{ t \in R_+ / \underset{|y| \leq t}{\inf} (\varepsilon_x u_0(x+y)) = 0 \right\} = \text{dist}(x, \Gamma)$$

is a viscosity solution of (4.61). Note that (4.65) is in fact a Hopf–Lax formula, i.e., an explicit expression of the solution. Let us write $\Omega_+ = \{x/u_0(x) > 0\}$, the outside (in the strict sense) of Γ, and $\Omega_- = \{x/u_0(x) < 0\}$, the inside (in the strict sense) of Γ $(R^2 = \Omega_+ \cup \Omega_- \cup \Gamma)$.

Proposition 4.3.3 *Let u_0 be a uniformly continuous function on R^2 and set $\Gamma = \{x/u_0(x) = 0\}$. Then u defined by (4.65) is a uniformly continuous function on $R^2 \times R_+$.*

Remark A consequence of the definition of t_x is that $x \mapsto t_x$ is continuous on R^2. ∎

Proof Let us prove that u given by (4.65) is uniformly continuous on $\Omega \times [0, T], \forall T > 0$. To do this, let us split the domain into the three cases below:

(i) $\Omega_+ \times [0, T], \forall T > 0$,

(ii) $\Omega_- \times [0, T], \forall T > 0$,

(iii) $\mathcal{V} \times [0, T], \forall T > 0$ (where $\mathcal{V}$ is a neighborhood in R^2 of $\{x/u_0(x) = 0\}$),

and prove that u is uniformly continuous on each of these subdomains.

(i) First case: Let us show that u is uniformly continuous on $\Omega_+ \times [0, T], \forall T > 0$. We first set $T > 0$. Let (x, t) and $(\hat{x}, \hat{t})$ in $\Omega_+ \times [0, T]$.

Step 1: Let us first assume that $\hat{t} \leq t_{\hat{x}}$ and $t \leq t_x$.

We denote by ρ the modulus of continuity of u_0, defined by

$$\rho(r) = \sup_{|x-y|\leq r} |u_0(x) - u_0(y)|.$$

We have

$$u(x,t) - u(\hat{x},\hat{t}) = \inf_{\{|y|\leq t\}} (u_0(x+y) + t) - \inf_{\{|y|\leq \hat{t}\}} (u_0(\hat{x}+y) + \hat{t}).$$
(4.66)

Since u_0 is continuous, there exists $\hat{b} \in B(\hat{x},\hat{t})$ (where $B(z,r)$ denotes the open ball in R^2 centered at z and of radius r) such that

$$u_0(\hat{b}) = \inf_{\{|y|\leq \hat{t}\}} (u_0(\hat{x}+y)).$$

Since $\hat{b} \in B(\hat{x},\hat{t})$, there exists $\lambda \in [0,1]$ such that $|\hat{x} - \hat{b}| = \lambda\hat{t}$. Let a be on $C(x,\lambda t)$, where $C(z,r)$ denotes the circle in R^2 centered at z and of radius r such that

$$\left\langle \hat{x} - \hat{b}, x - a \right\rangle = -|\hat{x} - \hat{b}||x - a|,$$

where $\langle .,. \rangle$ stands for the Euclidean scalar product. In particular, we have $a \in B(x,t)$ (since $|\lambda| \leq 1$). Then (4.66) becomes

$$u(x,t) - u(\hat{x},\hat{t}) = \inf_{\{|y|\leq t\}} (u_0(x+y) + t) - u_0(\hat{b}) - \hat{t}$$

$$\leq u_0(a) - u_0(\hat{b}) + t - \hat{t}$$

$$\leq \rho(|a - \hat{b}|) + t - \hat{t}.$$

But since

$$|a - \hat{b}| = |a - x + x - \hat{x} + \hat{x} - \hat{b}|$$

$$\leq \underbrace{|a - x + \hat{x} - \hat{b}|}_{=\lambda|t-\hat{t}|\leq|t-\hat{t}|} + |x - \hat{x}|,$$

and $\rho(r + s) \leq \rho(r) + \rho(s)$, we deduce that

$$u(x,t) - u(\hat{x},\hat{t}) \leq \rho(|x - \hat{x}|) + |t - \hat{t}| + \rho(|t - \hat{t}|).$$

In the same way we can show that

$$u(x,t) - u(\hat{x},\hat{t}) \geq - \left(\rho(|x - \hat{x}|) + |t - \hat{t}| + \rho(|t - \hat{t}|)\right),$$

and then

$$|u(x,t) - u(\hat{x},\hat{t})| \leq \rho(|x - \hat{x}|) + |t - \hat{t}| + \rho(|t - \hat{t}|).$$
(4.67)

Step 2: Let us now assume that $t \geq \hat{t} \geq \max(t_x, t_{\hat{x}})$.
Then we have

$$|u(x,t) - u(\hat{x},\hat{t})| = |u(x,t_x) - u(\hat{x},t_{\hat{x}})|$$

$$= |\text{dist}(x,\Gamma) - \text{dist}(\hat{x},\Gamma)|.$$

Hence, since the Euclidean distance is 1-Lipschitz,

$$|u(x,t) - u(\hat{x}, \hat{t})| \leq |x - \hat{x}|. \tag{4.68}$$

Step **3**: Let us now assume that $\hat{t} \geq t_{\hat{x}}$ and $t \leq t_x$.
Then we have

$$|u(x,t) - u(\hat{x}, \hat{t})| = |u(x,t) - u(\hat{x}, t_{\hat{x}})|.$$

The difference $|u(x,t) - u(\hat{x}, t_{\hat{x}})|$ can now be estimated as in Step 1.
So, using (4.67), we have

$$|u(x,t) - u(\hat{x}, t_{\hat{x}})| \leq \rho(|x - \hat{x}|) + |t - t_{\hat{x}}| + \rho(|t - t_{\hat{x}}|);$$

hence

$$|u(x,t) - u(\hat{x}, \hat{t})| \leq \rho(|x - \hat{x}|) + |t_x - t_{\hat{x}}| + \rho(|t_x - t_{\hat{x}}|).$$

But since

$$|t_x - t_{\hat{x}}| = |\mathrm{dist}(x, \Gamma) - \mathrm{dist}((\hat{x}, \Gamma)| \leq |x - \hat{x}|,$$

we have

$$|u(x,t) - u(\hat{x}, \hat{t})| \leq 2\rho(|x - \hat{x}|) + |x - \hat{x}|.$$

From Steps 1 to 3, we deduce the uniform continuity of u on $\Omega_+ \times [0,T]$, because u_0 uniformly continuous implies $\rho(|t - \hat{t}|) \to 0$ as $t \to \hat{t}$, and $\rho(|x - \hat{x}|) \to 0$ as $x \to \hat{x}$.

(ii) Second case: We can show with an identical proof that u is also uniformly continuous on $\Omega_- \times [0,T]$, $\forall T > 0$.

(iii) Third case: Let us show now that u is uniformly continuous in any neighborhood $\mathcal{V}$ in R^2 of $\{x/u_0(x) = 0\}$. In fact, thanks to the first two cases, we have to show only that u is continuous on $\mathcal{V} \times [0,T]$ (Heine's theorem will enable us to conclude that u is uniformly continuous on $\mathcal{V} \times [0,T]$).
Let us choose $T > 0$. Let $\varepsilon > 0$. Since u_0 is uniformly continuous, for all $\varepsilon > 0$, there exists $\delta > 0$ such that

$$(0 \leq r \leq \delta) \Rightarrow \left(\rho(r) \leq \frac{\varepsilon}{3} \right).$$

Moreover, there exists $\eta > 0$ such that

$$(\mathrm{dist}(x, \Gamma) \leq \eta) \Rightarrow \left(|u_0(x)| \leq \frac{\varepsilon}{3} \right).$$

Let $x \in R^2$ such that $\mathrm{dist}(x, \Gamma) \leq \min\left(\delta, \eta, \frac{\varepsilon}{3}\right)$. Two cases can occur.

Step **1**: If $t \leq t_x$, then

$$|u(x,t)| = |\inf_{\{|y| \leq t\}} (u_0(x+y) + t)| \leq \inf_{\{|y| \leq t\}} |u_0(x+y)| + t,$$

and since $t_x = \text{dist}(x, \Gamma)$ and $t \leq t_x$, we have

$$
\begin{aligned}
|u(x,t)| &\leq |u_0(x)| + \rho(t) + t \\
&\leq |u_0(x)| + \rho(t_x) + t_x \\
&\leq \frac{\varepsilon}{3} + \frac{\varepsilon}{3} + \frac{\varepsilon}{3} \\
&\leq \varepsilon.
\end{aligned}
$$

Step 2: If $t > t_x$, then

$$
|u(x,t)| = \text{dist}(x, \Gamma) \leq \frac{\varepsilon}{3} \leq \varepsilon.
$$

Thus, $u(x,t)$ is continuous on $\{x / u_0(x) = 0\} \times [0, T]$.

From the three cases, we conclude that u is uniformly continuous on $R^2 \times R_+$. ∎

We can now give the main result of this section.

Theorem 4.3.4 *Let u be defined by (4.65). Then u is the unique viscosity solution of (4.61) uniformly continuous on $R^2 \times [0, T], \forall T > 0$, and vanishing on $\Gamma, \forall t \in [0, T]$.*

Proof We split the proof into two parts.

Part 1: Existence of a solution.
Let us first prove that u defined by (4.65) is a viscosity subsolution of (4.61). From Proposition 4.3.3, we already know that u is uniformly continuous on $R^2 \times R_+$. Let us set $(x_0, t_0) \in R^2 \times R_+$, and let ϕ be of class C^2 such that $u - \phi$ has a local maximum in (x_0, t_0).

Step 1: Let us first assume $u_0(x_0) \neq 0$.
For instance, we assume $u_0(x_0) > 0$.

(i) If $t_0 > t_{x_0}$, we have $u(x_0, t_0) = \text{dist}(x_0, \Gamma)$ and $\frac{\partial u}{\partial t}(x_0, t_0) = 0$. Let us show that $F(x_0, D\phi(x_0, t_0)) \leq 0$, i.e.,

$$
-1 + |D\phi(x_0)| \leq 0.
$$

For the sake of clarity, we will not write the t-dependency, which plays no role here. For x small enough, we have (if $h \in [0, \text{dist}(x, \Gamma)]$):

$$
h + u(x_0 + x) - \phi(x_0 + x) \leq u(x_0) - \phi(x_0) + h,
$$

that is,

$$
u(x_0 + x) + h \leq u(x_0) - \phi(x_0) + \phi(x_0 + x) + h,
$$

which implies

$$
\inf_{|x|=h} (u(x_0 + x) + h) \leq u(x_0) - \phi(x_0) + \inf_{|x|=h} (\phi(x_0 + x) + h). \quad (4.69)
$$

Since the left-hand side of (4.69) is in fact equal to $u(x_0)$, thanks to Proposition 4.3.2, the inequality (4.69) becomes

$$\inf_{|x|=h} \left(\phi(x_0 + x) - \phi(x_0) \right) \geq -h.$$

But we have

$$\phi(x_0 + x) - \phi(x_0) = (D\phi(x_0), x) + \mathcal{O}(h^2).$$

Then

$$\inf_{|x|=h} \left(\phi(x_0 + x) - \phi(x_0) \right) = \inf_{|x|=h} \left(-|x| |D\phi(x_0)| \right) + \mathcal{O}(h^2)$$
$$= -h |D\phi(x_0)| + \mathcal{O}(h^2),$$

which leads to

$$|D\phi(x_0)| \leq 1 + \mathcal{O}(h). \tag{4.70}$$

So we can deduce from (4.70) that u is a viscosity subsolution of (4.61) in (x_0, t_0).

Remarks

- If $u_0(x_0) < 0$, it suffices to repeat the same proof as above by starting with the inequality

$$-h + u(x_0 + y) - \phi(x_0 + y) \leq u(x_0) - \phi(x_0) - h.$$

- We can also notice that in fact we have shown that the signed distance function is a viscosity solution of the stationary equation $\text{sign}(u_0) \left(|Du| - 1 \right) = 0$.

(ii) If $t_0 \leq t_{x_0}$, for h and y small enough ($h > 0$), we have

$$h + u(x_0 + y, t_0 - h) - \phi(x_0 + y, t_0 - h) \leq u(x_0, t_0) - \phi(x_0, t_0) + h,$$

which implies

$$\inf_{|y| \leq h} \left(u(x_0 + y, t_0 - h) + h \right)$$
$$\leq \inf_{|y| \leq h} \left(\phi(x_0 + y, t_0 - h) \right) - \phi(x_0, t_0) + h + u(x_0, t_0).$$

But

$$\inf_{|y| \leq h} \left(u(x_0 + y, t_0 - h) + h \right) = \inf_{|y| \leq h} \left(\inf_{|z| \leq t_0 - h} \left(u_0(x_0 + y + z) + t_0 \right) \right)$$
$$= \inf_{|w| \leq t_0} \left(u_0(x_0 + w) + t_0 \right)$$
$$= u(x_0, t_0),$$

thanks to (4.65). Hence

$$\phi(x_0, t_0) \leq h + \inf_{|y| \leq h} \left(\phi(x_0 + y, t_0 - h) \right),$$

and then

$$\phi(x_0, t_0) - \phi(x_0, t_0 - h) \leq \inf_{|y| \leq h} (\phi(x_0 + y, t_0 - h) - \phi(x_0, t_0 - h)) + h.$$

Since

$$\phi(x_0, t_0) - \phi(x_0, t_0 - h) = h \frac{\partial \phi}{\partial t}(x_0, t_0) + 0(h^2)$$

and

$$\phi(x_0 + y, t_0 - h) - \phi(x_0, t_0 - h) = (D\phi(x_0, t_0 - h), y) + 0(h^2),$$

we obtain

$$\inf_{|y| \leq h} (\phi(x_0 + y, t_0 - h) - \phi(x_0, t_0 - h))$$

$$= \inf_{|y| \leq h} (-|y| |D\phi(x_0, t_0 - h)|) + 0(h^2)$$

$$= -h|D\phi(x_0, t_0 - h)| + 0(h^2),$$

from which we deduce

$$\frac{\partial \phi}{\partial t}(x_0, t_0) \leq -|D\phi(x_0, t_0 - h)| + 0(h) + 1.$$

Since

$$D\phi(x_0, t_0 - h) = D\phi(x_0, t_0) + 0(h),$$

we have finally

$$\frac{\partial \phi}{\partial t}(x_0, t_0) + \underbrace{(|D\phi(x_0, t_0 - h)| - 1)}_{=F(x_0, D\phi(x_0, t_0))} \leq 0,$$

i.e., u is a viscosity subsolution of (4.61) in (x_0, t_0).

Remark If $u_0(x_0) < 0$, it suffices to start from

$$-h + u(x_0 + y, t_0 - h) - \phi(x_0 + y, t_0 - h) \leq u(x_0, t_0) - \phi(x_0, t_0) - h.$$

It can be shown by a similar proof that u is a viscosity supersolution of (4.61) in (x_0, t_0). So we have shown that u is a viscosity solution of (4.61) in the case $u_0(x_0) \neq 0$ (since, thanks to Proposition 4.3.3, u is also continuous).

Step **2**: Let us now assume $u_0(x_0) = 0$.
We have

$$F_*(x_0, D\phi(x_0, t_0)) = \min (0, |D\phi(x_0, t_0)| - 1, 1 - |D\phi(x_0, t_0)|)$$

and

$$F^*(x_0, D\phi(x_0, t_0)) = \max (0, |D\phi(x_0, t_0)| - 1, 1 - |D\phi(x_0, t_0)|).$$

Thus

$$F_*(x_0, D\phi(x_0, t_0)) \leq 0 \quad \text{and} \quad F^*(x_0, D\phi(x_0, t_0)) \geq 0,$$

and so u is still a viscosity solution of (4.61) in this case.

Part **2**: Uniqueness of the solution.
Let us recall that

$$\begin{cases} u_0(x) = 0 & \text{if } x \in \Gamma, \\ u_0(x) \geq 0 & \text{if } x \in \Omega_+, \\ u_0(x) \leq 0 & \text{if } x \in \Omega_-. \end{cases}$$

Let us consider the two following problems:

$$\begin{cases} \frac{\partial u}{\partial t} + |Du| - 1 = 0, \\ u(.,0) = u_0|_{\Omega_+}, \\ u(x,t) = 0 \text{ if } x \in \Gamma, \end{cases} \tag{4.71}$$

and

$$\begin{cases} \frac{\partial u}{\partial t} - (|Du| - 1) = 0, \\ u(.,0) = u_0|_{\Omega_-}, \\ u(x,t) = 0 \text{ if } x \in \Gamma. \end{cases} \tag{4.72}$$

Let us recall that u_0 is assumed to be uniformly continuous. From classical results on viscosity solutions (see Theorem 1 in [186]), (4.71) (respectively (4.72)) has at most one viscosity solution uniformly continuous on $\Omega_+ \times [0,T]$, $\forall T > 0$ (respectively on $\Omega_- \times [0,T]$, $\forall T > 0$). By reconsidering the arguments given for the existence, it can be shown that the solution of (4.71) (respectively of (4.72)) is the restriction of the function (4.65) to $\Omega_+ \times R_+$ (respectively to $\Omega_- \times R_+$). Since a viscosity solution of (4.61) has to be continuous, the only possible solution is the function (4.65).

The proof of Theorem 4.3.4 is thus complete. ∎

Remark on Theorem 4.3.4. Equation (4.61) is used to reinitialize a function as the signed distance function to a closed curve in R^2. This reinitialization is performed only in a narrow band in which Γ is embedded (see [223]). But according to the form of the solution of (4.61), the first reinitialized values of u are the closest ones to Γ. This explains why this reinitialization method is so fast. Moreover, an immediate consequence of the definition of t_x is that $u(x, t_x) = \varepsilon_x t_x$. Thus, if we choose the bandwidth around Γ on which we want a signed distance function, we know how many iterations of the equation (4.61) are necessary. ∎

To conclude this section, let us comment on the introduction of t_x in the definition of (4.65), which could appear arbitrary. Let us consider the function $v : R^2 \times R_+ \rightarrow R$ defined by

$$v(x,t) = \varepsilon_x \inf_{|y| \leq t} (\varepsilon_x u_0(x+y) + t). \tag{4.73}$$

We recall that $\varepsilon_x = \text{sign}(u_0(x))$. By reconsidering the arguments of the existence part of Theorem 4.3.4, we see that v is a viscosity solution of (4.61) on $\Omega_+ \times R_+$ and on $\Omega_- \times R_+$. But v is not necessarily a continuous function on $R \times R_+$, and thus is not a viscosity solution of (4.61) on $R \times R_+$.

The following 1-D example illustrates this remark. Let $\Omega = (-1, 1)$, and let us take as initial condition

$$u_0(x) = 2\left(|x| - 1\right).$$

In this case, it is easy to compute both functions u and v given by (4.65) and (4.73) respectively. We get

$$u(x,t) = \begin{cases} 2\left(|x| - 1\right) + t & \text{if } |x| \leq 1 \text{ and } t \leq 1 - |x| = t_x, \\ |x| - 1 & \text{if } |x| \leq 1 \text{ and } t \geq 1 - |x| = t_x, \\ 2\left(|x| - 1\right) - t & \text{if } |x| \geq 1 \text{ and } t \leq |x| - 1 = t_x, \\ |x| - 1 & \text{if } |x| \geq 1 \text{ and } t \geq |x| - 1 = t_x, \end{cases}$$

$$v(x,t) = \begin{cases} 2\left(|x| - 1\right) + t & \text{if } |x| \leq 1, \\ 2\left(|x| - 1\right) - t & \text{if } |x| \geq 1 \text{ and } t \leq |x|, \\ t - 2 & \text{if } |x| \geq 1 \text{ and } t \geq |x|. \end{cases}$$

The function u is continuous on $R \times R^+$, but this is not the case for v. One may just consider $v(1+\varepsilon, t)$ and $v(1-\varepsilon, t)$ for t large enough to be convinced.

Remark Let us consider the following PDE:

$$\begin{cases} \dfrac{\partial u}{\partial t} + \text{sign}(u)\left(|Du| - 1\right) = 0, \\ u(., 0) = u_0. \end{cases} \tag{4.74}$$

This PDE is very close to (4.61), but the discontinuity of the Hamiltonian $H(r, p) = \text{sign}(r)\left(|p| - 1\right)$ is now with respect to r. In fact, u defined by (4.65) is still a uniformly continuous viscosity solution of (4.74). It suffices to remark that for all $x \in R^2$, we have for all $t \in R_+$

$$\text{sign}(u(x, t)) = \text{sign}(u_0(x)),$$

and the arguments are then the same as those in the proof of Theorem 4.3.4 (Part 1). However, as far as we know, the uniqueness of a solution of (4.74) remains an open problem. ∎

4.3.5 Experimental Results

This section concerns the experimental results that can be obtained with (4.34). In fact, we will show only the result obtained on the "objects" image, which clearly illustrates the behavior of the method.

As far as the discretization is concerned, it is classical to use finite difference schemes. As is recalled in Appendix A, these schemes are well adapted to the structure of digital images, since we can associate with it a natural

Figure 4.13. Segmentation of the "objects" image using the geodesic active contour model (4.75). From top left to bottom right, different iterations (solution as time evolves) are displayed. Initialized by the boundaries of the images, the curve is shrinking and splitting until it isolates each object.

C++ regular grid. Now, returning to our problem we can rewrite (4.34) as the sum of three separate terms:

$$\frac{\partial u}{\partial t} = g(|\nabla I|)\,|\nabla u|\,\mathrm{div}\left(\frac{\nabla u}{|\nabla u|}\right) + \alpha g(|\nabla I|)\,|\nabla u| + \langle \nabla g, \nabla u \rangle. \qquad (4.75)$$

This reveals two kinds of terms:

- The first term in (4.75) acts as a parabolic term.

- The second and the third terms are hyperbolic terms. The second term describes motion in the normal direction to the front, while the third (linear) term corresponds to pure advection.

As one would imagine, this difference must be taken into account at the discrete level. So discretizing equation (4.75) is not straightforward. For a better understanding, we refer the reader to Appendix A, where the main ideas of finite differences are explained, and more precisely to Section A.3.4, where the discretization of (4.75) is presented. Given the discrete scheme, one can implement and test this approach. An example of a result is presented in Figure 4.13 for the "objects" image. During the evolution, the curve is shrinking, stopping as soon as it is close to an object boundary (high gradients) and splitting in order to detect the five objects. However, the following can be noticed:

- The interior of the objects is not segmented (for instance, the interior of the disk). Once the curve has detected a contour, it stops.

- Because of the level set description, we always have closed curves. This is not really a problem in this image, but one may think about situations where open contours are present.

We review in the next section recent work that considers these two issues.

4.3.6 About Some Recent Advances

GLOBAL STOPPING CRITERION

We have previously presented active contour models and snakes that use the gradient as a criterion to stop the curve. However, there are some objects whose boundaries cannot be defined or are badly defined through the gradient. This includes, for example, smeared boundaries and cognitive contours (boundaries of larger objects defined by grouping smaller ones; see G. Kanizsa [189]), as shown in Figure 4.14.

We present here a different active contour model, called "without edges" [99, 101]. The main idea is also to consider the information inside the regions, and not only at their boundaries. Let us describe the model. Let u_0 be the original image to be segmented, and let c denote the evolving curve and i_1, i_2 two unknown constants. In [99, 101] the authors introduced

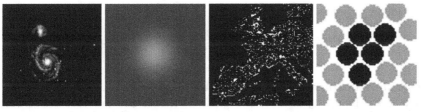

Smeared contours   Cognitive contours

Figure 4.14. The gradient is not always adapted to detect "boundaries".

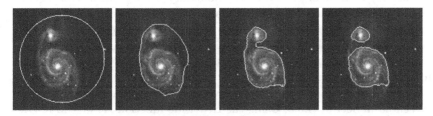

Figure 4.15. Smeared contours of a galaxy.

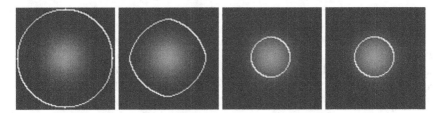

Figure 4.16. Contour of a blurred circular object.

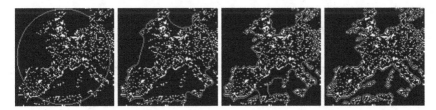

Figure 4.17. Cognitive contour for an image representing European night lights.

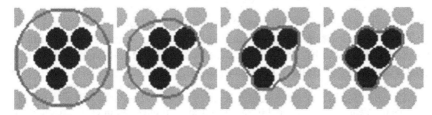

Figure 4.18. Cognitive contours. Another example that illustrates grouping based on chromatic identity.

the following minimization problem:

$$\inf_{i_1, i_2, c} F(i_1, i_2, c) = \mu |c| + \int_{\text{inside}(c)} |u_0 - i_1|^2 \, dx + \int_{\text{outside}(c)} |u_0 - i_2|^2 \, dx,$$

where μ is a positive parameter. This model looks for the best approximation of the image u_0 as a set of regions with only two different intensities (i_1 and i_2). Typically, one of the regions represents the objects to be detected (inside(c)), and the second region corresponds to the background (outside(c)). The snake c will be the boundary between these two regions. We see that this model is closely related to a binary segmentation. This model has many advantages. It allows one to detect both contours with or without gradient, and it automatically detects interior contours (think about the interior of the compact disk in the "objects" image), and it is robust in the presence of noise.

This approach has been implemented using the level set method that we presented in Section 4.3.3. Using the Heaviside function H, we can rewrite the energy F as

$$\tilde{F}(i_1, i_2, \phi) = \mu \int_{\Omega} |\nabla(H(\phi))| + \int_{\Omega} |u_0 - i_1|^2 H(\phi) \, dx + \int_{\Omega} |u_0 - i_2|^2 (1 - H(\phi)) \, dx,$$

where ϕ is the level set function. To find the minimum, we need to consider the problem that the functional $\tilde{F}$ is not Gâteaux differentiable with respect to the third variable. The reason is simply that the Heaviside function is not differentiable. It is then classical to regularize the problem by changing H into H_ε (a C^1-approximation of H) to be able to compute the derivatives.[3] So, to minimize $\tilde{F}$ with respect to i_1, i_2, and ϕ, we need to solve the equations

$$i_1 = \frac{\displaystyle\int_{\Omega} u_0 H(\phi) \, dx}{\displaystyle\int_{\Omega} H(\phi) \, dx}, \qquad i_2 = \frac{\displaystyle\int_{\Omega} u_0 (1 - H(\phi)) \, dx}{\displaystyle\int_{\Omega} (1 - H(\phi)) \, dx},$$

$$\frac{\partial \phi}{\partial t} = \delta_\varepsilon(\phi) \left(\mu \text{div}\left(\frac{\nabla \phi}{|\nabla \phi|} \right) - |u_0 - i_1|^2 + |u_0 - i_2|^2 \right),$$

where $\delta_\varepsilon = H'_\varepsilon$. Notice that we do not need H_ε for calculating i_1 and i_2. These equations can then be implemented using standard finite differences. We display in Figures 4.15–4.18 some numerical results from [99, 101] illustrating the possibilities of the approach.

---

[3] This problem is closely related to the classification problem from [290], presented in Section 5.4.2. In particular, the reader will find more details regarding the regularization of functionals.

From a theoretical point of view, we can prove the existence of minimizers of the energy $\tilde{F}$, but the convergence of the algorithm is an open problem.

To conclude, we mention that this kind of model is also used in other segmentation problems, like Mumford–Shah [100, 102], vector-valued images [96], texture [268, 270], and classification [290].

Toward More General Shape Representation

As seen in previous sections, level sets are a very convenient way to describe and implement the evolution of hypersurfaces, and this method has been extensively used in numerous applications. Unfortunately, this description has two main limitations. The first is that level sets do not permit the representation of shapes with a codimension different from one, or open shapes (see Figure 4.19). The second is that level sets do not permit one to describe the motion of self-intersecting interfaces.

point   region   open curve   self–intersecting curves

Figure 4.19. Examples of shapes in the 2-D case that cannot be described using level sets.

To overcome these difficulties, we mention some recent contributions in which an alternative to level sets is proposed [288, 169, 170]. Instead of considering the distance function $d(x, \Gamma) = |x - y|$ for representing and evolving objects, the idea is to use the vector distance function $u(x, \Gamma) = x - y$ (see Figure 4.20 for some examples). This vectorial function has remarkable properties, and related PDEs satisfied by $u(x)$ allow us to envisage more complicated motions than those treated with the classical level set method (see [169, 170]). This direction is very promising and should be investigated further, in particular from a numerical point of view.

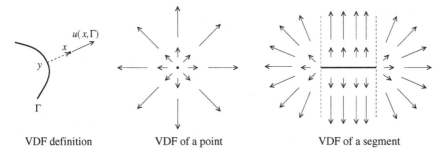

VDF definition   VDF of a point   VDF of a segment

Figure 4.20. Vector distance function (VDF): Definition and examples.

5
Other Challenging Applications

How to Read This Chapter

In this chapter we present some challenging applications that extend the ideas developed previously.

- In Section 5.1 we present the inpainting application, which consists in reinventing some missing parts in an image. We show how the two kinds of frameworks, namely variational formulations (Section 5.1.2) and PDE-based approaches (Section 5.1.3), can be successfully applied for this problem. In Section 5.1.4 we will discuss the limitations of such methods and mention some related work using other techniques.

- In Section 5.2 we show how an image can be decomposed as a sum of its regular geometric parts and its oscillating texture parts. This is based on recent ideas of Meyer, who introduced a suitable functional space (Section 5.2.2) and proposed a variational formulation to solve this problem (Section 5.2.3). The main part of this section is concerned with the numerical approximation of Meyer's model, which is described in Section 5.2.4. Some numerical results are given in Section 5.2.5, and we will conclude this part by mentioning some recent extensions in this area.

- In Section 5.3 we address the problem of processing sequences of images. As it is shown in Section 5.3.1, a sequence contains much information, such as motion, objects, depth. Our aim is to show how some of this information may be recovered from the sequence alone. We start by studying the problem of motion estimation. We show in Section 5.3.2 how it can be estimated using variational formulations. Another task that is becoming more and more important is the segmentation of the sequence, which consists in describing it by its different elements or objects (typically background and foreground). Little work has been done in this direction using PDEs, and we present in Section 5.3.3 an approach developed in [206] that has interesting mathematical properties. Finally, if image restoration is quite a well-known problem, sequence restoration is still an important challenge. In Section 5.3.4 we try to analyze the variety of problems involved, propose a classification of defects, and emphasize the importance of perception.

- In Section 5.4 we consider the problem of classification of aerial images, which presents some similarities with the segmentation problem (Section 5.4). We present two different and original approaches from Samson et al. [290, 291]. In the first one (Section 5.4.2), the evolution of a given number of curves (level sets) is used to describe a partition of the image. The difficulty comes from the opposite actions imposed on the curves (to cover the whole image without overlapping). In the second one (Section 5.4.3), an optimization problem is considered in which borrowed from the van der Waals, Cahn–Hilliard theory of phase transitions in mechanics. This is also the occasion to give a complete proof of a Γ-convergence theorem.

- In Section 5.5 we consider vector-valued images, e.g., color images. In fact, most algorithms and results given for scalar images can be extended for vector-valued images. However, extending an algorithm designed for scalar images is not generally as simple as applying the same algorithm to each component independently. In this section we present an elegant and unified presentation by Tschumperlé et al. [324] that is well adapted for restoration tasks. We first introduce in Section 5.5.2 the structure tensor as a way to estimate variations of a vector-valued image. Then we show how structure tensor properties can be used to propose variational and PDE-based approaches for vector-valued image restoration (Sections 5.5.3 and 5.5.4).

Although some theoretical results are given and proved, the goal of this chapter is to show the variety of domains for which PDEs can be useful. There are naturally many other applications of PDEs in image analysis that we do not present here. Some examples are diffusion on non-flat manifolds, shape from shading, image interpolation, shapes interpolation, stereovision. There are also many applications of PDEs in computer graphics; see for example [274, 326, 151].

5.1   Reinventing Some Image Parts by Inpainting

5.1.1   Introduction

In this section we intend to illustrate a useful and relatively recent application of PDEs in image processing called inpainting or filling in. The goal of inpainting is to restore a damaged or corrupted image in which part of the information has been lost. Such degradations of an image may have different origins (see Figure 5.1), such as image transmission problems and degradation in real images due to storage conditions or manipulation. Inpainting may also be an interesting tool for graphics people who need to remove artificially some parts of an image such as overlapping text or to implement tricks used in special effects. In any case, the restoration of missing parts has to be done so that the final image looks unaltered to an observer who does not know the original image.

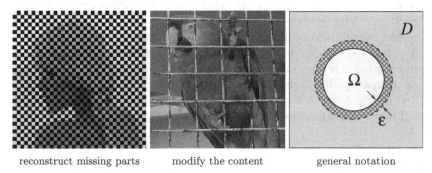

reconstruct missing parts   modify the content   general notation

Figure 5.1. Examples of filling in applications and general notation. In the first example, one would like to fill in black parts, while in the second one the objective is to remove the cage.

The problem can be described as follows: given a domain image D, a hole $\Omega \subset D$, and an intensity u_0 known over $D - \Omega$, we want to find an image u, an inpainting of u_0, that matches u_0 outside the hole and that has "meaningful" content inside the hole Ω (see Figure 5.1). This can be achieved by examining the intensity around Ω and propagating it in Ω. From a mathematical point of view this is an interpolation problem. Several kinds of approaches have been proposed to solve this problem. They differ by the application under consideration (for instance, inpainting of textured images or movies will require specific treatments) and by the mathematical tools used to model the problem (PDEs or others). In this section, we will mainly focus on geometric images, i.e., without fine texture content, for which PDE-based approaches are particularly suitable. As we will see, most inpainting PDEs methods are based on the simultaneous interpolation of isophotes and gray level intensities.

☞   For a general review on PDE-based inpainting methods see [49].

Section 5.1.2 reviews two variational approaches. In Section 5.1.3 three PDE-based models are presented. A discussion on limitations of the approaches, alternative methods, and psychophysics facts will be given in Section 5.1.4.

5.1.2 Variational Models

The Masnou and Morel Approach

The starting point of many contributions in inpainting was the pioneering work by Nitzberg, Mumford, and Shiota [255], who tried to identify occluding and occluded objects in a plane image, in order to compute the image depth map. To do this, the authors had the idea to apply principles of gestalt psychology. From perceptual experiments, see for example [189], it is known that our visual system detects occlusion at very low levels, and it is able to complete partially occluded boundaries following a principle of good continuation. When an object occludes another, the occluding and occluded boundaries form a particular configuration called a T-junction. If it is not too difficult to detect T-junctions, it may be much harder to perform good continuation between two T-junctions. Let us make things more precise and let us define materials that will be useful in the sequel.

Let us first define a T-junction properly using the notion of upper level set. If u is a Borel function, its upper level sets are defined by

$$\chi_\lambda u = \{x; u(x) \geq \lambda\}.$$

The family of $\{\chi_\lambda u\}$ gives a complete representation of u since we can recover it thanks to the formula

$$u(x) = \sup\{\lambda; x \in \chi_\lambda u\} \quad \text{for a.e. } x.$$

Upper level sets are invariant with respect to any increasing contrast change, and if u is a function of bounded variation, then almost every level set has finite perimeter. Level lines are defined as the boundaries of level sets. Now let us suppose that $u_0 \in BV(D - \bar{\Omega})$, where Ω is a hole in the image with smooth boundary. Then T-junctions are defined as the points where $\partial\Omega$ intersects the level lines of u_0. We shall say that two T-junctions are compatible if they are associated with the same level set and if the orientation of ∇u_0 is the same at both points.

The question now is, given two T-junctions T_1 and T_2 on $\partial\Omega$ with corresponding orientations θ_1 and θ_2 of ∇u (see Figure 5.2), how can we find a curve Γ lying in Ω, joining T_1 to T_2, and respecting at best the good continuation principle?

Inspired by the work by Nitzberg et al. [255], Masnou and Morel [230] proposed to search for Γ among the minimizers of the criterion

$$\int_\Gamma (\alpha + \beta\kappa^p)d\mathcal{H}^1 + (\theta_1, N_1) + (\theta_2, N_2),$$

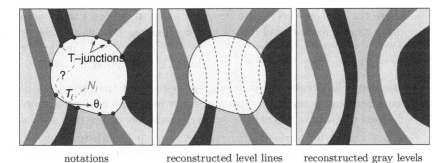

notations   reconstructed level lines   reconstructed gray levels

Figure 5.2. Illustration of Masnou and Morel's approach (adapted from [230]), where level lines are first reconstructed, and then gray levels.

where α, β are positive constants, $p \geq 1$, κ is the curvature of Γ, and the last two terms denote the angles between θ_i and N_i, the normal to Γ at T_i (for $i = 1, 2$). Γ is called a completion curve. Then by considering all the level sets and all the admissible T-junctions, the global energy to minimize is of the form

$$\int_{-\infty}^{+\infty} \sum_{\Gamma \in \mathcal{F}_\lambda} \left(\int_\Gamma (\alpha + \beta \kappa^p) d\mathcal{H}^1 + (\theta_1, N_1) + (\theta_2, N_2) \right) d\lambda, \qquad (5.1)$$

where $\mathcal{F}_\lambda$ denotes the family of completion curves associated with the level sets $\chi_\lambda u_0$. In [230] the existence of an optimal solution for any $p \geq 1$ is proven, provided that the restriction of u_0 to $\partial\Omega$ takes finitely many values. In practice (with $p = 1$), for computing the completion curves, the authors in [230] use an algorithm based on dynamic programming, and then fill in the space between these curves with the appropriate gray level, the one at the boundary of the hole. This method gives a reconstruction that is invariant by change of contrasts and is in accordance with the good continuation principle. However, due to the presence of the curvature term, this model tends to favor the creation of straight lines.

Remarks

- The model (5.1) is in fact a generalization of the Elastica model proposed in [255]. The main difference is that in [255] edges are considered rather than level lines, which yields a model dependent on the edge detector process.

- A slightly different model has been proposed by [16], where the authors skip the last two terms and consider the energy

$$\int_{-\infty}^{+\infty} \sum_{\Gamma \in \mathcal{F}_\lambda} \left(\int_\Gamma (\alpha + \beta \kappa^p) d\mathcal{H}^1 \right) d\lambda, \qquad (5.2)$$

where the elements of F_λ are now the union of a completion curve and a small piece of level line of u_0 for a domain Ω^+ slightly bigger than Ω. In [16] the authors have proven the existence of a minimizer for this new energy for any $p > 1$ without the restriction of finiteness.

- One can reformulate the problem by seeking not the level lines of the interpolated image u, but u itself, which will lead to a more classical problem of the calculus of variations. Observing that $\frac{\nabla u}{|\nabla u|}$ is orthogonal to the surface $\partial\{u \geq \lambda\}$ at every point where $|\nabla u| > 0$ and using a change of variables formula, if $u \in C^2(\bar{\Omega})$, one may rewrite (5.2), with $\alpha = \beta = 1$,

$$F(u) = \int_{\bar{\Omega}} |\nabla u| \left(1 + \operatorname{div}\left(\frac{\nabla u}{|\nabla u|}\right)^p\right) \, dx, \qquad (5.3)$$

with the convention that the integrand is 0 whenever $|\nabla u| = 0$. Of course, in order to get the existence of a minimizer for $F(u)$ we need to enlarge the space of admissible functions, and consider a relaxed version of $F(u)$ defined on $L^1(D)$ by

$$RF(u, D) = \inf\left\{\underline{\lim}\, F(u_h), u_h \to u \ \text{in} \ L^1(D), h \to +\infty\right\}.$$

In [16] the authors have proven that if u_0 is in $BV(D - \bar{\Omega})$ then the problem

$$\min\left\{RF(u, \bar{\Omega}), \ u = u_0 \ \text{on} \ D - \bar{\Omega}\right\}$$

has at least one solution in $BV(D)$.

■

The Ballester et al. Approach

In a series of papers, Ballester et al. [34, 35, 36] proposed to tackle the inpainting problem by considering a joint interpolation of vector fields and gray levels. Their approach is also inspired by the Elastica model and is closely related to the previous variational approach by Masnou et al. Let us fix the notation. As before, D is the image domain and Ω the hole to fill in. Again let us introduce Ω^+ such that $\bar{\Omega} \subset \Omega^+ \subset D$ and $B = \Omega^+ - \bar{\Omega}$. The set B will be called the band around Ω.

In the model proposed by Ballester et al. there are two unknowns: the orthogonal direction of level lines θ and the gray levels u of the image. The vector field θ will be then related to the image u according to the constraints

$$\theta \cdot \nabla u = |\nabla u| \quad \text{and} \quad |\theta| \leq 1.$$

Thus when $|\nabla u| \neq 0$, we have $\theta = \frac{\nabla u}{|\nabla u|}$. Note that θ is always defined even if the gradient vanishes. To fill in the hole Ω, the information is the value

u_0 of u in the band B and the vector field θ_0 of normals to the level lines of u_0 in B, i.e. θ_0 satisfies $\theta_0(x) \cdot \nabla u_0(x) = |\nabla u_0(x)|$ and $|\theta_0(x)| \leq 1$. We will assume that $\theta_0(x)$ has a trace on $\partial\Omega^+$. The problem of inpainting can be formulated as follows: given the pair of functions (u_0, θ_0), is it possible to extend it from the band B to a pair of functions (u, θ) defined on Ω? Inspired from the Elastica model Ballester et al. [34, 35, 36] proposed to search for (u, θ) as a minimizer of the constrained problem

$$\min \int_\Omega |\operatorname{div}(\theta)|^p (a + b\nabla k \star u) \, dx, \tag{5.4}$$

$$|\theta| \leq 1, \ \theta \cdot \nabla u - |\nabla u| = 0 \text{ in } \Omega, \tag{5.5}$$

$$u = u_0 \text{ in } B, \ \theta \cdot N = \theta_0 \cdot N \text{ on } \partial\Omega, \tag{5.6}$$

where $p > 1, a > 0, b \geq 0, k$ denotes a regularizing kernel of class C^1 with $k(x) > 0$ a.e., and $N(x)$ denotes the outer unit normal at $x \in \partial\Omega$. Let us remark that since $|\nabla k \star u| = |\nabla(k \star u)|$, the second term in (5.4) is nothing other than the norm of the gradient of the regularized image.

Remark The model (5.4)–(5.6) can be viewed as another relaxed formulation of functional (5.3). ∎

In order to get the existence of a solution, let us define suitable functional spaces for the pair (u, θ). We will suppose that

- u belongs to $BV(\Omega, u_0) = \{u \in BV(\Omega); \ u = u_0 \text{ in } B\}$ and $|u| \leq |u_0|_{L^\infty(B)}$,

- θ belongs to $W^{1,p}(\operatorname{div}, \Omega) = \{\theta \in L^p(\Omega); \ \operatorname{div}(\theta) \in L^p(\Omega); \ 1 \leq p < \infty\}$.

Following [19], let us recall that if $u \in BV(\Omega) \cap L^q(\Omega)$ and if $\theta \in L^\infty(\Omega)$, $\operatorname{div}(\theta) \in L^p(\Omega)$, $(1/p + 1/q = 1)$, then the product $\theta \cdot \nabla u$ is well-defined in the sense of measure, as well as the trace on $\partial\Omega$ of the normal component of θ, and a Green formula remains valid:

$$\int_\Omega u \operatorname{div} \theta \, dx + \int_\Omega (\theta \cdot \nabla u) \, dx = \int_{\partial\Omega} \theta \cdot N \, u \, d\mathcal{H}^1.$$

We can also remark that the constraint (5.5) can be incorporated into the energy (5.4) by using a Lagrange multiplier $\alpha \in \mathbb{R}$:

$$\min \int_\Omega |\operatorname{div}(\theta)|^p (a + b\nabla k \star u) \, dx + \alpha \int_\Omega (|\nabla u| - \theta \cdot \nabla u) \, dx. \tag{5.7}$$

Note that since $(|\nabla u| - \theta \cdot \nabla u)$ is always nonnegative, it suffices to impose

$$\int_\Omega (|\nabla u| - \theta \cdot \nabla u) \, dx = 0.$$

But thanks to the Green formula, the last term in (5.7) can be rewritten

$$\int_\Omega (|\nabla u| + \mathrm{div}(\theta)u)\ dx,$$

where we have omitted the boundary term since it is fixed. In [35] the authors have proven the existence of a minimizer (u, θ) with $\theta \in W^{1,p}(\mathrm{div}, \Omega)$, $p > 1$, $u \in BV(\Omega, u_0)$, and satisfying the constraints (5.5)–(5.6) and $|u| \leq |u_0|_{L^\infty(B)}$.

In practice, the datum u_0 is decomposed in B into level sets and we get a family of binary images $u_{0,\lambda} = \chi_{u_0 \geq \lambda}$, $\lambda = 0, 1, 2, \ldots, 255$. Then the algorithm is applied for each element of this family, resulting in an interpolated image u_λ with level sets $X_\lambda u$. Finally, the total image is reconstructed using the formula $u(x) = \sup \{\lambda \in \{0, 1, 2, \ldots, 255\} ; x \in X_\lambda u\}$. For each λ a minimizer of (5.4) is found by using a steepest-descent method involving two coupled second-order evolution equations.

An example of result is given in Figure 5.3, where some text can be removed automatically. Interestingly, this approach has also been extended to the 3-D case, when one wants to repair some surfaces. A result is shown in Figure 5.4 and we refer the interested reader to [49] for more details.

The Chan and Shen Total Variation Minimization Approach

In a series of papers, Chan and Shen [103, 95, 97] have introduced an approach based on total variation. In its simplest formulation, it can be seen as a simplification of the criterion (5.1), where one sets $\beta = 0$. It gives rise to the minimization of the criterion

$$F(u) = \int_D |Du|\ dx, \quad u = u_0 \text{ in } D - \Omega.$$

A proof of the existence of a minimizer under mild assumptions is given in [95].

Denoising capabilities can be added to the known data by replacing the constraint $u = u_0$ in $D - \Omega$ by a denoising one

$$\frac{1}{|D - \Omega|} \int_{D-\Omega} (u - u_0)^2\ dx = \sigma^2,$$

where σ^2 is the variance of the white (Gaussian) noise. The authors proposed to minimize in $BV(D)$ the unconstrained formulation

$$\lambda \int_{D-\Omega} (u - u_0)^2\ dx + \int_D |Du|.$$

We consider a slightly modified version of the criterion and show the existence of a solution. We suppose that $u_0 \in L^2(D - \Omega)$ and define $\bar{u}_0$ by extending u_0 by 0 in Ω. Then $\bar{u}_0 \in L^2(D)$. We also notice that if $f \in L^2(D)$, $|\chi_{D-\Omega} f|_{L^2(D)} \leq |f|_{L^2(D)}$, the multiplication by $\chi_{D-\Omega}$ is continuous in $L^2(\Omega)$ and the application $f \to \chi_{D-\Omega} f$ does not annihilate the

Original image   Result

Figure 5.3. Example of result obtained with the approach [36]. The overlapping text in the left-hand side image is the region to be filled in.

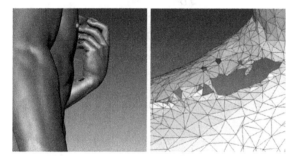

Original surface   Detail with meshes   Result

Figure 5.4. Example of result obtained with the approach [49] on 3-D surfaces (data from the Digital Michelangelo Project, Stanford University [213]).

constants. Therefore

$$\lambda \int_{D-\Omega} (u - u_0)^2 \, dx = \lambda \int_D (\chi_{D-\Omega} u - \bar{u}_0)^2 \, dx.$$

Let ϕ be a function satisfying conditions (3.15)–(3.17) and let Ψ be defined by $\Psi(\xi) = \phi(|\xi|)$. Then the functional

$$E(u) = \lambda \int_D (\chi_{D-\Omega} u - \bar{u}_0)^2 \, dx + \int_D \Psi(Du) \qquad (5.8)$$

has a minimizer in $BV(D)$. We are indeed exactly in the hypotheses of Theorem 3.2.2 with $R = \chi_{D-\Omega}$. Notice also that if ϕ is assumed to be only convex instead of strictly convex, the proof of Theorem 3.2.2 provides nevertheless the existence of a minimizer.

When u_0 is bounded a.e. in D, i.e., $m \le u_0 \le M$ a.e. in D, we have the following lemma.

Lemma 5.1.1 *Let u be a solution of (5.8), and let us suppose that u_0 satisfies $m \le u_0 \le M$ a.e. in D. Let $\varphi_{\alpha,\beta}$ be the cut-off function defined by*

$$\varphi_{m,M}(x) = \begin{cases} m & \text{if } x \leq m, \\ x & \text{if } m \leq x \leq M, \\ M & \text{if } x \geq M. \end{cases} \tag{5.9}$$

Then we have

$$E(\varphi_{m,M}(u)) \leq E(u).$$

Proof From Lemma 5.3.1 we deduce that

$$\int_D \Psi(D\varphi_{m,M}(u)) \leq \int_D \Psi(Du). \tag{5.10}$$

Define D_m (respectively D_M) as the subset of $D - \Omega$ where $u(x) < m$ (respectively $u(x) > M$) a.e. and $D_c = D - \Omega - D_m - D_M$. We compute

$$\int_{D-\Omega} (\varphi_{m,M}(u) - u_0)^2 \, dx - \int_{D-\Omega} (u - u_0)^2 \, dx$$

$$= \int_{D-\Omega} (\varphi_{m,M}(u) + u - 2u_0)(\varphi_{m,M}(u) - u) \, dx$$

$$= \int_{D_m} (m + u - 2u_0)(m - u) \, dx + \int_{D_M} (M + u - 2u_0)(M - u) \, dx.$$

The part on D_c vanishes since $u = \varphi_{m,M}(u)$ in D_c. In D_m, $m - u > 0$ a.e., and $m + u < 2m < 2u_0$ a.e from which the integral on D_m is negative. By similar arguments, the integral on D_M is also negative and we conclude that

$$\int_{D-\Omega} (\varphi_{m,M}(u) - u_0)^2 \, dx \leq \int_{D-\Omega} (u - u_0)^2 \, dx. \tag{5.11}$$

From (5.10) and (5.11) we get that $E(\varphi_{m,M}(u)) \leq E(u)$. ■

A consequence of Lemma 5.1.1 is that we can constrain the solution of (5.8) to satisfy $m \leq u \leq M$ without loss of generality.

As noticed in [98], the major drawback of the TV inpainting model is that it does not restore correctly a single object when its disconnected parts are separated far apart by the inpainting domain (see Figures 5.5 (a)–(c)). The main reasons are that TV models do not respect the connectivity principle highlighted by psychological experience [189] and that they do not encourage the connection of slim broken objects. In fact, in TV models, the geometry of objects is not necessarily taken into account.

5.1.3 PDE-Based Approaches

In this section we review two PDE-based models that do not explicitly come from a variational principle. The first one, by Bertalmio et al. [50],

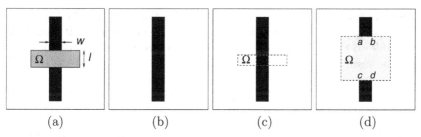

Figure 5.5. Results from TV inpainting algorithms (adapted from [103]). (a) Image to be inpainted; (b) answer from most humans; (c) TV inpainting for $w > l$; (d) TV inpainting for $w \ll l$.

is inspired by basic techniques used by art conservators to inpaint when restoring real paintings. The resulting mathematical model is a third-order PDE that propagates the level lines arriving at the hole. The second model, by Chan and Shen [98], comes from restoration problems and also involves third-order equations. Let us note that the two models described below are based on heuristic arguments and to the best of our knowledge, they are not totally justified from a mathematical point of view. Of course, other PDE-based approaches for inpainting have been proposed [49], and we also refer to Section 5.5 for color image inpainting.

The Bertalmio et al. Approach

The basic idea is still the same, namely to extend the structure present in the area surrounding the hole Ω. By structure we mean level lines arriving at the boundary of the hole, and the gray levels matching those of $\partial\Omega$.

In [50], the authors proposed an alternative method using transport equations. So which information will be propagated and in which direction needs to be defined. To answer to the first question, we can think about some edge detector, for example a function of the Laplacian [229], or Canny's edge detector [82, 81]. To simplify we will consider the Laplacian itself. Concerning the second question, the most natural direction to choose is the direction of the isophotes, i.e., the direction orthogonal to the gradient of the image: $\nabla u^\perp$. Then the authors model the propagation of the Laplacian in the direction of the level lines by solving the PDE

$$\frac{\partial u}{\partial t}(x,y,t) = \nabla(\Delta u(x,y,t)).\nabla u^\perp(x,y,t), \quad \forall(x,y) \in \Omega, \qquad (5.12)$$

$$u(x,y,t) = u_0(x,y) \text{ for all } (x,y) \in \partial\Omega \text{ and for all } t \geq 0,$$

$$u(x,y,0) = u_0^{ext}(x,y),$$

where $u_0^{ext}(x,y)$ is any continuous extension of the datum u_0 (which is known only in $D - \Omega$).

Remark Note that in terms of transport, (5.12) means that the Laplacian of u is propagated along the isophotes of u. In our opinion, it would be preferable to transport gray levels of u in the direction of the isophotes of Δu. The resulting PDE is

$$\frac{\partial u}{\partial t} = \nabla(\Delta u)^{\perp} \cdot \nabla u.$$

However, we can remark that at convergence (i.e., when $\frac{\partial u}{\partial t} = 0$), the solution satisfies the same equation since for any vectors V and W, we have $V^{\perp} \cdot W = -V \cdot W^{\perp}$. ∎

The authors in [50] have also remarked that in order to ensure a correct evolution of the direction field, a diffusion process has to be interleaved with the image inpainting process. This trick is made every two or three steps and implies a curving of level lines, preventing them from crossing each other. Any anisotropic diffusion process preserving sharpness can be used. In [50] the authors use the equation

$$\frac{\partial u}{\partial t}(x,y,t) = g_{\varepsilon}(x,y) \; \kappa(x,y,t) \; |\nabla u|(x,y,t), \; \forall(x,y) \in \Omega_{\varepsilon},$$

where κ is the Euclidean curvature, Ω_{ε} a small dilatation of Ω, and g_{ε} a smooth function in Ω_{ε} such that $g_{\varepsilon}(x,y) = 0$ on $\partial\Omega_{\varepsilon}$, and $g_{\varepsilon}(x,y) = 1$ in Ω.

The Chan and Shen Curvature-Driven Diffusion Approach

This model, proposed by Chan and Shen in [103], follows their previous work on total variation minimization inpainting described previously. The authors start from the PDE inpainting model based on the Euler–Lagrange equation associated with the total variation energy,

$$\frac{\partial u}{\partial t} = \operatorname{div}\left(\frac{\nabla u}{|\nabla u|}\right) + \lambda_e(u - u_0) \text{ in } D, \tag{5.13}$$

where $\lambda_e(x) = \lambda(1 - \chi_{\Omega}(x))$ is an extended Lagrange multiplier, χ_{Ω} is the characteristic function of the hole Ω, and λ is a constant. Note that since the fidelity attach term is written in the equation (5.13), then this approach also performs denoising in $D-\Omega$. Instead, when $u = u_0$ is imposed in $D-\Omega$, we have a pure inpainting approach: the image is not modified in $D - \Omega$.

In [103] the authors proposed a PDE model involving the curvature κ of the isophotes. In (5.13) the diffusion coefficient, namely $\frac{1}{|\nabla u|}$, depends only on the contrast or the strength of the isophotes; thus the infinite curvature at corners a, b, c, and d was not taken into account (see Figures 5.5 (a)–(d)). That is why the authors called their model curvature-driven diffusion

(CDD). The proposed PDE is

$$\frac{\partial u}{\partial t} = \text{div}\left(\frac{g(\kappa)\ \nabla u}{|\nabla u|}\right) \quad \text{in} \ \ \Omega, \tag{5.14}$$

$$u = u_0 \ \ \text{in} \ \ D - \Omega, \tag{5.15}$$

where g is an increasing function such that $g(0) = 0$ and $g(+\infty) = +\infty$. Typically we can choose $g(s) = s^p$ with $p \geq 1$. If we apply this approach to the case (d) in Figure 5.5, then we obtain a result similar to Figure 5.5 (b).

Unfortunately in most cases the available part of the image u_0 is noisy and the inpainting algorithm is sensitive to the noise. So it is necessary to incorporating to the model a denoising process. The simplest way to do it is to combine inpainting and TV denoising in a single equation:

$$\frac{\partial u}{\partial t} = \text{div}\left(\frac{G(x,|\kappa|)\ \nabla u}{|\nabla u|}\right) + \lambda_e(u - u_0) \quad \text{in} \ \ D, \tag{5.16}$$

where

$$(G(x,s),\lambda_e) = \begin{cases} (1,\lambda) & \text{if } x \in D - \Omega, \\ (g(s),0) & \text{otherwise, with } g \text{ is as in (5.14)} \\ & \text{and } \lambda \text{ as in (5.13).} \end{cases}$$

5.1.4 Discussion

In this section we have analyzed several variational or PDE-based models for inpainting. In the examples shown, we have mainly considered geometric images, where the main difficulty was to prolong level lines and preserve the saliency when edges have to be extended.

One limitation of such methods appears when lost regions should contain texture informations. If we consider the upper left-hand image of Figure 5.6, or if we imagine more complex textures, it seems clear that previous methods will fail in this case. In order to avoid this difficulty, one possibility is to process independently the geometric and texture components of the image, using suitable tools, which has been proposed by several authors. In the variational framework, Bertalmio et al. [51] propose to use some recent ideas by Meyer [235]. Given an image f, we can find u and v such that

$$f = u + v,$$

where u is well-structured and models homogeneous regions, and v contains oscillating patterns like texture and noise. We refer the reader to Section 5.2, where such methods are analyzed in detail. An example of decomposition is shown in the second row of figure 5.6. Then the geometric part can be restored using some standard inpainting approaches (for instance [50]) and the texture part can be recovered using texture synthesis algorithms (for instance [143]). The reconstructed image is obtained by summing again both components (see the upper right-hand image of Figure 5.6).

Another difficulty is related to the size of the region to be filled in. Here again, it is unlikely that only diffusing some information in the hole will give satisfying results. Reinventing the content is also here very closely related to texture synthesis (see for example [143]). In recent work by Criminisi et al. [120], the authors proposed an algorithm that allows one to fill in an image progressively by looking for closest prototypes in a dictionary. More precisely, given a hole, a set of patterns around the hole is extracted to build a dictionary. Then, given a set of pixels, this set is filled in with the best-matching pattern. Note that the interest in considering patterns instead of working at the pixel level is that it speeds up the process and also adds some coherence in the reconstruction. Although the image filling in itself is not related to PDE-based inpainting, there is an underlying "priority" map, which drives the order of the filling process, that is diffused from the boundary inward and makes use of both geometric and photometric gradient-based information. An example of a result is shown in Figure 5.7.

The approach [120] does not have any direct link with PDEs but this idea of looking for the best match between two patterns in order to select an intensity can be found in the nonlocal filter model by Buades et al. [77]. As explained in Section 3.3.4, the model (3.119) uses the similarity between two pixels x and y defined by the similarity of the intensity gray level between their neighborhoods. The new intensity in x will be estimated thanks to values y such that their corresponding neighborhoods best match.

When we have a sequence of images where some images have been degraded, filling in these parts is also an interesting and challenging application. Naturally, such data have some specificities in such a way that spatial and temporal dimensions should be exploited differently. For such sequences, time naturally plays an important role since some parts of the degraded area may be visible in neighboring times. Adding an extra dimension brings some difficulties, and we refer the reader to Section 5.3.4 for more information.

To conclude, one may wonder whether inpainting could also be used for another purpose than to restore images. For example, in [211] the authors showed how inpainting can be used to drive a segmentation algorithm. In this approach, the object to segment was seen as a foreground signal partially occluding a background, and the principle was to optimize a "background disocclusion" criterion, where disocclusion was performed by any inpainting algorithm. This kind of idea might be another track to follow.

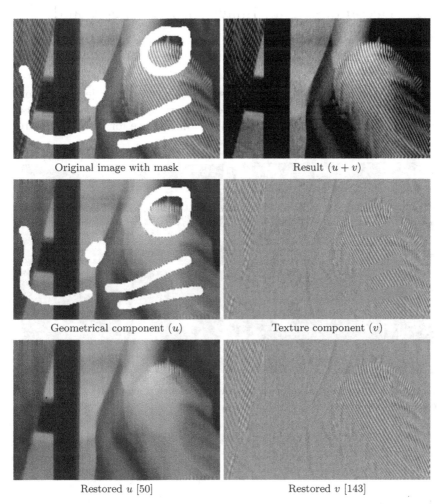

Original image with mask   Result $(u + v)$

Geometrical component (u) Texture component (v)

Restored u [50]   Restored v [143]

Figure 5.6. Example of result obtained with the approach [51] on real textured images.

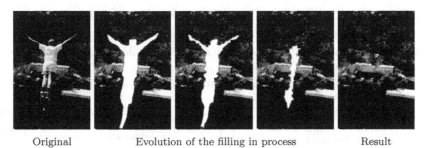

Original   Evolution of the filling in process   Result

Figure 5.7. Example of result obtained with the approach [120]

5.2 Decomposing an Image into Geometry and Texture

5.2.1 Introduction

In this section we present contributions [235, 333] concerning decomposition models. These models aim at characterizing strongly oscillating patterns. Such patterns can be encountered in real images; see for instance Figure 5.8, where they can be either texture or noise.

Figure 5.8. Sample image with piecewise smooth parts but also strongly oscillatory components.

The problem we address in this section is to decompose a given image f as the sum of two components

$$f = u + v,$$

so that

- u is well-structured and models homogeneous regions,

- v contains oscillating patterns such as texture and noise.

It will be shown how a variational approach can solve this problem, after we have introduced a suitable functional space to handle oscillating patterns in an image. The origins of these ideas is a remarkable book by Meyer [235], in which the author examined in detail the restoration model proposed in [284]:

$$\inf_{(u,v)\in BV(\Omega)\times L^2(\Omega)/f=u+v} \left(\int_\Omega |Du| + \frac{1}{2\lambda} |v|^2_{L^2(\Omega)} \right). \tag{5.17}$$

Meyer has observed that if f is a characteristic function and if f is sufficiently small with respect to a suitable norm (to be clarified in the sequel;

see Lemma 5.2.4) then the Rudin–Osher–Fatemi model [284] gives $u = 0$ and $v = f$, which is not what one would expect ($u = f$ and $v = 0$).

5.2.2 A Space for Modeling Oscillating Patterns

Meyer has introduced in [235] the following space:

Definition 5.2.1 $G(R^2)$ *is the Banach space composed of distributions f that can be written*

$$f = \partial_1 g_1 + \partial_2 g_2 = \mathrm{div}(g),$$

with g_1 and g_2 in $L^\infty(R^2)$. The space $G(R^2)$ is endowed with the following norm:

$$|v|_{G(R^2)} = \inf \left\{ |g|_{L^\infty(R^2)} = \operatorname*{ess\,sup}_{x \in R^2} |g(x)| \; / v = \mathrm{div}(g), \; g = (g_1, g_2), \right.$$
$$\left. g_1 \in L^\infty(R^2), \; g_2 \in L^\infty(R^2), |g(x)| = \sqrt{(|g_1|^2 + |g_2|^2)(x)} \right\}.$$

The idea of Meyer was to decompose an image f into a sum $u + v$ with u in $BV(R^2)$ and v in $G(R^2)$. Though $BV(R^2)$ has no simple dual space, Meyer showed that $G(R^2)$ is the dual of the closure in $BV(R^2)$ of the Schwartz class. So, it is closely related to the dual of $BV(R^2)$. This is an argument for decomposing a function on $BV(R^2) + G(R^2)$.

Before going further, since the domain of an image is bounded, it is more natural to work on a bounded subset of R^2 than in the whole R^2. So, we need to adapt the definition of $G(R^2)$ to this situation. We follow here the presentation given in [21]. Let us denote by Ω a bounded connected open set of R^2 with a Lipschitz boundary. The definition of $G(\Omega)$ is as follows:

Definition 5.2.2 $G(\Omega)$ *is the subspace of $W^{-1,\infty}(\Omega)$ defined by*

$$G(\Omega) = \left\{ v \in L^2(\Omega) \; / \; v = \mathrm{div}(\xi) \; , \; \xi \in L^\infty(\Omega, R^2) \; , \; \xi \cdot N = 0 \; on \; \partial\Omega \right\}. \tag{5.18}$$

$G(\Omega)$ *can be endowed with the norm*

$$|v|_{G(\Omega)} = \inf \left\{ |\xi|_{L^\infty(\Omega, R^2)} \; / v = \mathrm{div}(\xi) \; , \; \xi \cdot N = 0 \; on \; \partial\Omega \right\}. \tag{5.19}$$

Remark In Definition 5.2.2, since $\mathrm{div}(\xi) \in L^2(\Omega)$ and $\xi \in L^\infty(\Omega, R^2)$, we can define $\xi \cdot N$ on $\partial\Omega$ (see [216]).   ∎

The following lemma is straightforward.

Lemma 5.2.1 *If $v \in G(\Omega)$, then there exists $\xi \in L^\infty(\Omega, R^2)$, with $v = \mathrm{div}(\xi)$ and $\xi \cdot N = 0$ on $\partial\Omega$, and such that $|v|_G = |\xi|_{L^\infty(\Omega, R^2)}$.*

The main properties of $G(\Omega)$ are given below in Proposition 5.2.1 and Lemmas 5.2.2 and 5.2.3.

Proposition 5.2.1

$$G(\Omega) = \left\{ v \in L^2(\Omega) \ / \ \int_\Omega v = 0 \right\}. \tag{5.20}$$

Proof Let us denote by $H(\Omega)$ the right-hand side of (5.20). We split the proof into two steps.

Step **1**: Let v be in $G(\Omega)$.
Then from (5.18) it is immediate that $\int_\Omega v = 0$, i.e., $v \in H(\Omega)$.

Step **2**: Let v be in $H(\Omega)$.
Then from [70] (Theorem 3), there exists $\xi \in C^0(\bar{\Omega}, R^2) \cap W^{1,2}(\Omega, R^2)$ such that $v = \mathrm{div}(\xi)$ and $\xi = 0$ on $\partial\Omega$. In particular, we have $\xi \in L^\infty(\Omega, R^2)$ and $\xi \cdot N = 0$ on $\partial\Omega$. Thus $v \in G(\Omega)$.   ∎

Remark (About the second step of the proof of Proposition 5.2.1). Let us emphasize how powerful the result by Bourgain and Brezis in [70] is. It deals with the limit case v in $L^q(\Omega)$, $q = 2$, when the dimension of the space is $N = 2$. The classical method for tackling the equation

$$\mathrm{div}(\xi) = v \text{ with } \xi \cdot N = 0 \text{ on } \partial\Omega$$

consists in solving the problem

$$\Delta u = v \text{ with } \frac{\partial u}{\partial N} = 0 \text{ on } \partial\Omega,$$

and in setting $\xi = \nabla u$. If v is in $L^q(\Omega)$ with $q > 2$, this problem admits a unique solution (up to a constant) in $W^{2,q}(\Omega)$. Moreover, thanks to standard Sobolev embeddings (see [148, 161]), $\xi = \nabla u$ belongs to $L^\infty(\Omega, R^2)$. If $q = 2$, the result is not true and the classical approach does not work. So the result by J. Bourgain and H. Brezis [70] is very sharp.   ∎

The next lemma was stated in [235]. Using approximation results with $C_c^\infty(\Omega)$ functions [15], the proof is straightforward.

Lemma 5.2.2 *Let* $u \in BV(\Omega)$ *and* $v \in G(\Omega)$. *Then*

$$\int_\Omega uv \leq \int_\Omega |Du| \ |v|_{G(\Omega)}.$$

As a consequence of the previous proposition we have the following corollary.

Corollary 5.2.1 *For all $u \in BV(\Omega)$ and $v \in G(\Omega)$, we have*

$$\int_\Omega |Du| = \sup \left\{ \int_\Omega uv \; ; \; v \in G(\Omega) \; , \; |v|_{G(\Omega)} \leq 1 \right\},$$

$$|v|_{G(\Omega)} = \sup \left\{ \int_\Omega uv \; ; \; u \in BV(\Omega) \; , \; \int_\Omega |Du| \leq 1 \right\}.$$

The next result explains why the chosen norm in $G(\Omega)$ is well adapted to capturing oscillating signals in a minimization process. It is due to Meyer in the case $\Omega = R^2$, and it has been extended in the case of a bounded domain [21].

Lemma 5.2.3 *Let Ω be a Lipschitz bounded open set, and let f_n, $n \geq 1$, be a sequence of functions in $L^q(\Omega) \cap G(\Omega)$ with the following two properties:*

(i) *There exist $q > 2$ and $C > 0$ such that $|f_n|_{L^q(\Omega)} \leq C$.*

(ii) *The sequence f_n converges to 0 in the distributional sense.*

Then $|f_n|_G$ converges to 0 as n goes to infinity.

Example We show here an example of a sequence having a fixed L^q-norm and a $G(\Omega)$-norm converging to zero. Let $\Omega = [-\pi, \pi]$ and f_n a sequence of functions defined by

$$f_n(x, y) = \cos(nx) + \cos(ny).$$

After some calculus, we obtain that

$$|f_n|_{L^3(\Omega)} = \left(\frac{104\pi}{3} \right)^{\frac{1}{3}}.$$

Then, we observe that

$$f_n(x, y) = \mathrm{div}(\xi_n(x, y)) \quad \text{with} \quad \xi_n(x, y) = \frac{1}{n}(\sin(nx), \sin(ny)),$$

with $\xi_n \in L^\infty(\Omega, R^2)$, $\xi_n \cdot N = 0$ on $\partial\Omega$, and $|\xi_n|_{L^\infty} \leq \frac{2}{n} \to 0$ as $n \to \infty$. So we have

$$|f_n|_{G(\Omega)} = \inf_\xi \{ |\xi|_{L^\infty} , f_n = \mathrm{div}(\xi) \} \leq |\xi_n|_{L^\infty} \to 0.$$

∎

Proof of Lemma 5.2.3. Let us consider a sequence $f_n \in L^q(\Omega) \cap G(\Omega)$ satisfying assumption (i) and let us define the Neumann problem

$$\begin{cases} \Delta u_n = f_n & \text{in } \Omega, \\ \frac{\partial u_n}{\partial N} = 0 & \text{on } \partial\Omega. \end{cases} \tag{5.21}$$

We recall that since $f_n \in G(\Omega)$, we also have $\int_\Omega f_n \, dx = 0$. We know (see [173, 237, 123]) that problem (5.21) admits a solution $u_n \in W^{2,q}(\Omega)$. From [237, 236], we also know that there exists a constant $B > 0$ such that $|u_n|_{W^{2,q}(\Omega)} \leq B|f_n|_{L^q(\Omega)}$, and since we assume that $|f_n|_{L^q(\Omega)} \leq C$, we get

$$|u_n|_{W^{2,q}(\Omega)} \leq BC. \tag{5.22}$$

Since $q > 2$ and Ω is bounded, we know (see [2]) that there exists $\theta \in (0,1)$ such that $W^{2,q}(\Omega)$ is compactly embedded in $C^{1,\theta}(\Omega)$. We set $g_n = \nabla u_n$. We have $|g_n|_{W^{1,q}(\Omega)} \leq |u_n|_{W^{2,q}(\Omega)} \leq BC$. And it is also standard that $W^{1,q}(\Omega)^2$ is compactly embedded in $C^{0,\theta}(\Omega)^2$.

Hence, up to an extraction, we get that there exist u and $g \in C^{0,\theta}$ such that $u_n \to u$ and $g_n \to g$ (for the $C^{0,\theta}$ topology). It is then standard to pass to the limit in (5.21) to deduce that $g_n \to 0$ uniformly (we recall that $g_n = \nabla u_n$). The previous reasoning being true for any subsequence extracted from u_n, we conclude that the whole sequence ∇u_n is such that $\nabla u_n \to 0$ as $n \to +\infty$ in $L^\infty(\Omega, R^2)$-strong, i.e., $g_n = \nabla u_n \to 0$ in $L^\infty(\Omega, R^2)$-strong. Since $f_n = \mathrm{div}(g_n)$, we easily deduce that $|f_n|_G \to 0$. ∎

Lemma 5.2.3 says that a sequence of oscillating functions (e.g., $f_n \to 0$ in the distributional sense) having bounded L^q-norm tends to zero in $G(\Omega)$ without converging strongly to zero in $L^q(\Omega)$. Thus the L^2-norm in problem (5.17) is not well adapted to capturing oscillations in a signal, which has motivated Meyer to propose a new space.

5.2.3 Meyer's Model

Using this new functional space $G((\Omega))$, we have the following result:

Lemma 5.2.4 *Let (u,v) be the unique solution of the Rudin–Osher–Fatemi model (5.17), referred to in the sequel as the ROF model. Then*

$$|f|_{G(\Omega)} \leq \frac{1}{2\lambda} \tag{5.23}$$

is equivalent to $u = 0$ and $v = f$.

Proof We divide the proof into two steps.

Step 1: Let us suppose $u = 0$.
Then for all functions h in $BV(\Omega)$ we have

$$\lambda|f|^2_{L^2(\Omega)} \leq \lambda|f - h|^2_{L^2(\Omega)} + \int_\Omega |Dh|, \tag{5.24}$$

which is equivalent to

$$\int_\Omega fh \, dx \le \frac{1}{2\lambda}\left(\lambda|h|^2_{L^2(\Omega)} + \int_\Omega |Dh|\right). \qquad (5.25)$$

By changing h into $h\varepsilon$, $\varepsilon > 0$, and letting $\varepsilon \to 0$, we get

$$\int_\Omega fh \, dx \le \frac{1}{2\lambda}\int_\Omega |Dh| \text{ for all } h \text{ in } BV(\Omega), \qquad (5.26)$$

and with Corollary 5.2.1, we conclude that $|f|_{G(\Omega)} \le \frac{1}{2\lambda}$.

Step 2: If $|f|_{G(\Omega)} \le \frac{1}{2\lambda}$, then for all h in $BV(\Omega)$,

$$\int_\Omega fh \, dx \le \frac{1}{2\lambda}\int_\Omega |Dh|, \qquad (5.27)$$

which clearly implies (5.25) and (5.24), which means that $u = 0$ is a solution of the ROF model. ∎

This result shows that an oscillating image that has a small G-norm will be treated by the ROF model according to our terminology, as texture or noise. This is not what we could expect if f were a pure geometric image, for example a characteristic function. This observation led Meyer to propose the following model:

$$\inf_{(u,v)\in BV(\Omega)\times G(\Omega)/f=u+v}\left(\int_\Omega |Du| + \alpha\,|v|_{G(\Omega)}\right), \qquad (5.28)$$

where α is a weighting parameter. We remark that if f is in $L^q(\Omega)$, $q \ge 2$, we can always decompose f into the form $f = u+v$. In fact, we can choose

$$u = \int_\Omega f \text{ and } v = f - \int_\Omega f,$$

so that u is in $BV(\Omega)$ (since Ω is bounded) and v is in $G(\Omega)$, thanks to Proposition 5.2.1. Moreover, using standard arguments, it easy to show that Meyer's model has a solution. However, the uniqueness is still an open problem. In the next section, it is discussed how to compute numerically a solution of (5.28).

5.2.4 An Algorithm to Solve Meyer's Model

The main difficulty in solving Meyer's problem numerically is that it involves a term coming from an L^∞-norm, which is not easy to handle in practice. To overcome this difficulty, it is classical to introduce a new related problem that in some sense approximates the original one. In this section we follow the approach given in [32], which essentially consists in replacing $|v|_{G(\Omega)}$ in (5.28) with a constraint $|v|_{G(\Omega)} \le \mu$. More precisely, we will consider the problem

$$\inf_{\substack{(u, v) \in BV(\Omega) \times G(\Omega) \\ |v|_{G(\Omega)} \leq \mu}} \left(\int_\Omega |Du| + \frac{1}{2\lambda} |f - u - v|^2_{L^2(\Omega)} \right), \qquad (5.29)$$

where λ, μ are two nonnegative parameters. We first show that if λ and μ are fixed, then the problem (5.29) admits a unique solution, which is numerically computed using a projection-based algorithm. The convergence of the algorithm is rigorously proved. Then we examine the asymptotic case λ goes to zero. We find a limit variational problem whose minimizers are in some sense solutions of the original problem.

Prior Numerical Contribution

Vese and Osher [333] were the first authors to numerically tackle Meyer's model. They propose to solve the problem

$$\inf_{(u,v) \in BV(\Omega) \times G(\Omega)} \left(\int_\Omega |Du| + \frac{1}{2\lambda} |f - u - v|^2_{L^2(\Omega)} + \mu |v|_{G(\Omega)} \right). \qquad (5.30)$$

To compute the solution, they replace the term $|v|_{G(\Omega)}$ by $|\sqrt{g_1^2 + g_2^2}|_p$, where $v = \text{div}(g)$. Then they formally derive the Euler–Lagrange equations from (5.30). For numerical reasons, the authors use the value $p = 1$ (they claim that they made experiments for $p = 1, \ldots, 10$ and that they did not see any qualitative difference). They report good numerical results. See also [263] for another related model concerning the case $\lambda = 0$ and $p = 2$.

The Aujol et al. Approach

We present here the approach developed in [32]. The proposed functional to approximate Meyer's model is similar to the one given by [333], but the algorithm for finding a minimizer is quite different. Moreover, this approach is completely justified from a mathematical point of view. Let us first consider the following functional defined on $BV(\Omega) \times G(\Omega)$:

$$F_{\lambda,\mu}(u, v) = \begin{cases} \int_\Omega |Du| + \frac{1}{2\lambda} |f - u - v|^2_{L^2(\Omega)} & \text{if } v \in G_\mu(\Omega), \\ +\infty & \text{if } v \in G(\Omega) \backslash G_\mu(\Omega), \end{cases} \qquad (5.31)$$

where

$$G_\mu(\Omega) = \{ v \in G(\Omega) / |v|_G \leq \mu \}. \qquad (5.32)$$

We recall that $|v|_G$ is defined by (5.19). The parameter μ plays the same role as the one in problem (5.30). $F_{\lambda,\mu}(u, v)$ is finite if and only if (u, v)

belongs to $BV(\Omega) \times G_\mu(\Omega)$, and the problem we want to solve is

$$\inf_{(u,v) \in BV(\Omega) \times G(\Omega)} F_{\lambda,\mu}(u,v). \qquad (5.33)$$

Later in this section, we will show that the unique solution of (5.33), (u_λ, v_λ), tends (up to a subsequence) to a solution of Meyer's model as λ goes to 0. Before doing that we give now a precise description of the algorithm for solving (5.33). We describe it in the discrete case, but it is exactly the same in the continuous setting. We choose here the same notation as in [91] (see also Section 3.2.4). The image is a two-dimensional vector of size $N \times N$. We denote by X the Euclidean space $R^{N \times N}$, and $Y = X \times X$. The space X will be endowed with the scalar product $(u,v)_X = \sum_{1 \le i,j \le N} u_{i,j} v_{i,j}$ and the norm $|u|_X = \sqrt{(u,u)_X}$. To define a discrete total variation, we introduce a discrete version of the gradient operator. If $u \in X$, the gradient ∇u is a vector in Y given by $(\nabla u)_{i,j} = ((\nabla u)^1_{i,j}, (\nabla u)^2_{i,j})$, with

$$(\nabla u)^1_{i,j} = \begin{cases} u_{i+1,j} - u_{i,j} & \text{if } i < N, \\ 0 & \text{if } i = N, \end{cases} \qquad (\nabla u)^2_{i,j} = \begin{cases} u_{i,j+1} - u_{i,j} & \text{if } j < N, \\ 0 & \text{if } j = N. \end{cases}$$

The discrete total variation of u is then defined by

$$J_d(u) = \sum_{1 \le i,j \le N} |(\nabla u)_{i,j}|. \qquad (5.34)$$

We also introduce a discrete version of the divergence operator. We define it by analogy with the continuous setting by $\text{div} = -\nabla^*$ where ∇^* is the adjoint of ∇, that is, for every $p \in Y$ and $u \in X$

$$(-\text{div}(p), u)_X = (p, \nabla u)_Y.$$

It is easy to check that

$$(\text{div } p)_{i,j} = (\text{div } p)^1_{i,j} + (\text{div } p)^2_{i,j} \qquad (5.35)$$

with

$$(\text{div } p)^1_{i,j} = \begin{cases} p^1_{i,j} - p^1_{i-1,j} & \text{if } 1 < i < N, \\ p^1_{i,j} & \text{if } i = 1, \\ -p^1_{i-1,j} & \text{if } i = N, \end{cases}$$

$$(\text{div } p)^2_{i,j} = \begin{cases} p^2_{i,j} - p^2_{i,j-1} & \text{if } 1 < j < N, \\ p^2_{i,j} & \text{if } j = 1, \\ -p^2_{i,j-1} & \text{if } j = N. \end{cases}$$

From now on, we will use these discrete operators. We are now in a position to introduce the discrete version of the space G.

Definition 5.2.3

$$G^d = \{v \in X \; / \; \exists g \in Y \; such \; that \; v = \text{div}(g)\}, \qquad (5.36)$$

and if $v \in G^d$,

$$|v|_{G^d} = \inf \left\{ |g|_\infty \ / \ v = \mathrm{div}(g), \right.$$
$$\left. g = (g^1, g^2) \in Y, |g_{i,j}| = \sqrt{(g_{i,j}^1)^2 + (g_{i,j}^2)^2} \right\}, \qquad (5.37)$$

where $|g|_\infty = \max_{i,j} |g_{i,j}|$. Moreover, we will define

$$G_\mu^d = \left\{ v \in G^d \ / \ |v|_{G^d} \le \mu \right\}, \qquad (5.38)$$

or in other words,

$$G_\mu^d = \left\{ v \in X / \exists g \in Y \ \text{such that} \ |g|_\infty \le \mu \ \text{and} \ v = \mathrm{div}(g) \right\}. \qquad (5.39)$$

The next proposition gives a very simple characterization of G^d.

Proposition 5.2.2 *The space G^d can be identified with the following subspace:*

$$X_0 = \left\{ v \in X \ / \ \sum_{i,j} v_{i,j} = 0 \right\}.$$

Proof We split our proof into two steps.

Step 1: Let us assume that $v \in G^d$.
Therefore, there exists $g \in Y$ such that $v = \mathrm{div}(g)$. But

$$\sum_{i,j} (\mathrm{div}(g))_{i,j} = (-\nabla^* g, 1)_Y = (g, \nabla 1)_X = 0,$$

i.e., $v \in X_0$. Hence $G^d \subset X_0$.

Step 2: Conversely, let $v \in X_0$.
Since the kernel of ∇ is the set of constant images, i.e., the vectors $x \in X$ such that $x_{i,j} = x_{i',j'}$ for all i, j, i', j', it is clear that a discrete Poincaré inequality holds:

$$\left| x - \frac{1}{N^2} \sum_{i,j} x_{i,j} \right|_X \le c |\nabla x|_Y.$$

Hence one shows easily that the problem $\min_{x \in X} A(x)$, with $A(x) = |\nabla x|^2 + 2(x, v)$, has a solution. This solution satisfies $A'(x) = 0$, that is, $-2\,\mathrm{div}(\nabla x) + 2v = 0$. Hence $v = \mathrm{div}(\nabla x) \in G^d$, and we conclude that $X_0 \subset G^d$. ∎

The discretized functional associated with (5.31), defined on $X \times X$, is given by

$$F_{\lambda,\mu}(u, v) = \begin{cases} J_d(u) + \frac{1}{2\lambda} |f - u - v|_X^2 & \text{if } v \in G_\mu^d, \\ +\infty & \text{if } v \in X \backslash G_\mu^d. \end{cases} \qquad (5.40)$$

The problem we want to solve is

$$\inf_{(u,v)\in X\times X} F_{\lambda,\mu}(u,v). \tag{5.41}$$

A way for numerically solving (5.41) is to use an alternate projection algorithm. We first remark that since J_d defined by (5.34) is homogeneous of degree one (i.e., $J_d(\lambda u) = \lambda J_d(u)$ $\forall u$ and $\lambda > 0$), it is then standard to show that J_d^* (the Legendre–Fenchel transform of J_d) is the indicator function of some closed convex set [144], which turns out to be the set G_1^d defined by (5.38):

$$J_d^*(v) = \chi_{G_1^d}(v) = \begin{cases} 0 & \text{if } v \in G_1^d, \\ +\infty & \text{otherwise.} \end{cases} \tag{5.42}$$

Hence $F_{\lambda,\mu}(u,v)$ can be written as

$$F_{\lambda,\mu}(u,v) = \frac{1}{2\lambda}|f - u - v|_X^2 + J_d(u) + J_d^*\left(\frac{v}{\mu}\right). \tag{5.43}$$

With this formulation, we see the symmetric roles played by u and v. And the problem we want to solve is

$$\inf_{(u,v)\in X\times X} F_{\lambda,\mu}(u,v). \tag{5.44}$$

To solve (5.44), we consider the following two problems:

(i) v being fixed, we search for u as a solution of

$$\inf_{u\in X}\left(J_d(u) + \frac{1}{2\lambda}|f - u - v|_X^2\right). \tag{5.45}$$

(ii) u being fixed, we search for v as a solution of

$$\inf_{v\in G_\mu^d}|f - u - v|_X^2. \tag{5.46}$$

From Proposition 3.2.1 in Section 3.2.4, we know that the solution of (5.45) is given by

$$\hat{u} = f - v - P_{G_\lambda^d}(f - v),$$

and the solution of (5.46) is simply given by

$$\hat{v} = P_{G_\mu^d}(f - u).$$

From (5.45) and (5.46), we deduce the alternate projection algorithm (see Table 5.1).

Some mathematical properties of the algorithm (5.47)–(5.49) are described below. The first result is about the existence and uniqueness of a solution.

Proposition 5.2.3 *There exists a unique pair* $(\hat{u}, \hat{v}) \in X \times G_\mu^d$ *minimizing* $F_{\lambda,\mu}$ *on* $X \times X$.

For $(u^0, b^0) = (0,0)$ given

- Step 1:

$$v_{n+1} = P_{G_\mu^d}(f - u_n).$$ (5.47)

- Step 2:

$$u_{n+1} = f - v_{n+1} - P_{G_\lambda^d}(f - v_{n+1}).$$ (5.48)

- Go back to the first step until convergence. The stopping test is defined by

$$\max(|u_{n+1} - u_n|, |v_{n+1} - v_n|) \leq \varepsilon.$$ (5.49)

Table 5.1. Alternate projection algorithm.

Proof We split the proof into two steps.

Step 1: Existence

- We first remark that the set $X \times G_\mu^d$ is convex, and then that $F_{\lambda,\mu}$ is convex on $X \times G_\mu^d$. We thus deduce that $F_{\lambda,\mu}$ is convex on $X \times X$.

- It is immediate to see that $F_{\lambda,\mu}$ is continuous on $X \times G_\mu^d$. We then deduce that $F_{\lambda,\mu}$ is lower semicontinuous on $X \times X$.

- Let $(u,v) \in X \times G_\mu^d$. We have $|v|_{G^d} \leq \mu$. Moreover, since X is of finite dimension, there exists $g \in Y$ such that $v = \mathrm{div}(g)$ and $|g|_{L^\infty} = |v|_{G^d} \leq \mu$. We deduce from (5.35) that (N^2 is the size of the image)

$$|v|_X \leq 4\mu N^2.$$ (5.50)

We recall that $X \times X$ is endowed with the Euclidean norm:

$$|(u,v)|_{X \times X} = \sqrt{|u|_X^2 + |v|_X^2}.$$

Thus, if $|(u,v)|_{X \times X} \to +\infty$, then we get from (5.50) that $|u|_X \to +\infty$. We therefore deduce, since f is fixed and since (5.50) holds, that

$$|f - u - v|_X^2 \to +\infty.$$

And since $F_{\lambda,\mu}(u,v) \geq \frac{1}{2\lambda}|f - u - v|_2^2$, we get $F_{\lambda,\mu}(u,v) \to +\infty$. Hence we deduce that $F_{\lambda,\mu}$ is coercive on $X \times G_\mu^d$. We therefore conclude that $F_{\lambda,\mu}$ is coercive on $X \times X$.

We deduce the existence of a minimizer $(\hat{u}, \hat{v})$.

Step **2**: Uniqueness

To get the uniqueness, we first remark that $F_{\lambda,\mu}$ is strictly convex on $X \times G_\mu^d$, as the sum of a convex function and a strictly convex function, except in the direction $(u, -u)$. Hence it suffices to check that if $(\hat{u}, \hat{v})$ is a minimizer of $F_{\lambda,\mu}$, then for $t \neq 0$, $(\hat{u} + t\hat{u}, \hat{v} - t\hat{u})$ is not a minimizer of $F_{\lambda,\mu}$. The result is obvious if $\hat{v} - t\hat{u} \in X \backslash G_\mu^d$. Let us show that if $\hat{v} - t\hat{u} \in G_\mu^d$ then the result is still true. In fact, if $\hat{v} - t\hat{u} \in G_\mu^d$, we have

$$F_{\lambda,\mu}(\hat{u} + t\hat{u}, \hat{v} - t\hat{u}) = F_{\lambda,\mu}(\hat{u}, \hat{v}) + (|1 + t| - 1)J_d(\hat{u}). \qquad (5.51)$$

To obtain a contradiction, let us assume that there exists $\hat{t} \neq \{-2, 0\}$ such that $\hat{v} - \hat{t}\hat{u} \in G_\mu^d$ and

$$F_{\lambda,\mu}(\hat{u} + \hat{t}\hat{u}, \hat{v} - \hat{t}\hat{u}) \leq F_{\lambda,\mu}(\hat{u}, \hat{v}). \qquad (5.52)$$

Since $(\hat{u}, \hat{v})$ minimizes $F_{\lambda,\mu}$, (5.52) is an equality. From (5.51), we deduce that $(|1 + \hat{t}| - 1)J_d(\hat{u}) = 0$. Since $\hat{t} \neq \{-2, 0\}$, we get that $J_d(\hat{u}) = 0$. Therefore there exists $\gamma \in R$ such that for all (i, j), $\hat{u}_{i,j} = \gamma$.

- If $\gamma = 0$, then $\hat{u} = 0$. Thus $(\hat{u} + \hat{t}\hat{u}, \hat{v} - \hat{t}\hat{u}) = (\hat{u}, \hat{v})$.

- If $\gamma \neq 0$, then $\hat{v} - \hat{t}\hat{u}$ cannot belong to G_μ^d since its mean is not zero (see Proposition 5.2.2). This contradicts our assumption.

There remains to check what happens in the case $\hat{t} = -2$. In this case, by convexity, we get that if $t \in (-2, 0)$, then $\hat{v} - t\hat{u} \in G_\mu^d$ and

$$F_{\lambda,\mu}(\hat{u} + t\hat{u}, \hat{v} - t\hat{u}) \leq F_{\lambda,\mu}(\hat{u}, \hat{v}).$$

Thus we get (5.52), and the proof is complete.   ∎

Next, we show that the alternate projection algorithm gives asymptotically as $n \to \infty$ the solution of the discrete problem (5.44).

Proposition 5.2.4 *The sequence $F_{\lambda,\mu}(u_n, v_n)$ built in Table 5.1 converges to the minimum of $F_{\lambda,\mu}$ on $X \times X$.*

Proof We first remark that as we solve successive minimization problems, we have

$$F_{\lambda,\mu}(u_n, v_n) \geq F_{\lambda,\mu}(u_n, v_{n+1}) \geq F_{\lambda,\mu}(u_{n+1}, v_{n+1}). \qquad (5.53)$$

In particular, the sequence $F_{\lambda,\mu}(u_n, v_n)$ is nonincreasing. Since it is bounded from below by 0, it thus converges in R. We denote by m its limit. We want to show that

$$m = \inf_{(u,v) \in X \times X} F_{\lambda,\mu}(u, v). \qquad (5.54)$$

Without any restriction, we can assume that $\forall n, (u_n, v_n) \in X \times G_\mu^d$. Since $F_{\lambda,\mu}$ is coercive and since the sequence $F_{\lambda,\mu}(u_n, v_n)$ converges, we deduce that the sequence (u_n, v_n) is bounded in $X \times G_\mu^d$. We can thus extract a subsequence (u_{n_k}, v_{n_k}) that converges to $(\hat{u}, \hat{v})$ as $n_k \to +\infty$, with $(\hat{u}, \hat{v}) \in X \times G_\mu^d$. Moreover, we have, for all $n_k \in N$ and all v in X,

$$F_{\lambda,\mu}(u_{n_k}, v_{n_k+1}) \leq F_{\lambda,\mu}(u_{n_k}, v), \tag{5.55}$$

and for all $n_k \in N$ and all u in X,

$$F_{\lambda,\mu}(u_{n_k}, v_{n_k}) \leq F_{\lambda,\mu}(u, v_{n_k}). \tag{5.56}$$

Let us denote by $\tilde{v}$ a cluster point of (v_{n_k+1}). Considering (5.53), we get (since $F_{\lambda,\mu}$ is continuous on $X \times G_\mu^d$)

$$m = F_{\lambda,\mu}(\hat{u}, \hat{v}) = F_{\lambda,\mu}(\hat{u}, \tilde{v}). \tag{5.57}$$

By passing to the limit in (5.47), we get $\tilde{v} = P_{G_\mu^d}(f - \hat{u})$. But from (5.57), we know that $|f - \hat{u} - \hat{v}| = |f - \hat{u} - \tilde{v}|$. By uniqueness of the projection, we conclude that $\tilde{v} = \hat{v}$. Hence $v_{n_k+1} \to \hat{v}$. By passing to the limit in (5.55) ($F_{\lambda,\mu}$ is continuous on $X \times G_\mu^d$), we have, for all v,

$$F_{\lambda,\mu}(\hat{u}, \hat{v}) \leq F_{\lambda,\mu}(\hat{u}, v). \tag{5.58}$$

And by passing to the limit in (5.56), for all u,

$$F_{\lambda,\mu}(\hat{u}, \hat{v}) \leq F_{\lambda,\mu}(u, \hat{v}). \tag{5.59}$$

Inequalities (5.58) and (5.59) can be rewritten as

$$F_{\lambda,\mu}(\hat{u}, \hat{v}) = \inf_{v \in X} F_{\lambda,\mu}(\hat{u}, v), \tag{5.60}$$

$$F_{\lambda,\mu}(\hat{u}, \hat{v}) = \inf_{u \in X} F_{\lambda,\mu}(u, \hat{v}). \tag{5.61}$$

But from Definition 5.43 of $F_{\lambda,\mu}(u, v)$, (5.61) is equivalent to (see [144])

$$0 \in -f + \hat{u} + \hat{v} + \lambda \partial J_d(\hat{u}), \tag{5.62}$$

and (5.60) to

$$0 \in -f + \hat{u} + \hat{v} + \lambda \partial J_d^* \left(\frac{\hat{v}}{\mu}\right). \tag{5.63}$$

The subdifferential of $F_{\lambda,\mu}$ at $(\hat{u}, \hat{v})$ is given by

$$\partial F_{\lambda,\mu}(\hat{u}, \hat{v}) = \frac{1}{\lambda} \begin{pmatrix} -f + \hat{u} + \hat{v} + \lambda \partial J_d(\hat{u}) \\ -f + \hat{u} + \hat{v} + \lambda \partial J_d^*\left(\frac{\hat{v}}{\mu}\right) \end{pmatrix}. \tag{5.64}$$

Thus, according to (5.62) and (5.63), we have

$$\begin{pmatrix} 0 \\ 0 \end{pmatrix} \in \partial F_{\lambda,\mu}(\hat{u}, \hat{v}), \tag{5.65}$$

which is equivalent to $F_{\lambda,\mu}(\hat{u}, \hat{v}) = \inf_{(u,v) \in X^2} F_{\lambda,\mu}(u, v) = m$. Hence the whole sequence $F_{\lambda,\mu}(u_n, v_n)$ converges to m, the unique minimum of $F_{\lambda,\mu}$ on $X \times G_\mu^d$. We deduce that the sequence (u_n, v_n) converges to $(\hat{u}, \hat{v})$, the minimizer of $F_{\lambda,\mu}$, as n tends to $+\infty$. ∎

STUDY OF THE ASYMPTOTIC CASE

We examine now the role played by the parameter λ and in particular the asymptotic behavior of problem (5.44) as λ goes to zero. In fact, we are going to show that problem (5.44) tends to the following limit problem:

$$\inf_{(u,v)\in X\times X/f=u+v} J_d(u) + J_d^* \left(\frac{v}{\mu}\right). \tag{5.66}$$

Since problem (5.66) is convex, coercive, and continuous on G^d, it is then straightforward to show that it admits at least one solution (the uniqueness is an open problem). Problem (5.66) is the limit problem of (5.44) in the following sense.

Proposition 5.2.5 *Let us assume that problem (5.66) has a unique solution $(\hat{u}, \hat{v})$. We fix $\mu = |\hat{v}|_G$ in (5.66). Let us denote by (u_λ, v_λ) the unique solution of problem (5.40). Then (u_λ, v_λ) converges to $(\hat{u}, \hat{v}) \in X \times X$ as λ goes to 0.*

Remark In the case that the solution of problem (5.66) is not unique, the result of Proposition 5.2.5 does not hold. We can just show that any cluster point of $(u_{\lambda_n}, v_{\lambda_n})$ is a solution of problem (5.66). ∎

Proof Since (u_λ, v_λ) is the solution of problem (5.40), we have $v_\lambda \in G_\mu^d$, i.e., $|v_\lambda|_{G^d} \leq \mu$. As we saw in the proof of Lemma 5.2.3, this inequality implies

$$|v_\lambda|_X \leq 4\mu N^2. \tag{5.67}$$

Since (u_λ, v_λ) is the solution of problem (5.40), we have

$$F_{\lambda,\mu}(u_\lambda, v_\lambda) \leq F_{\lambda,\mu}(f, 0), \tag{5.68}$$

which means that

$$F_{\lambda,\mu}(u_\lambda, v_\lambda) \leq J_d(f). \tag{5.69}$$

The left-hand side of (5.69) is given by

$$F_{\lambda,\mu}(u_\lambda, v_\lambda) = J_d(u_\lambda) + \frac{1}{2\lambda} |f - u_\lambda - v_\lambda|_X^2 + J_d^* \left(\frac{v_\lambda}{\mu}\right)$$

$$= J_d(u_\lambda) + \frac{1}{2\lambda} |f - u_\lambda - v_\lambda|_X^2.$$

Hence $J_d(u_\lambda) + \frac{1}{2\lambda} |f - u_\lambda - v_\lambda|_X^2 \leq J_d(f)$, and

$$|f - u_\lambda - v_\lambda|^2 \leq 2\lambda J_d(f). \tag{5.70}$$

Since $|v_\lambda|_X$ is bounded (from (5.67)), we conclude that if $\lambda \in [0; 1]$, u_λ is bounded by a constant $C > 0$ that does not depend on λ.

Consider a sequence (λ_n) that goes to 0 as $n \to +\infty$. Then, up to an extraction (since $(u_{\lambda_n}, v_{\lambda_n})$ is bounded in $X \times X$), there exists $(u_0, v_0) \in$

$X \times X$ such that $(u_{\lambda_n}, v_{\lambda_n})$ converges to (u_0, v_0). By passing to the limit in (5.70), we get $|f - u_0 - v_0|_X = 0$, i.e., $f = u_0 + v_0$.

To conclude the proof of the proposition, there remains to show that (u_0, v_0) is a solution of problem (5.66). We first notice that as $\forall \lambda > 0$, and since $|v_\lambda|_{G^d} \leq \mu$, we get $|v_0|_{G^d} \leq \mu$. Let $(u, v) \in X \times X$ be such that $f = u + v$. We have

$$J_d(u) + J_d^* \left(\frac{v}{\mu}\right) + \frac{1}{2\lambda} \underbrace{|f - u - v|^2}_{=0}$$

$$\geq J_d(u_{\lambda_n}) + J_d^* \left(\frac{v_{\lambda_n}}{\mu}\right) + \frac{1}{2\lambda_n} |f - u_{\lambda_n} - v_{\lambda_n}|^2$$

$$\geq \underbrace{J_d(u_{\lambda_n}) + J_d^* \left(\frac{v_{\lambda_n}}{\mu}\right)}_{\to J_d(u_0) + J_d^* \left(\frac{v_0}{\mu}\right)}.$$

Hence (u_0, v_0) is a solution of problem (5.66). And since we have assumed that problem (5.66) has a unique solution, we deduce that $(u_0, v_0) = (\hat{u}, \hat{v})$, i.e., (u_0, v_0) is the solution of problem (5.66).   ∎

BACK TO MEYER'S MODEL

Now let us examine the link between the previous results and Meyer's model. Is Meyer's model solved? The answer will be given in Proposition 5.2.6 and Proposition 5.2.9. We first recall the discrete version of Meyer's problem:

$$\inf_{(u,v) \in X \times G^d / f = u + v} H_\alpha(u, v), \tag{5.71}$$

with

$$H_\alpha(u, v) = (J_d(u) + \alpha |v|_{G^d}). \tag{5.72}$$

Proposition 5.2.6 *Let us fix $\alpha > 0$ in problem (5.71). Let $(\hat{u}, \hat{v})$ be a solution of problem (5.71). We fix $\mu = |\hat{v}|_{G^d}$ in (5.66). Then*

(i) *$(\hat{u}, \hat{v})$ is also a solution of problem (5.66).*

(ii) *Conversely, any solution $(\tilde{u}, \tilde{v})$ of (5.66) (with $\mu = |\hat{v}|_{G^d}$) is a solution of (5.71).*

Proof We split the proof into two steps.

Step 1: Proof of (i)
Since $(\hat{u}, \hat{v})$ is a solution of (5.71) and $|\hat{v}|_{G^d} = \mu$, then $\hat{u}$ is solution of

$$\inf_{u \in X / u = f - v, |v|_{G^d} = \mu} J_d(u) + \alpha\mu,$$

i.e., $\hat{u}$ is solution of

$$\inf_{u \in X/u=f-v, |v|_{G^d}=\mu} J_d(u). \tag{5.73}$$

Since we have the inclusion

$$\{u \in X/u = f - v, |v|_{G^d} = \mu\} \subset \{u \in X/u = f - v, |v|_{G^d} \leq \mu\},$$

we have

$$\inf_{u \in X/u=f-v, |v|_{G^d}=\mu} J_d(u) \geq \inf_{u \in X/u=f-v, |v|_{G^d} \leq \mu} J_d(u).$$

To obtain a contradiction, let us assume that

$$\inf_{u \in X/u=f-v, |v|_{G^d}=\mu} J_d(u) > \inf_{u \in X/u=f-v, |v|_{G^d} \leq \mu} J_d(u). \tag{5.74}$$

Thus, there exists $v' \in X$ such that $|v'|_{G^d} < \mu$ and

$$J_d(f - v') < \inf_{u \in X/u=f-v, |v|_{G^d}=\mu} J_d(u). \tag{5.75}$$

Setting $u' = f - v'$, we have $J_d(u') + \alpha |v'|_{G^d} < J_d(u') + \alpha\mu$. But since $(\hat{u}, \hat{v})$ is a solution of (5.71),

$$J_d(\hat{u}) + \alpha |\hat{v}|_{G^d} \leq J_d(u') + \alpha |v'|_{G^d} < J_d(u') + \alpha\mu. \tag{5.76}$$

Hence (we recall that $|\hat{v}|_{G^d} = \mu$), we get from (5.76) that $J_d(\hat{u}) < J_d(u')$. This contradicts (5.75). We conclude that (5.74) cannot hold. Hence

$$\inf_{u \in X/u=f-v, |v|_{G^d}=\mu} J_d(u) = \inf_{u \in X/u=f-v, |v|_{G^d} \leq \mu} J_d(u).$$

From (5.73), we see that $\hat{u}$ is a solution of $\inf_{u \in X/u=f-v, |v|_{G^d} \leq \mu} J_d(u)$, i.e., $\hat{u}$ is a solution of

$$\inf_{u \in X/u=f-v} J_d(u) + J_d^* \left(\frac{v}{\mu} \right).$$

Hence $(\hat{u}, \hat{v})$ is also a solution of (5.66).

Step **2**: Proof of (ii)
Let us now consider $(\tilde{u}, \tilde{v})$, a solution of (5.66). We can repeat the computations we made in Step 1. We get that $\tilde{u}$ is a solution of

$$\inf_{u \in X/u=f-v, |v|_{G^d}=\mu} J_d(u) + \alpha\mu.$$

We therefore have $J_d(\tilde{u}) + \alpha\mu = J_d(\hat{u}) + \alpha |\hat{v}|_{G^d}$. But since $(\tilde{u}, \tilde{v})$ is a solution of (5.66), we have $|\tilde{v}|_{G^d} \leq \mu$. Hence $J_d(\tilde{u}) + \alpha |\tilde{v}|_{G^d} \leq J_d(\hat{u}) + \alpha |\hat{v}|_{G^d}$. And since $(\hat{u}, \hat{v})$ is a solution of (5.71), we get that

$$J_d(\tilde{u}) + \alpha |\tilde{v}|_{G^d} = J_d(\hat{u}) + \alpha |\hat{v}|_{G^d}.$$

Thus we conclude that $(\tilde{u}, \tilde{v})$ is a solution of (5.71). ■

In fact, we can say more about the link between Meyer's problem (5.71) and our limit problem (5.66). When α is fixed, let us define

$$Z_\alpha = \left\{ v_\alpha, \ v_\alpha \text{ is a solution of the problem } \inf_{v \in G^d} H_\alpha(f - v, v) \right\},$$

$$S_\alpha = \left\{ |v_\alpha|_{G^d}, \ v_\alpha \text{ is a solution of the problem } \inf_{v \in G^d} H_\alpha(f - v, v) \right\}.$$

We know that Z_α and S_α are not empty. Then we consider the two multi-applications

$$Y: \quad R_+ \to \mathbb{P}(G^d),$$
$$\alpha \mapsto Z_\alpha,$$

and

$$T: \quad R_+ \to \mathbb{P}(R_+),$$
$$\alpha \mapsto S_\alpha,$$

where $\mathbb{P}(G^d)$ (respectively $\mathbb{P}(R_+)$) stands for the set of subsets of G^d (respectively R_+). We want to show a kind of reciprocal result to Proposition 5.2.6, i.e., that, for a certain range of μ, there exists α such that $\mu \in T(\alpha)$. The two following propositions summarize the main properties of the multi-application T. We give it without proofs and we refer the interested reader to [32] for details.

Proposition 5.2.7   (i) T is a nonincreasing multi-application.

(ii) $Y(0) = \{f - \bar{f}\}$ and $T(0) = |f - \bar{f}|_{G^d}$ (where $\bar{f}$ stands for the mean value of f over Ω).

(iii) If α goes to $+\infty$, then $Y(\alpha)$ (respectively $T(\alpha)$) goes to $\{0\}$ (respectively $\{0\}$) (with respect to the Hausdorff metric).

Remark In fact, we have $T: R_+ \to \left[0, |f - \bar{f}|_{G^d}\right]$. In particular, T has uniformly bounded values:

(i) $T(0) = \{f - \bar{f}\}$.

(ii) If $\alpha > 0$, then if $v_\alpha \in T(\alpha)$, we have

$$|v_\alpha|_{G^d} \leq |f - \bar{f}|_{G^d}. \tag{5.77}$$

■

Proposition 5.2.8 T is u.s.c. (upper semicontinuous) (i.e., T has a closed graph and convex compact values).

From previous propositions we deduce that the multi-application $T_\mu = T - \mu$ is such that for all μ in $(0, |f - \bar{f}|_{G^d})$, there exists α with $0 \in T_\mu(\alpha)$, which is equivalent to the following statement:

Proposition 5.2.9 *For all μ in $(0, |f - \bar{f}|_{G^d})$, there exists α in R_+ such that there exists (u, v) in $X \times G^d$ with $|v|_{G^d} = \mu$ and solving Meyer's model (5.71).*

Remark This proposition completes the result of Proposition 5.2.6. It completely closes the link between Meyer's problem (5.71) and the limit problem (5.66).   ∎

We end this section on image decomposition models by giving some experimental results.

5.2.5   Experimental Results

In this section, we compare the results obtained by the ROF model (5.17) and Meyer's (5.28). The latter is approximated by the Aujol et al. algorithm described in Section 5.2.4. In order to shorten notation, we will use the abbreviations ROF and A²BC for the respective methods. In the case of the ROF problem, the parameter λ corresponds to the one in (5.17), and in the case of Meyer's problem, the parameters λ and μ correspond to those in (5.43). The parameters of both algorithms are tuned so that $|v_{\text{ROF}}|_{L^2}$ is equivalent to $|v_{A^2BC}|_{L^2}$, so that v contains the same quantity of information.

DENOISING CAPABILITIES

In this experiment, we start from a synthetic image provided by the GdR-PRC ISIS (http://www-isis.enst.fr/) that has been corrupted by a Gaussian additive noise of variance $\sigma = 50$ (see Figure 5.9). The interest in such an image is that besides the noise, it is a pure geometric image.

The second (respectively third) row of Figure 5.9 shows the results obtained using the A²BC (respectively ROF) model. Both u and v components are displayed. We can observe that v_{ROF} still contains some geometric information. To see it more clearly, the top right-hand image represents the difference between v_{A^2BC} and v_{ROF} using the following color code. The value of a pixel in position (i, j) is 255.0 (i.e., white) if $v_{A^2BC}(i,j) > v_{\text{ROF}}(i,j)$, 127.0 (i.e., gray) if $v_{A^2BC}(i,j) = v_{\text{ROF}}(i,j)$, and 0.0 (i.e., black) if $v_{A^2BC}(i,j) < v_{\text{ROF}}(i,j)$.

One observes on the "difference image" that $v_{A^2BC}(i,j) > v_{\text{ROF}}(i,j)$ in the darkest regions of the original image, and that $v_{A^2BC}(i,j) < v_{\text{ROF}}(i,j)$ in the lightest regions. This means that the v component in the ROF model depends more on the mean gray level value of the original image than in the case of Meyer's model. For instance, let us have a look at the dark circle on upper left side of the orginal image: the mean value of the pixels corresponding to this circle is -1.0 in v_{A^2BC} and -4.2 in v_{ROF}. Both components v tend to have a negative mean because in the original image the circle is a dark component.

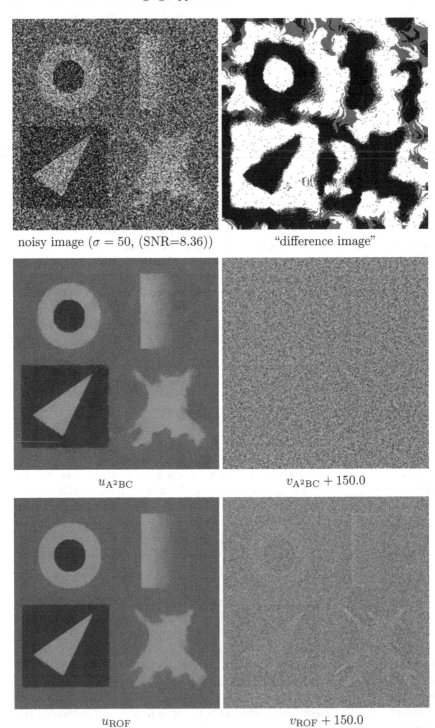

Figure 5.9. Comparison of reconstruction algorithms on a noisy geometric image. Parameters for A²BC: ($\lambda = 0.1$, $\mu = 70$) and ROF: ($\lambda = 70$).

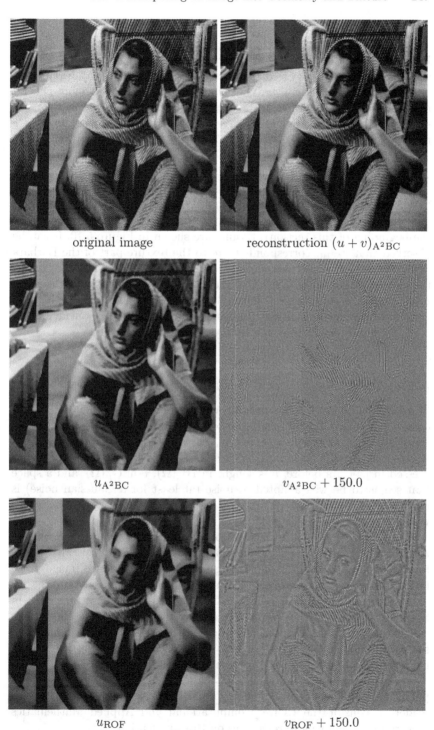

Figure 5.10. Comparison of reconstruction algorithms on a real image with texture. Parameters for A^2BC: ($\lambda = 1.0$, $\mu = 100$) and ROF: ($\lambda = 70$).

Remark In this section we have shown some denoising results on a simple geometric image, in order to demonstrate the capabilities of this framework. In fact, this approach can be very efficient in dealing with more general noise classes. For example, in [32], the authors show an application for synthetic aperture radar (SAR) images that are strongly corrupted by a noise called speckle [214].   ∎

DEALING WITH TEXTURE

In this experiment we test the ROF and A^{2}BC algorithms on a real image having some texture and geometric parts (see for example Figure 5.8 at the beginning of this section). Results are shown in Figure 5.10: it can be observed that v_{A^2BC} corresponds more to the texture part of the Barbara image than v_{ROF}, which also contains some geometric content. For example, the face of Barbara and the leg of the table appear much more clearly in v_{ROF} than in v_{A^2BC}. This confirms that the Aujol et al. approximation gives a better decomposition of an image into a BV component and an oscillatory component than the ROF model.

5.2.6   *About Some Recent Advances*

Of course, a $(u + v)$ decomposition cannot be satisfying, since in the v component noise and texture are mixed. It would be better to get a decomposition of the type $u + v + w$, where u is the geometric part, v is only the texture part, and w the noise. This question has been addressed in [30]. As before, u is sought in $BV(\Omega)$, v in $G_\mu(\Omega)$, and a space that seems to be well adapted to noise (at least for a Gaussian noise) is the Besov space $B_{-1}^\infty(L^\infty(\Omega))$ consisting of (generalized) functions whose wavelet coefficients are in l^∞.

Apart from the intrinsic interest of such decomposition algorithms in restoration, there are may other applications in which they can be useful. For example, in image inpainting we can decompose an image f (to be inpainted) as the sum $(u + v)$ and then apply separately specific inpainting for the geometric component u (for example those presented in Sections 5.1.2 and 5.1.3)), while the missing texture parts are filled in using texture synthesis methods (see Section 5.1.4). This approach has been successfully applied in [51] (see also [30]). Another interesting application is image compression, see for example [31].

Since the seminal work by Meyer [235], a huge amount of work has been published on decomposition models. In particular, we invite the interested reader to consult the UCLA Computational and Applied Mathematics reports web site for more references on this application.

5.3 Sequence Analysis

5.3.1 Introduction

When dealing with digital sequences, we already have to deal with all the characteristics of static digital images. As is mentioned in Section 1.2, we have to consider problems like low resolution, low contrasts, and the same variety of gray-level information like graduated shading, sharp transitions, and fine elements (see Figure 1.3). It is naturally more complex because of the temporal dimension, which intrinsically contains a large amount of information such as motion, depth, and the different objects in the scene. To illustrate this, let us comment on the real sequence presented in Figure 5.11. From this sequence we may note the following:

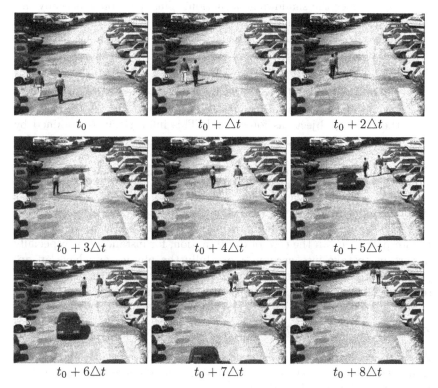

Figure 5.11. The "street" sequence shows different ranges of motion, spurious motions (from the bushes or from the shadows), occlusions, the effect of noise, etc. Notice that the noise visible in the sequence has been added afterwards (it is a Gaussian additive noise of variance $\sigma = 20$).

- As one can observe, this sequence has a static background, and three objects are moving (two people and a car) with motions of different ranges. This velocity is naturally linked with the sampling in time of the sequence (the number of frames per second), or equivalently, the time Δt between the acquisition of two consecutive frames. Naturally, the more images we have, the better the motion can be understood, because the variations between two consecutive images will be small.

- The natures of the motions are also quite different. The car has a smooth trajectory, the two people can change direction quickly, and the bushes on the bottom right-hand corner have a "random" motion (due to the wind).

- As we will discuss in the next section, the motion that we perceive is based on intensity variations. Thus we can already wonder which interpretation of the motion may be recovered. For instance, should we consider the shadows as a motion? This question is actually unclear, and we see that to answer it we need all our "life experience" and knowledge about the sequence to know the difference between a real object and its shadow. Another concern will be the sensitivity of motion perception with respect to noise.

- There is also partially contained in the sequence the notion of depth. For instance, one can distinguish elements from the background and the moving objects as foreground. This information is obtained by analyzing the occlusions, and we will see in Section 5.3.3 how this can be used. More generally, if we have a sequence taken with a moving camera, we can imagine that more precise depth information may be recovered from the scene.

Beyond these remarks, we can also imagine that analyzing a weather forecast sequence or a soccer game may involve different problems and models to interpret them. In the case of cloud motion, for instance, it is especially important to take into account more fully the physics and have an appropriate notion of smoothness for the motion, while for the soccer game sequence motions are discontinuous and occlusions are very important.

5.3.2   The Optical Flow: An Apparent Motion

As soon as we consider a sequence, there is the idea of motion. Coming from displacements in the physical world, we can observe only a projection of it. This is illustrated in Figure 5.12.

Unfortunately, we are not able to measure the *2-D motion field* (the projection on the image plane of the 3-D velocity of the scene). As is mentioned in the title of this section, what we are able to perceive is just an apparent motion, also called the *optical flow*. By apparent, we mean that this 2-D motion is observable only through intensity variations. Unfortunately, the optical flow and the 2-D motion field are in general quantitatively different, unless very special conditions are satisfied. We refer to the discussion of

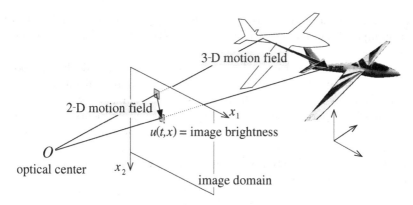

Figure 5.12. Simplified illustration of a camera (the pinhole camera model) .

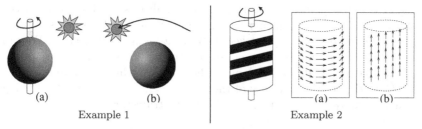

Figure 5.13. The optical flow is not always the motion field. Example 1 (see also [183]): In (a), no motion is perceived because intensity remains constant, while in (b) a static sphere is illuminated by a moving source, producing intensity variations. Example 2: The barber's pole. (a) shows the flow field and (b) the optical flow, which is the perceived motion.

Verri and Poggio [329, 330] for more details. Just to illustrate this difference, we show in Figure 5.13 two examples.

Still, if the optical flow and the 2-D motion field are quantitatively different, they often share the same qualitative properties, for instance motion discontinuities. The optical flow is then a rich source of information about the 3-D kinematic behavior of objects (see, for instance, [156]) or the geometric structure of the world. It is also used in many other applications: segmentation, time to collision, earth sciences, and so on.

In the last decade numerous methods have been proposed to compute optical flow, and this is still an active field of research. Three different strategies can be distinguished. *Correlation-based* techniques compare parts of the first image with parts of the second in terms of the similarity in brightness patterns in order to determine the motion vectors. Correlation is generally used to aid the matching of image features or to find image motion once features have been determined by alternative methods. *Feature-based* approaches aim at computing and analyzing the optic flow at a small number of well-defined image features (such as corners, edges, blobs) in a scene.

Gradient-based methods (also called differential techniques) make use of spatiotemporal partial derivatives to estimate the image flow at each point in the image.

There is also a wide range of methodologies: wavelets, Markov random fields, Fourier analysis, and, naturally, partial differential equations. In [41], Barron, Fleet, and Beauchemin present the main different classes of techniques and perform numerical quantitative experiments to compare them. We also mention to the interested reader other interesting reviews: [259, 238, 328, 264, 157, 24].

We now focus on differential techniques for which variational formulations can be proposed.

THE OPTICAL FLOW CONSTRAINT (OFC)

One of the first points to clarify and to formalize is the link between the intensity variations and the motion. A common and widely used assumption is that the intensity of a point keeps constant along its trajectory. We can consider it as reasonable for small displacements for which changes of the light source are small, and as long as there is no occlusion. More precisely, let $u(t, x)$ denote the intensity of the pixel $x = (x_1, x_2)$ at time t. Starting from a point x_0 at the time t_0, we define the trajectory

$$t \mapsto (t, x(t))$$

such that

$$u(t, x(t)) = u(t_0, x_0) \quad \forall t, \tag{5.78}$$
$$(t_0, x(t_0)) = (t_0, x_0). \tag{5.79}$$

By formally differentiating (5.78) with respect to t, we obtain at $t = t_0$,

$$\frac{dx}{dt}(t_0) \cdot \nabla u(t_0, x_0) + \frac{\partial u}{\partial t}(t_0, x_0) = 0. \tag{5.80}$$

So we will search the optical flow as the velocity field $\sigma(x_0) = \frac{dx}{dt}(t_0)$ satisfying (5.80).

To summarize, for a given sequence $u(t, x)$ and a time of observation t_0, the aim is to find the instantaneous apparent velocity $\sigma(x)$ such that

$$\sigma(x_0) \cdot \nabla u(t, x_0) + u_t(t, x_0) = 0. \tag{5.81}$$

This equation is called the *optical flow constraint* (OFC). [1]

---

[1] Observe that (5.81) is just an approximation to the first order of (5.78) and is valid only for small time differences. This will be commented on next.

☞ *Unfortunately, one scalar equation is not enough for finding both com-*
ponents of the velocity field. It gives only the component in the direction
of ∇u, that is, the normal to the isophotes of the images. It is called the
normal flow. *This problem is usually called the* aperture problem.

SOLVING THE APERTURE PROBLEM

As we just saw, equation (5.81) is not sufficient for computing the optical
flow. Several ideas have been proposed to overcome this difficulty.

- Use second-order derivative constraints. For instance, one could
  impose the conservation of $\nabla u(t, x)$ along trajectories; that is,

$$\frac{d\nabla u}{dt}(t, x) = 0.$$

  This is a stronger restriction than (5.81) on permissible motion
  fields. This implies that rigid deformations are not considered. This
  condition can be rewritten in the following form:

$$\begin{bmatrix} u_{x_1 x_1} & u_{x_2 x_1} \\ u_{x_1 x_2} & u_{x_2 x_2} \end{bmatrix} \begin{pmatrix} \sigma_1 \\ \sigma_2 \end{pmatrix} + \begin{pmatrix} u_{x_1 t} \\ u_{x_2 t} \end{pmatrix} = \begin{pmatrix} 0 \\ 0 \end{pmatrix}. \qquad (5.82)$$

  These equations can be used alone or together with the optical flow
  constraint. Several possibilities are then proposed (see [265, 318].
  However, this kind of method is often sensitive to noise because we
  need to compute second-order derivatives.

- Another possibility is to solve the problem using a weighted least
  squares approach [220, 221]. The central point of this method is a
  model of constant velocities in a small spatial neighborhood. For
  instance, to compute the velocity σ at point x_0, the idea is to minimize

$$\inf_z \int_{B(x_0,r)} w^2(x) (z \cdot \nabla u + u_t)^2 \, dx,$$

  where $B(x_0, r)$ is the ball of center x_0 and radius r (the neighborhood), and $w(x)$ is a window function that gives more influence to
  the constraint at the center of the neighborhood than at the periph-
  ery. This approach, which gives good results, has been extended, and
  it is still often used to compute the optical flow. However, it is local,
  and there is no notion of global regularity for the resulting flow.

- One may also use parametric models of velocity that respect as much
  as possible the optical flow constraint. In the affine case, one looks
  for σ such that

$$\sigma(x) = \sigma_\theta(x) = \begin{pmatrix} \theta_1 + \theta_2 x_1 + \theta_3 x_2 \\ \theta_4 + \theta_5 x_1 + \theta_6 x_2 \end{pmatrix},$$

where the unknown parameter vector $\theta \in R^6$ is determined by minimizing

$$E(\theta) = \int_\Omega \phi(\sigma_\theta \cdot \nabla u + u_t) \, dx,$$

where ϕ is a suitable given function. In general, this minimization leads to nonquadratic (and possibly nonconvex) optimization problems that we can solve by half-quadratic techniques if ϕ satisfies the hypotheses of Section 3.2.4. Naturally, other models may be proposed [185, 258].

• Regularizing the velocity field is another possibility. The idea is to consider a minimization problem of the form

$$\inf_\sigma \, (A(\sigma) + S(\sigma)), \tag{5.83}$$

where $A(\sigma)$ is the fidelity term, for instance based on (5.81) or (5.78), and $S(\sigma)$ is the smoothing term.

Horn and Schunck [184] (see also [295]) were among the first to propose solving the following problem:

$$\inf_\sigma \underbrace{\int_\Omega (\sigma \cdot \nabla u + u_t)^2 \, dx}_{A(\sigma)} + \alpha \underbrace{\sum_{j=1}^2 \int_\Omega |\nabla \sigma_j|^2 \, dx}_{S(\sigma)}, \tag{5.84}$$

where α is a constant. However, this kind of penalty term introduced by Tikhonov and Arsenin [317] is well known to smooth isotropically without taking into account the discontinuities (see Figure 5.14 and also Section 3.2.2, where a similar term arises for image restoration). Unfortunately, as it has been mentioned, the discontinuities of the optical flow field are a very important cue for sequence analysis.

Since then, much research has been carried out to compute discontinuous optical flow fields by changing the smoothing term $S(\sigma)$ (see for example [339, 340] for a taxonomy of optical flow regularizers). We describe below some of the most significant:

– Modifying the Horn and Schunck functional was pioneered by Black et al. [54, 57]. The idea was to change the regularization term into

$$\sum_{j=1}^2 \int_\Omega \phi(|\nabla \sigma_j|) \, dx, \tag{5.85}$$

where the function ϕ would permit noise removal and edge conservation. Some examples include Cohen [114], Kumar, Tannenbaum, and Balas [207] with the L^1 norm (i.e., the total variation, $\phi(s) = s$) and Aubert, Deriche et al. [133, 24, 25]

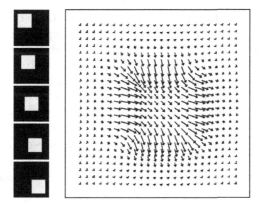

Figure 5.14. The method of Horn and Schunck applied to a synthetic example. Notice that this example is very artificial since we do not have any texture information on the background and on the moving object. One may observe that the discontinuities near the edges are lost.

and Blanc-Féraud, Barlaud et al. [60] with the ϕ functions traditionally used for image restoration to preserve discontinuities. This was also proposed at the same time in a statistical framework where ϕ functions are called robust estimators (see, for instance, [55, 234]).

- Suter [313], Gupta and Prince [178], and Guichard and Rudin [177] add some penalty terms based on the divergence and the rotational of the flow field:

$$\int_\Omega \varphi(\text{div}(\sigma), \text{rot}(\sigma)) \, dx, \qquad (5.86)$$

where several possibilities for φ may be proposed. For instance, in [177] the authors choose $\varphi(\text{div}(\sigma), \text{rot}(\sigma)) = |\text{div}(\sigma)|$, which is adapted to rigid 2-D objects with a "2-D" motion.

- Nagel and Enkelmann [248, 146] propose an oriented smoothness constraint in which smoothness is not imposed across steep intensity gradients (edges) in an attempt to handle occlusions. So the penalty term is of the form

$$\int_\Omega \frac{1}{|\nabla u|^2 + 2\lambda^2} \, \text{trace}\left((\nabla\sigma)^T D(\nabla u)(\nabla\sigma)\right) \, dx$$

with

$$D(\nabla u) = n \, n^T + \lambda^2 \, \text{Id} \quad \text{and} \quad n = \begin{pmatrix} u_{x_2} \\ -u_{x_1} \end{pmatrix}, \qquad (5.87)$$

where λ is a constant. The idea is to attenuate the blurring of the flow across the boundaries of the intensity when $|\nabla u| \gg \lambda$.

In this case the smoothing is essentially in the direction tangent
to the isophotes. Otherwise, the smoothing is isotropic. This is a
major difference with (5.85) and (5.86), since the characteristics
of the smoothing depend here on the intensity and not on the
motion itself. However, one can wonder about the action of this
term (5.87) for highly textured scenes where the gradient varies
considerably and is not very representative of object boundaries.

- Nési [253] adapts the formulation of Horn and Schunck, intro-
ducing the length of the discontinuity set of σ (noted $|S_\sigma|$).
We recall that this kind of idea was introduced by Mumford
and Shah [247] for image segmentation (see Chapter 4). The
regularization term is of the form

$$\sum_{j=1}^{2} \int_\Omega |\nabla \sigma_j|^2 \, dx + \alpha |S_\sigma|,$$

where α is a constant. Numerically, as for the image segmentation problem, the main difficulty is to approximate the last term.
One possible solution is to use the notion of Γ-convergence (see
Section 2.1.4). We introduce a sequence of functionals such that
the sequence of minimizers converges to the unique minimum
of the initial functional. Typically, the way to approximate the
regularization term is (see Section 4.2.4 and [17] for more detail)

$$\sum_{j=1}^{2} \int_\Omega z^2 |\nabla \sigma_j|^2 \, dx + \alpha \int_\Omega \left(\frac{|\nabla z|^2}{k} + \frac{k(1-z)^2}{4} \right) \, dx,$$

where z is an additional function and k is a parameter that is
destined to tend to infinity. The function z can be considered as
a control variable that is equal to zero near discontinuities and
close to 1 in homogeneous regions.

OVERVIEW OF A DISCONTINUITY-PRESERVING VARIATIONAL APPROACH

Although many models have been proposed for finding the optical flow,
little work has considered its mathematical analysis. We summarize below
some results presented in [24, 25, 203], where the smoothing term is the
same as in the restoration problem of Section 3.2.3.

Given a sequence $u(t, x)$ we search for the velocity field σ that realizes
the minimum of the energy:

$$E(\sigma) = \underbrace{\int_\Omega |\sigma \cdot Du + u_t|}_{A(\sigma)} + \alpha^s \underbrace{\sum_{j=1}^{2} \int_\Omega \phi(D\sigma_j)}_{S(\sigma)} + \alpha^h \underbrace{\int_\Omega c(|Du|)|\sigma|^2 \, dx}_{H(\sigma)}, \quad (5.88)$$

where α^s, α^h are positive constants. From now on, unless specified otherwise all derivatives are written in a formal setting (in the distributional sense). Since we look for discontinuous optical flows, the suitable theoretical background to study this problem will be $\mathbf{BV}(\Omega)$, the space of bounded variation (see Section 2.2). The energy is compounded of three terms:

- $A(\sigma)$ is the "L^1"-norm of the OFC (5.81). In fact, it is formal and has to be interpreted as a measure.

- $S(\sigma)$ is the smoothing term. As for image restoration (see Section 3.2), one would like to find conditions on ϕ such that discontinuities may be kept. We recall the assumptions of Section 3.2.3:

  ϕ is a strictly convex, nondecreasing function from R^+ to R^+, with $\phi(0) = 0$ (without loss of generality), $\qquad$ (5.89)

  $$\lim_{s \to +\infty} \phi(s) = +\infty, \qquad (5.90)$$

  There exist two constants $c > 0$ and $b \geq 0$ such that $cs - b \leq \phi(s) \leq cs + b \ \forall s \geq 0.$ $\qquad$ (5.91)

  Notice that these conditions will guarantee the well-posedness of the theoretical problem. Regarding the edge preservation properties, other qualitative conditions should be added (see Section 3.2.2, conditions (3.9) and (3.12)). Under these assumptions, $S(\sigma)$ has to be interpreted as a convex function of measures (see Section 2.2.4). We recall that this term is l.s.c. for the BV$-$w* topology.

- $H(\sigma)$ is related to homogeneous regions. The idea is that if there is no texture, that is, no gradient, there is no way to estimate correctly the flow field. Then one may force it to be zero. This is done through a weighted L^2-norm, where the function c is such that

  $$\lim_{s \to 0} c(s) = 1 \quad \text{and} \quad \lim_{s \to +\infty} c(s) = 0.$$

  Without any loss of generality, we assume that

  $$x \to c(|Du|(x)) \in C^\infty(\Omega), \qquad (5.92)$$

  There exists $m_c > 0$ such that $c(|Du|(x)) \in [m_c, 1]$ for all $x \in \Omega$. $\qquad (5.93)$

  Although this term may be criticized from a modeling point of view, it is necessary for the coercivity of the functional. Let us also remark that this term is well-defined on $\mathbf{BV}(\Omega)$ thanks to the inclusion of $\mathbf{BV}(\Omega)$ into $\mathbf{L}^2(\Omega)$ ($N = 2$).

Now that we have stated the problem, let us consider its theoretical study. Up to now nothing has been told about the regularity of the data u. Interestingly, this will influence very importantly the nature of the problem. As a first example, let us consider that the data is Lipschitz in space and time [24, 203]:

$$u \in W^{1,\infty}(R \times \Omega). \tag{5.94}$$

The derivatives in $A(\sigma)$ are then functions, and $A(\sigma)$ is simply the L^1-norm of the OFC. Notice that this assumption is realistic from a numerical point of view because a presmoothing is usually carried out to diminish noise effects, remove small amounts of temporal aliasing, and improve the subsequent derivative estimates.[2] Then, we have the following result:

Theorem 5.3.1 [24] *Under the hypotheses* (5.89)–(5.91), (5.92)–(5.93), *and* (5.94), *the minimization problem*

$$\inf_{\sigma \in \mathbf{BV}(\Omega)} E(\sigma) = \int_{\Omega} |\sigma \cdot \nabla u + u_t| \, dx + \sum_{j=1}^{2} \int_{\Omega} \phi(D\sigma_j) + \int_{\Omega} c(|Du|)|\sigma|^2 \, dx \tag{5.95}$$

admits a unique solution in $\mathbf{BV}(\Omega)$.

In this case, the proof follows from classical arguments. According to (5.91)–(5.93), the functional E is coercive on $BV(\Omega)$. Thus, we can uniformly bound the minimizing sequences and extract a convergent subsequence for the BV$-$w* topology. Since E is lower semicontinuous (l.s.c.) for this topology, we easily deduce the existence of a minimum.

Now the problem is to get an approximation of the solution. As in the case of image restoration, an algorithm based on Γ-convergence and half-quadratic minimization can be proposed (see Section 3.2.4 and [24]). We show in Figure 5.15 a typical result where one can observe the qualitative differences with the original Horn and Schunck's model.

Unfortunately, the regularity assumption (5.94) on the data u may not always be verified, and one may wonder what would be the problem if we assumed only that u is a function of bounded variation. This is considered in [25, 203], and the fact that u may have jumps induces nontrivial theoretical questions. The first difficulty is to give a sense to the first term $A(\sigma)$. It is now a measure. To be more explicit, one needs to find an integral representation of this term.[3] The precise assumptions on u are as follows:

$$u \in SBV(R \times \Omega) \cap L^{\infty}(R \times \Omega), \tag{5.96}$$

There exist $h_1 \in L^1(\Omega)$ and $h_2 \in L^1_{\mathcal{H}^1}(S_u)$
such that $u_t = h_1 \, dx + h_2 \mathcal{H}^1|_{S_u}$, $\tag{5.97}$

---

[2] As mentioned in [41], differential techniques are naturally very sensitive to the quality of the estimation of the spatiotemporal derivative. Along the same lines, from a discrete point of view the method of numerical differentiation is very important.

[3] By integral representation, we mean finding a measure μ and a function $h \in L^1_\mu(\Omega)$ such that $A(\sigma) = \int_\Omega h \, d\mu$.

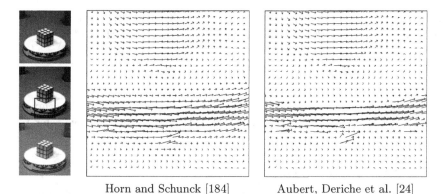

Horn and Schunck [184]   Aubert, Deriche et al. [24]

Figure 5.15. Example of result obtained on the rotating cube sequence. A close-up of the lower part of the plate is displayed to highlight the qualitative differences. One can observe that the diffusion is controlled in the right-hand case.

where $SBV(R \times \Omega)$ is the space of special functions of bounded variation (no Cantor part) and $\mathcal{H}^1$ is the one-dimensional Hausdorff measure. Notice that (5.97) means that the measure u_t is absolutely continuous with respect to $|Du|$. This is physically correct, since when there is no texture (no gradient) no intensity variation should be observed. Then, under these assumptions, it can be established that the energy E defined in (5.88) can be rewritten as

$$E(\sigma) = \int_\Omega |\sigma \cdot \nabla u + h_1| \, dx + \int_{S_u} |\overset{\bullet}{\sigma} \cdot n_u (u^+ - u^-) + h_2| \, ds$$

$$+ \sum_{j=1}^2 \int_\Omega \phi(D\sigma_j) + \int_\Omega c(|Du|)|\sigma|^2 \, dx,$$

where S_u is the jump set of u and $\overset{\bullet}{\sigma}$ is the precise representation of σ (see Section 2.2.3). We recall that $\overset{\bullet}{\sigma}$ belongs to the same class as σ but is now defined $\mathcal{H}^1$-a.e. by

$$\overset{\bullet}{\sigma}(x) = \lim_{r \to 0} \frac{1}{|B(x,r)|} \int_{B(x,r)} \sigma(y) dy,$$

where $B(x,r)$ is the ball of center x and radius r. We also recall that

$$\overset{\bullet}{\sigma}(x) = \sigma(x) \; dx \text{ a.e. and } \overset{\bullet}{\sigma}(x) = \frac{\sigma^+(x) + \sigma^-(x)}{2} \quad \mathcal{H}^1 \text{ a.e. on } S_\sigma.$$

So, we can observe that the OFC is now split into two parts: an absolutely continuous part and a length part on the jump set of u. To study the existence of a solution for this problem the difficulty is that the functional is no longer l.s.c. for the $BV - w^*$ topology because of the term on S_u. Basically, the problem is that the trace function is l.s.c. for the strong topology of $BV(\Omega)$ (even continuous; see Section 2.2.3) but not for the BV–w* topology. If σ^n is a minimizing sequence of E such that $\sigma^n \xrightarrow[BV-w^*]{} \sigma$, then in general

$$\mathrm{tr}(\sigma^n) \to \nu \in \mathcal{M}(\Omega) \quad \text{and} \quad \nu \neq \mathrm{tr}(\sigma),$$

where tr(.) is the trace operator. It is then necessary to compute the relaxed functional. This leads to long and technical calculations and we refer to [25] for more details.

ALTERNATIVES TO THE OFC

Is the OFC unavoidable? Even if it is widely used to compute optical flow, several reasons may invite us to look for something different.

The first is that one may wonder about the validity of using the OFC (5.81) in case of large displacements. If we denote by Δt the time interval between two consecutive images (in the discrete temporal case), the differentiation (5.78) is valid only for "small" Δt, or equivalently for "small" displacements. To deal with large displacements there are two possibilities:

- One can use a multiresolution approach with computations at each resolution level according to a coarse to fine strategy (see, for instance, [146, 234]). It is then a modified OFC that is considered at each resolution level.

- One may prefer to keep the conservation equation (5.78) without differentiating it, i.e., the displacement frame difference. So the fidelity term in (5.83) is of the form [248, 177, 9]

$$A(\sigma) = \int_\Omega \left(u(t + \Delta t, x + \sigma \Delta t) - u(t, x) \right)^2 \, dx,$$

which is now nonlinear with respect to σ. It is then more similar to a correlation problem. If this approach gives satisfying numerical results, it is not clear from a theoretical point of view how to give sense to a possibly discontinuous function u depending on another discontinuous function σ.

The second is related to the a priori knowledge that we may have about the origins of the sequence. Typical examples include fluid flow estimation [341, 115] and weather forecast sequence analysis [47, 349] (see also [136, 3, 249, 296, 297]). For instance, for fluid motion, one should impose a conservation of mass

$$\mathrm{div}(\rho V) + \frac{\partial \rho}{\partial t} = 0,$$

where $\rho(t, x)$ is the density of the fluid at position $X \in R^3$ and time t, and $V(t, x) \in R^3$ is the velocity. Now the problem is to know the link between the density ρ and the image brightness. For example, in the case of a 2-D transmittance image of a 3-D fluid flow the intensities are proportional to the density integrated along the path of impinging energy. Then, it can be shown [152, 341] that the mass conservation equation implies that

$$\operatorname{div}(u\sigma) + u_t = 0,$$

which is different from the classical optical flow constraint.

The third is that we may simply want to relax the brightness consistency assumption, which is true when the scene surface is Lambertian and is either stationary as the camera moves or moves parallel to the image plane. It is also a good approximation as soon as the surface has rich texture, such that the brightness change due to shadings or surface lighting conditions is negligible relative to that due to the motion effects. Unfortunately, this is not the case in many applications [250], and one needs to relax this assumption. For instance, the models proposed by Negahdaripour and Yu [252] or Mattavelli and Nicoulin [233] permit an affine variation of the intensity during time (and not a conservation). We also refer the interested reader to [181, 56, 179, 251] for other possibilities.

✴ *The optical flow problem is not near to being completely solved. The choice of the data term is not yet well established, and there is not a unique possibility. As far as the regularity is concerned, when presenting the short overview of possible regularization terms, we may have noticed the following:*

- *Most of them are* intrinsic: *The diffusion is controlled only by the flow itself.*

- *Few of them are* extrinsic: *The diffusion is controlled by the intensity image itself (see (5.87)).*

It is not clear which solution is actually "the best" in terms of modeling and numerical results.

Another point that would need further development is a better understanding of the link between u and σ in terms of their discontinuity sets. If it seems clear that the set S_σ should be contained in S_u, this should be taken into account in the model, and one should do a finer analysis of these relationships.

Finally, one cannot ignore the coupling between the fidelity term and the regularization term.

5.3.3 Sequence Segmentation

INTRODUCTION

As suggested in the introduction, another important task in sequence analysis is segmentation. Here segmentation means finding the different objects in a scene, and this is naturally in relation to velocity estimation or optical flow. Two kinds of approaches based on motion estimation can be distinguished: Either they detect flow discontinuities (local operators) or they extract patches of self-consistent motion (global measurements). In any case, this is dependent on the quality of the flow that can be obtained. As suggested in the previous section, this estimation may be hard or impossible to obtain (for instance, it is necessary to have a reasonable time sampling, with a limited amount of noise). Unfortunately, this is not always the case for many real applications. Just think about video surveillance, where sensors are often of poor quality and for which low image rates are usually considered because of storage capacity.

To avoid these difficulties, another possibility is to consider that the sequence is compounded of layers, typically a background and a foreground. To make this point more clear, let us focus on the case of sequences with static background (see, for example, Figure 5.11 or the synthetic sequence in Figure 5.16). The idea is that the background has a "persistence," and then the objects can be seen as occluding it, "being on top of it" (see Figure 5.17). Naturally, this means that only the objects with a different color from that of the background can be detected. However, as this is not motion based, an object stopping for a while will still be detected.

This idea of comparing a reference image with the current image is very intuitive but not always applicable in real applications, especially when noise is present. This is illustrated for the sequence presented in Figure 5.18. If we simply compute the difference between one image of the sequence and the ideal background (see Figure 5.19), we obtain an image that enhances the objects but also the noise. A threshold can be used to turn this image into a binary one. Examples are shown in Figure 5.20. It can be observed that the choice of this threshold is not an easy task: Either noise is kept or objects are partially lost. Another important question is how a reference image (i.e., the background image) can be obtained. A simple idea is to compute the temporal mean of the sequence[4] (see Figure 5.20). It can be noticed that some shadows of the objects are still present, and thus making the difference with an image of the sequence will not be satisfying. This effect is even stronger given that the sequence is short. It is then necessary to have a robust technique to estimate the background, and classical approaches compute statistical background models [150, 172, 343, 269, 267].

Something important that comes out of this discussion is that in the case of noisy sequences obtaining a reference image is as difficult as segmenting the sequence. In fact,

---

[4]In this example, where the sequence is given by 5 images denoted by $(N_i)_{i=1,\ldots,5}$, the temporal mean is $M(x) = \dfrac{1}{5}\sum_{i=1}^{5} N_i(x)$, $x \in \Omega$.

Figure 5.16. Example of synthetic sequence of 5 images with 3 moving objects.

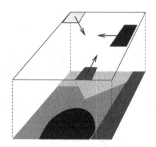

Figure 5.17. Interpretation of the sequence presented in Figure 5.16 as two layers: background and foreground.

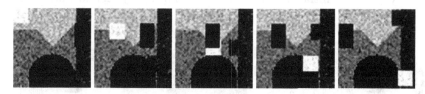

Figure 5.18. Synthetic sequence with Gaussian additive noise ($\sigma = 20$).

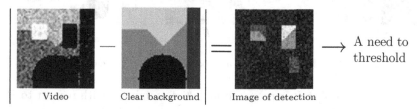

Figure 5.19. Example of background subtraction for the second image (the resulting image needs to be thresholded to extract the objects).

Image of detection with different thresholds   Temporal mean

Figure 5.20. Difficulties of background subtraction method.

☞ *Having a reference image means that we have previously extracted the objects. Conversely, being able to extract the objects means having somewhere a reference image.*

This remark suggests strong links in the estimation of background and of moving objects. It should then be more efficient to estimate both at the same time.

A VARIATIONAL FORMULATION

According to the previous discussion, let us present the variational approach proposed by Kornprobst, Deriche, and Aubert [206]. Let $N(t,x)$ denote the given noisy sequence ($t \in [0, t_{\max}], x \in \Omega$), for which the background is assumed to be static. We look simultaneously for (see Figure 5.21):

- The restored background $B(x)$.

- The sequence $C(t,x)$ that indicates the moving regions. Typically, $C(t,x) = 0$ if the pixel x belongs to a moving object, and 1 otherwise.

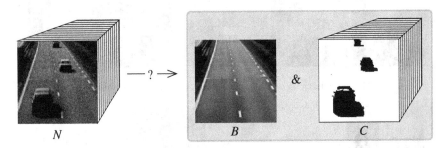

Figure 5.21. Objectives of the approach [206] on a typical example.

To solve this problem it is proposed in [206] to minimize with respect to B and C

$$\underbrace{\iint_V \left[C^2(B-N)^2 + \alpha_c(C-1)^2 \right] dx\, dt}_{A(\sigma)} + \underbrace{\alpha_b^r \int_\Omega \phi_1(DB) + \alpha_c^r \iint_V \phi_2(DC)}_{S(\sigma)},$$

(5.98)

where $V = [0, t_{\max}] \times \Omega$ and α_c, α_b^r, α_c^r are positive constants. As usual, all the derivatives are written in a formal setting (in the distributional sense). The energy is compounded of two kinds of terms:

- $A(\sigma)$ realizes the coupling between the two variables B and C. The second term forces the function C to be equal to 1, which corresponds to the background. However, if the current image N is too different from the background (meaning that an object is present), then the first term will be too high, which will force C to be 0.

- $S(\sigma)$ is the smoothing term. As for image restoration (see Section 3.2) one would like to find conditions on $(\phi_i)_{i=1,2}$ such that discontinuities may be kept (for the background as well as for the images of detection). We recall the assumptions of Section 3.2.3:

$$\phi_i \text{ is a strictly convex, nondecreasing function from } R^+ \text{ to } R^+, \text{ with } \phi_i(0) = 0 \text{ (without loss of generality),} \tag{5.99}$$

$$\lim_{s \to +\infty} \phi_i(s) = +\infty, \tag{5.100}$$

$$\text{There exist two constants } c > 0 \text{ and } b \geq 0 \text{ such that} \\ cs - b \leq \phi_i(s) \leq cs + b \;\; \forall s \geq 0. \tag{5.101}$$

As noted previously, these conditions will guarantee the well-posedness of the theoretical problem. Regarding the edge preservation properties, other qualitative conditions should be added (see Section 3.2.2, conditions (3.9) and (3.12)). Under these assumptions, $S(\sigma)$ has to be interpreted as a convex function of measures (see Section 2.2.4). We recall that this term is l.s.c. for the BV$-$w* topology. In the sequel, since there is no reason to choose $\phi_1 \neq \phi_2$, we will simply set $\phi = \phi_1 = \phi_2$.

MATHEMATICAL STUDY OF THE TIME-SAMPLED ENERGY

In fact, we do not have a continuum of images. A sequence is represented by a finite number of images. From the continuum $[0, t_{\max}]$ we have T images, $N_1(x), \ldots, N_T(x)$. For the theoretical framework it is assumed that

$$N_h \in BV(\Omega) \cap L^\infty(\Omega) \;\; \forall \; h = 1, \ldots, T. \tag{5.102}$$

The bounds of the gray-value levels over the sequence are denoted by

$$\begin{cases} m_N = \underset{h \in [1,\ldots,T], x \in \Omega}{\text{ess} - \inf} N_h(x), \\ M_N = \underset{h \in [1,\ldots,T], x \in \Omega}{\text{ess} - \sup} N_h(x). \end{cases} \tag{5.103}$$

where ess $-$ inf (respectively ess $-$ sup) is the essential infimum (respectively essential supremum), that is, the infimum up to Lebesgue measurable sets.

Again, the unknowns are B (the image of the restored background) and $C_1, \ldots, C_T$ (the T images of detection). The suitable functional space for studying this problem is the space of bounded variation (see Section 2.2), since both B and C_h are likely to have discontinuities across some contours. The time-discretized version of (5.98) is then to search for the solution of

$$\underset{(B,C_1,\ldots,C_T) \in BV(\Omega)^{T+1}}{\inf} E(B, C_1, \ldots, C_T) \;\; \text{with} \tag{5.104}$$

$$E(B, C_1, \ldots, C_T) = \tag{5.105}$$

$$\sum_{h=1}^{T} \int_{\Omega} \left[C_h{}^2 (B - N_h)^2 + \alpha_c (C_h - 1)^2 \right] dx + \alpha_b^r \int_{\Omega} \phi(DB) + \alpha_c^r \sum_{h=1}^{T} \int_{\Omega} \phi(DC_h).$$

We are now interested in proving existence and uniqueness for this problem. Before going further let us point out two main difficulties:

- The functional E is degenerate because of the first term. As a consequence, applying the direct method of the calculus of variations as described in Section 2.1.2 does not give any result: If we choose a minimizing sequence $(B_n)_{n \in N}$, $(C_n^h)_{n \in N}$, we can easily bound the sequence $(C_n^h)_{n \in N}$ thanks to the second term, but nothing can be said about $(B_n)_{n \in N}$ because the functions $(C_n^h)_{n \in N}$ have no lower bound.

- Though the functional E is convex with respect to each variable, it is nonconvex globally. This is naturally an issue as far as uniqueness is concerned.

To overcome the first difficulty, let us consider the problem (5.104)–(5.105) set over the constrained space

$$\overline{\mathcal{B}}(\Omega) = \Big\{ (B, C_1, \ldots, C_T) \in BV(\Omega)^{T+1} \text{ such that } m_N \le B \le M_N$$

$$\text{and } 0 \le C_h \le 1 \ \forall \, h \Big\}.$$

One may observe that these constraints are quite natural: It is not expected that B would have values never reached in the original sequence, and considering the functions C_h bounded is a priori reasonable. Since the variables are now uniformly bounded, it is now clear that the problem

$$\inf_{(B, C_1, \ldots, C_T) \in \overline{\mathcal{B}}(\Omega)} E(B, C_1, \ldots, C_T) \tag{5.106}$$

admits a solution in $\overline{\mathcal{B}}(\Omega)$. However, this result is not satisfying as soon as we are interested in finding a numerical solution: The optimality conditions are now inequalities instead of equations, and one would need to use Lagrange multipliers.

☞ *Interestingly, if there is a solution of the problem (5.104), then the solution belongs to $\overline{\mathcal{B}}(\Omega)$. As a consequence, we can prove the existence of a solution for the problem (5.104).*

The following technical lemma is helpful in proving the main result:

Lemma 5.3.1 *Let $u \in BV(\Omega)$, ϕ a function satisfying hypotheses (5.99), (5.100), (5.101), and $\varphi_{\alpha,\beta}$ the cut-off function defined by*

$$
\varphi_{\alpha,\beta}(x) = \begin{cases} \alpha & \text{if } x \leq \alpha, \\ x & \text{if } \alpha \leq x \leq \beta, \\ \beta & \text{if } x \geq \beta. \end{cases} \tag{5.107}
$$

Then we have

$$
\int_{\Omega} \phi(D\varphi_{\alpha,\beta}(u)) \leq \int_{\Omega} \phi(Du).
$$

Proof This lemma is very intuitive, but the proof requires some attention. Let us first recall the Lebesgue decomposition of the measure $\phi(Du)$:

$$
\int_{\Omega} \phi(Du) = \underbrace{\int_{\Omega} \phi(|\nabla u|)\, dx}_{\text{term 1}} + \underbrace{\int_{S_u} |u^+ - u^-| \mathcal{H}^{N-1}}_{\text{term 2}} + \underbrace{\int_{\Omega - S_u} |C_u|}_{\text{term 3}}.
$$

We are going to show that cutting the function u using the function $\varphi_{\alpha,\beta}$ permits us to reduce each term. To simplify the notation we will write $\hat{u}$ instead of $\varphi_{\alpha,\beta}(u)$.

Term 1: Let $\Omega_c = \{x \in \Omega;\ u(x) \leq \alpha \text{ or } u(x) \geq \beta\}$ and $\Omega_i = \Omega - \Omega_c$. Thanks to [196], we have $\int_{\Omega_i} \phi(|\nabla \hat{u}|)\, dx = \int_{\Omega_i} \phi(|\nabla u|)\, dx$. Consequently,

$$
\int_{\Omega} \phi(|\nabla \hat{u}|)\, dx = \int_{\Omega_i} \phi(|\nabla u|)\, dx + \underbrace{\int_{\Omega_c} \phi(|\nabla \hat{u}|)\, dx}_{(\equiv 0)} \leq \int_{\Omega} \phi(|\nabla u|)\, dx. \tag{5.108}
$$

Term 2: Using results proved in [11], we know that

$$
S_{\hat{u}} \subset S_u \quad \text{and} \quad \hat{u}^+ = \varphi_{\alpha,\beta}(u^+),\ \hat{u}^- = \varphi_{\alpha,\beta}(u^-).
$$

Since $\varphi_{\alpha,\beta}$ is Lipschitz continuous with a constant equal to 1, we then have

$$
\int_{S_{\hat{u}}} |\hat{u}^+ - \hat{u}^-| \mathcal{H}^{N-1} \leq \int_{S_{\hat{u}}} |u^+ - u^-| \mathcal{H}^{N-1} \leq \int_{S_u} |u^+ - u^-| \mathcal{H}^{N-1}. \tag{5.109}
$$

Term 3: We need to understand the Cantor part of the distributional derivative of the composed function $\varphi_{\alpha,\beta}(u)$. Vol'pert [334] first proposed a *chain rule formula* for functions $v = \varphi(u)$ for $u \in BV(\Omega)$ and when φ is continuously differentiable. Ambrosio and Dal Maso [14] gave extended results for functions φ uniformly Lipschitz continuous. Since u is scalar, it is demonstrated in [14] that we can write

$$
C(\varphi_{\alpha,\beta}(u)) = \varphi'_{\alpha,\beta}(\tilde{u})C(u) \quad |Du|\text{-a.e. on } \Omega - S_u, \tag{5.110}
$$

where $\tilde{u}$ is the approximate limit (see Section 2.2.3) of u defined by

$$\lim_{r \to 0^+} \frac{1}{|B(x,r)|} \int_{B(x,r)} |u(y) - \tilde{u}(x)| dy = 0,$$

where $B(x,r)$ is the closed ball with center x and radius r. Moreover, we have

$$\int_{\Omega - S_{\hat{u}}} |C_{\hat{u}}| = \int_{\Omega - S_u} |C_{\hat{u}}| + \int_{S_u/S_{\hat{u}}} |C_{\hat{u}}|, \qquad (5.111)$$

where the last term over $S_u/S_{\hat{u}}$ is zero.[5] Then, using the chain rule formula (5.110), we have

$$\int_{\Omega - S_{\hat{u}}} |C_{\hat{u}}| \leq \underbrace{|\varphi'_{\alpha,\beta}|_{L^\infty}}_{(\leq 1)} \int_{\Omega - S_u} |C_u| \leq \int_{\Omega - S_u} |C_u|. \qquad (5.112)$$

The inequalities (5.108), (5.109), and (5.112) conclude the proof.   ∎

Then, the following result can be established:

Theorem 5.3.2 *The problem (5.104) admits a solution on $BV(\Omega)^{T+1}$. If moreover*

$$\alpha_c \geq 3(M_N - m_N)^2, \qquad (5.113)$$

then the solution is unique.

Proof Existence is proven showing that minimizing (5.105) over $\overline{B}(\Omega)$ is equivalent to the same problem posed over $BV(\Omega)^{T+1}$, that is, without any constraint (this is a direct consequence of Lemma 5.3.1), for which we apply the direct method of the calculus of variations.

As far as uniqueness is concerned, the difficulty comes from the nonconvexity of the function

$$(B, C_1, \ldots, C_T) \to \sum_{h=1}^{T} C_h^2 (B - N_h)^2 + \alpha_c \sum_{h=1}^{T} (C_h - 1)^2.$$

However, if α_C is large enough, it can be proved that this functional is strictly convex over $\overline{B}$, which permits us to conclude the proof (see [206] for more details).   ∎

Remark The condition (5.113) is, in fact, quite natural. It means that the background must be taken sufficiently into account.   ∎

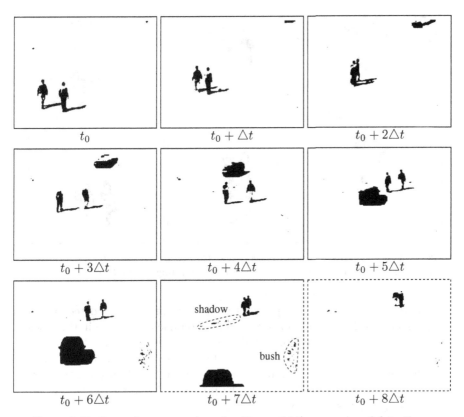

t_0	$t_0 + \triangle t$	$t_0 + 2\triangle t$
$t_0 + 3\triangle t$	$t_0 + 4\triangle t$	$t_0 + 5\triangle t$
$t_0 + 6\triangle t$	$t_0 + 7\triangle t$	$t_0 + 8\triangle t$

shadow

bush

Figure 5.22. "street" sequence (see also Figure 5.11): sequence of detections.

original video   temporal mean   restored background

Figure 5.23. "Street" sequence (see also Figure 5.11). Close-ups of the same area (top right-hand corner) for the initial sequence, the temporal mean of the sequence, and the restored background.

EXPERIMENTS

An important consequence of Theorem 5.3.2 is that it permits us to consider the minimization problem over all $BV(\Omega)^{T+1}$ without any constraint. Con-

---

[5]We recall that for any $v \in BV(\Omega)$ and any set S of Hausdorff dimension at most $N - 1$, we have $C_v(S) = 0$.

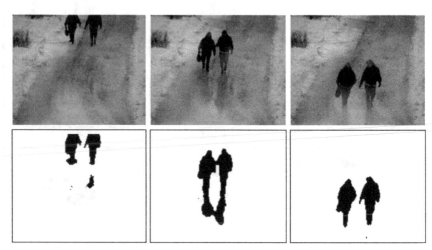

Figure 5.24. "Walking in Finland" sequence, from University of Oulu, 55 images. Three images of the sequence N and the corresponding C functions.

sequently, from a numerical point of view, the difficulties and techniques are the same as for the previous variational approaches studied in this book for image restoration (Section 3.2.4) or optical flow (Section 5.3.2).

Although optimality equations may be written, they remain hard to handle (see Section 3.2.4). Similarly, an algorithm based on Γ-convergence and half-quadratic minimization can be proposed. We refer to [206], where convergence results are proved.

To illustrate this approach, let us present some results from real sequences. We first show some results on the "street" sequence presented in Section 5.3.1 (Figure 5.11). Figure 5.22 shows an example of detection. Notice that the motions of the bush and the shadow are detected. We illustrate in Figure 5.23 the restoration aspect. The temporal mean of the sequence is still noisy (because the motion of the persons is taken into account and the noise is not of zero mean), while the restored background is of very good quality.

We show in Figure 5.24 another real sequence, where the reflections on the ground are detected as "motion."

We also recall that this method permits us to detect objects by comparison with a reference image. As it is not based on motion, a person stopping for some time and walking again will be detected. We refer the interested reader to [206] for more results and for quantitative experiments on the noise influence with respect to the results.

Finally, let us mention that this approach can be extended in the case of video-streams instead of batch (that is postprocessing a set of given images). This is the case of most real applications, for instance in video-surveillance applications. There is a continuous flux of images to be analyzed. In this case, one can update only the last image C_i and consider all the others as fixed. If computations have been done for $t = 1, \ldots, T$ and that a new image N_{T+1} is available, it is enough to minimize

$$\tilde{E}(B, C_{T+1}) = E(B, \overline{C_1}, \ldots, \overline{C_T}, C_{T+1}),$$

where $\overline{C_i}$ are the detections previously computed. We refer to [203] for more details.

5.3.4 Sequence Restoration

This section concerns the problem of restoration as applied to sequences. Unlike static images, very little research has been carried out on this subject. The aim of this section is essentially to show the difficulties and specificities of this problem by focusing on the problem of the restoration of degraded movies. If sound track restoration is now fairly well resolved as a problem, it is not the same case as far as images are concerned.

There are two main reasons for this lack of research. The first is economic. The decision to restore a movie may come either from political institutions or from companies in the broadcasting market. In both cases, budgets for this are limited, since these investments do not always pay. The second is simply because it is a very difficult problem.

To begin with, there are the technical problems associated with the storage and processing of the data. With 24 or 25 frames per second (US) size 1920×1080 (new high-definition progressive video format), we let the reader compute for himself the memory size necessary for the storage of a 90- or 120-minute movie.

Then, by simply trying to define what movie restoration is, this becomes quite impossible because of the numerous types of degradation. First, defects may affect the base of the film or the emulsion side or both. They may be mechanical, due to lack of precautions while handling the film or caused by cameras, printers, developing tanks, and all the various equipment used. It is the case for scratches (on one frame, caused by manual handling), vertical scratches on many contiguous frames (caused by mechanical devices), dirt or dust spots, hairs, emulsion tearing off, water marks. They may also be chemical, degradation of the nitrate base for old films or of the acetate base for more recent ones, or some kind of mushrooms, or irregular shadows. They may also be a combination of mechanical and chemical, like the vertical blue scratches caused by obstructed orifices in the development tank. The splices that tie pieces of film together may be deteriorated, and the film may have shrunk because of too much drying. The original film may be missing, leaving only a copy that may exhibit some or all of the preceding defects plus some photographed defects. That is, dust spots and scratches that were present on the original have been transferred to the copy in a blurred form, which renders their detection less evident. During the transfer operation a badly adjusted piece of equipment may leave a "hot spot" due to improper focusing of the illumination. Some of these degradations are illustrated in Figures 5.25 to 5.29.[6]

☛ *From this quick overview alone, it is clear that movie restoration cannot be defined as a unique and simple problem.*

Faced with this variety of defects, it may be interesting to classify them. As presented in a report by the "Commission Supérieure Technique de l'image et du son" [135], they can be characterized by the amount of human interaction needed to detect and correct them. This can be summarized in Table 5.2.

		Detection	
		Interactive	Automatic
Correction	Interactive	Defects that might be identified as an element of the image and correction of which could not be done without contextual information.	Emulsion tearing off or spots of any kind, nonstationary and wide (mushrooms, irregular shades)
	Automatic	Static defects of small areas (hairs, static spots), stable irregularities, "hot spots," contrast, color or sharpness differences in the sequence.	Defects whose mathematical model cannot be confused with an image element (dusts, scratches, small instabilities in the position or the exposition)

Table 5.2. Proposal of a classification for defects [135].

Let us comment Table 5.2:

- In the left-hand column can be found everything that cannot be detected automatically, that is, elements that could be identified as image elements like shadow effects or reflections that may be wanted in the scenario. The discrimination of such defects seems to be beyond reach for a long time to come. Any correction requiring an esthetic choice, like color grading, will also come into this category. Conversely, the right-hand column presents all the defects that can be detected automatically.

- On the upper line are all the defects that cannot be corrected automatically. This may be due to:
  - The impossibility of measuring a variation precisely, that is, to define a norm.

---

[6]Original images from Figures 5.27 to 5.29 are in color.

Figure 5.25. By courtesy of "association des frères Lumière." These images show general intensity variations throughout the sequence (observe the posters at the right-hand side of the images, or the background brightness). These variations are due to the nonuniform aperture time, since this was manually controlled.

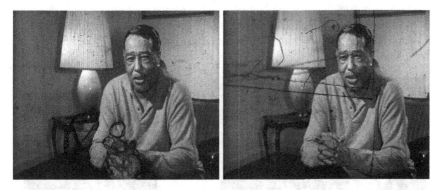

Figure 5.26. Images of Duke Ellington, "document amateur," by courtesy of "Cinémathèque de la Danse (Paris)." In the left-hand image, the film has burned. In the right-hand image, there are many spots and rays.

Figure 5.27. Example of chemical stain. By courtesy of DUST Restauration. This figure contains color: please visit the book web site `http://www-sop.inria.fr/books/imath` for the colored version.

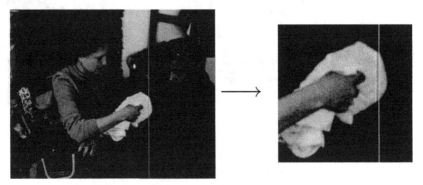

Figure 5.28. In this image there is an example of a ray, and there is also clearly a problem with the grain. By courtesy of DUST Restauration. This figure contains color: please visit the book web site `http://www-sop.inria.fr/books/imath` for the colored version.

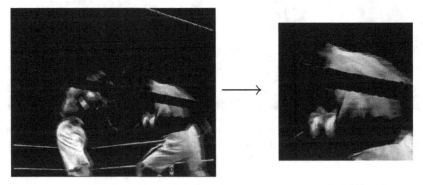

Figure 5.29. Example of interlaced video. By courtesy of DUST Restauration. This figure contains color: please visit the book web site `http://www-sop.inria.fr/books/imath` for the colored version.

- The impossibility of defining an automatic rule for the correction.
- The definitive loss of information that cannot be reconstructed without human intervention.

Interestingly, this degree of interaction/automation can also be found in the different systems that are currently proposed:

- Paint system is the standard fully manual system, which is also widely used in the special effects industry. Still, it does not allow us to correct all the problems, like some colorimetry defects, for instance. If subtle color variation occurs in a sequence, it is impossible for the human eye to perceive it and correct it frame by frame, though this is perfectly visible at the projection rate.

- A different strategy is to focus on degradation, to detect and to correct it as automatically as possible, but with a manual preselection. For each kind of degradation one needs to develop a suitable tool. Some such systems are Limelight and Da Vinci. Limelight is a Eureka project (1993 to 1997) whose purpose was the development of a prototype restoration system. Dedicated software was developed by the University of La Rochelle, in France, and by the Joanneum research, in Graz, Austria. A combination of mathematical morphology and temporal comparison with adjacent frames is used to detect dirt and dust. Da Vinci provides digital film, HDTV, and SDTV color enhancement technology and is now distributing "Revival," software developed initially within the Singapore University. This software also focuses on degraded areas of the picture.

- Some fully automatic systems also exist and some even operate in real time. They are based on a temporal diffusion process of the sequence. Such methods have been developed using wavelets or Markov random fields (see, for instance, [199, 200]). Within this framework, we can mention the system Archangel proposed by Snell & Wilcox. This system follows the research effort of BBC Research, Snell & Wilcox, INA, the University of Cambridge, and the University of Delft inside the Eureka project AURORA. Hardware embeds all the necessary memory to allow temporal filtering on many adjacent frames. However, even if the temporal filtering is adaptive, since it acts on the entire image domain it is unavoidable to have some blur effect, loss of grain, and loss of resolution for fine textures.

So we can see that from fully manual to fully automatic, nobody can pretend to have a "perfect" restoration tool. But what would "perfect" restoration mean?

✳ *This is one of the main and probably most difficult questions: How can a restoration be judged? Clearly, one cannot be satisfied with the results obtained on one image, even if it looks good. One needs to see the result at a rate of 24 frames per second. A typical example is the correction of vertical scratches. Imagine that a ray is well corrected for each frame except at some locations, which differ from frame to frame. The result is that the ray will still be visible when the sequence is played. This example clearly shows the temporal aspect but also the importance of perception. In other words, the main objective of movie restoration is not to reconstruct but to focus on perception: It will be sufficient for a defect to be no longer perceived even if it has not been completely removed. More than any other domain in computer vision, movie restoration should benefit from advances and studies of human perception (see [189], for instance).*

Could PDEs help with video restoration? In fact, very little research has been carried out up to now. Few approaches have been developed (see, for instance, [174, 241, 219]), and by observing the success of PDEs in image restoration, it will certainly be very interesting to continue work in this direction, taking perception more into account. When restoration is understood as reconstructing some missing parts, previously detected either automatically or manually, then the problem becomes very similar to inpainting as described in Section 5.1 for still images. In the rest of this section we investigate how the PDE methodology developed for still images can be extended to video inpainting.

PRINCIPLES OF VIDEO INPAINTING

Let us consider the restoration of videos with severe degradations, severe meaning that the initial content is lost over a large number of pixels. Examples of such videos are shown in Figures 5.26 and 5.28: the images contain many spots, rays, dust, etc. Now, if we assume that the areas that are degraded are known, then the problem becomes an inpainting problem as described in Section 5.1.

Using the same notation as in Section 5.1, the problem can be described as follows: given a spatiotemporal domain D, a hole $\Omega \subset D$, and a video $u_0(t,x)$ known over $D - \Omega$, we want to find a video u that matches u_0 outside the hole Ω and that has "meaningful" content inside.

A simple solution for video inpainting is to inpaint each damaged frame independently, as proposed for example by Bertozzi et al. [48]. Unfortunately, such an approach often leads to several artifacts such as flicker, since it does not take into account the adjacent frames.

☞ *In fact, taking into account the adjacent frames is the key to success for video inpainting. In most cases the missing information at a given time can be found in a spatiotemporal neighborhood.*

We present in the rest of this section two kinds of methods to take into account the interframe correlation.

TOTAL VARIATION (TV) MINIMIZATION APPROACH

A possible solution to take into account the spatiotemporal structure of the input is to consider a video as a 3-D image, i.e., a volume [271], so that we can apply 2-D image algorithms extended to the 3-D case.

For example, we propose in this section to extend the total variation (TV) minimization of Chan and Shen [97] (see Section 5.1.2). An inpainting of ⬚ C++ the video u_0 can be obtained as a minimizer of the energy

$$E(u) = \lambda \int_{D-\Omega} (u - u_0)^2 \, dx \, dt + \int_D |\nabla_{xt} u| \, dx \, dt, \tag{5.114}$$

where $\nabla_{xt} u$ is the spatiotemporal gradient of u. Formally, a minimizer of $E(u)$ will satisfy the following Euler–Lagrange equation:

$$2\lambda_e (u - u_0) - \operatorname{div}_{xt} \left(\frac{\nabla_{xt} u}{|\nabla_{xt} u|} \right) = 0, \tag{5.115}$$

where $\lambda_e(t, x) = \lambda(1 - \chi_\Omega(t, x))$, which provides a simple algorithm for the inpainting.

In order to estimate the quality of inpainting, we consider a classical sequence called "Mobile and Calendar," which was used for restoration by Kokaram in [200], but also by numerous other authors in the video compression community.

It is a real sequence that has been artificially degraded by generating blotches randomly (number, position, size) on each frame. Four of the most corrupted frames are displayed in the first column of Figure 5.30, and the result of TV inpainting is presented in the second column. Although seemingly quite satisfactory, the restored video contains a lot of flicker at the inpainted regions, and some close-ups shown in Figure 5.31 also reveal several spatial defects: oversmoothing and incorrect filling in are observed. Note that original and reconstructed videos are available online, from the book web site http://www-sop.inria.fr/books/imath.

So, extending 2-D algorithms to 3-D was a natural idea, but this has two main limitations. First, spatial and temporal dimensions are of different natures. Second, the local frame contents at adjacent temporal positions may be somewhat decorrelated due to the apparent motion. Consequently, video inpainting should use motion estimation. This is the subject of the next paragraph.

MOTION COMPENSATED (MC) INPAINTING

The key to video inpainting is the motion. Recent techniques, whether stochastic or deterministic, use motion information, either directly or indi-

Degraded video   TV inpainting   MC inpainting

Figure 5.30. Example of video restoration on a degraded video (left-hand column). The original sequence is the classical "Mobile and Calendar" sequence. Some frames of the video are shown with the parts to restore in red. Results of TV and MC inpainting are shown in the middle and right-hand columns. Note that this figure as well as the associated videos are available online on the book web site: http://www-sop.inria.fr/books/imath. We refer to Figure 5.31 for a close-up of some parts of this sequence.

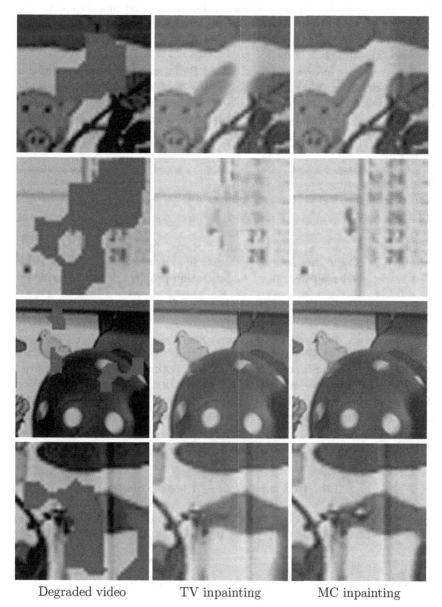

Degraded video   TV inpainting   MC inpainting

Figure 5.31. Close-up of the results presented in Figure 5.30. One may observe
on the still images the better performance of the MC inpainting compared to the
TV inpainting. This figure as well as the associated videos are available online
on the book web site: http://www-sop.inria.fr/books/imath.

rectly (see for example [201, 65, 129, 327, 202] for stochastic approaches). They are classically based on motion adaptibility and motion compensation. Motion adaptivity approaches take motion into account implicitly by estimating the presence or absence of motion, and then adapting their behavior accordingly.

However, as we have seen in Section 5.3.2, motion estimation is a difficult task in general. In the inpainting case, it is even more difficult since some intensity values are missing. When the missing areas are relatively small, motion trajectories can be estimated by some form of interpolation. However, when they are large, this may fail, which leads to an chicken-and-egg problem.

➤ *Chicken-and-egg problem: intensity values are needed to recover the motion trajectories and motion trajectories are needed to recover the missing intensity values.*

The chicken-and-egg problem can be solved by estimating jointly the intensity u and the corresponding motion field $\sigma = (\sigma_1, \sigma_2)^T$. One possible idea is to consider a constrained minimization problem of the form

$$\inf_{u,\sigma} \ (R(u) + A(u,\sigma) + S(\sigma)), \tag{5.116}$$

$$u = u_0 \ \text{in} \ D - \Omega,$$

where $R(u)$ is a spatial image smoothness term, $A(u,\sigma)$ encodes both a data fidelity term for the motion field and a smoothness term for intensities along motion trajectories, and $S(\sigma)$ is a smoothing term for the motion field. Note that when u is fixed, the part $A(u,\sigma) + S(\sigma)$ corresponds to the generic model for flow recovery (5.83), and we refer to Section 5.3.2 for more details.

For an inpainting denoising problem, one may consider an unconstrained problem of the form

$$\inf_{u,\sigma} \ (F(u,u_0) + R(u) + A(u,\sigma) + S(\sigma)), \tag{5.117}$$

where $F(u,u_0)$ is a fidelity term for the intensity values. For example, an energy of the form (5.117) was proposed by Chanas and Cocquerez [105, 113],

$$E_c(u,\sigma) = \int_{D-\Omega} (u - u_0)^2 \ dx \, dt + \lambda_1 \int_D \phi_1(|\nabla_x u|) \ dx \, dt \tag{5.118}$$

$$+ \lambda_2 \int_D |\sigma \cdot \nabla_x u + u_t| \ dx \, dt + \lambda_3 \sum_{j=1}^{2} \int_D \phi_2(|\nabla_x \sigma_i|) \ dx \, dt,$$

where ∇_x is the spatial gradient operator, and the ϕ_i's are strictly convex functions with linear growth. The optical flow part was adapted from (5.88) [24].

Based on the generic form (5.117), Lauze and Nielsen [210, 209] proposed a constrained minimization using work of Brox et al. [76] for motion recovery. The authors proposed to minimize the following energy compounded of three terms:

$$E(u, \sigma) = \int_D \phi_1(|\nabla_x u|) \, dx \, dt + \lambda_1 \int_D \phi \left(\left(\frac{\partial u}{\partial \Sigma} \right)^2 + \gamma \left| \frac{\partial \nabla_{xt} u}{\partial \Sigma} \right|^2 \right) \, dx \, dt$$

$$\tag{5.119}$$

$$+ \lambda_2 \int_D \phi(|\nabla_{xt} \sigma_1|^2 + |\nabla_{xt} \sigma_2|^2) \, dx \, dt,$$

with the constraint $u = u_0$ in $D - \Omega$, where $\partial/\partial\Sigma$ is the directional derivative along the spatiotemporal trajectory $\Sigma = (\sigma_1, \sigma_2, 1)^T$, and $\phi(s) = \sqrt{s + \varepsilon^2}$. The first term is a smoothing term on the reconstructed gray levels. The second term is an extended optical flow constraint as proposed by Brox et al. [76]. It combines the brightness consistency as well as a preservation of the gradient. The third term is a smoothing term on the optical flow; it is rotationally invariant.

MC inpainted frames are presented in the last column of Figure 5.30. Although the results of TV and MC inpainting look similar in that figure, the resulting video exhibits almost no flicker in the case of MC inpainting, while as mentioned above, TV produces a strong flicker. We remind the reader that the videos are also available on the book web site, which make it possible to evaluate the quality of the different reconstructions. Several close-ups in the last column of Figure 5.31 also illustrate the high quality achieved in the reconstruction with the MC inpainting: oversmoothing is strongly reduced and small details are recovered.

5.4   Image Classification

5.4.1   Introduction

In this section we present two supervised classification models for satellite images. We show in Figure 5.32 some typical images to be analyzed. Classification aims at finding in the image some classes that have been previously defined, in terms of intensity. This intensity usually corresponds to different ground natures. This kind of technique is especially useful in studying forest evolution, ground conditions, city development, etc.

The classification problem is closely related to the segmentation one, in the sense that we want to get a partition composed of homogeneous regions. The main difference is that the number of classes and their characteristics are fixed. Many models can be found in the field of stochastic approaches, with the use of Markov random field (MRF) theory [53, 350]. Hereafter, we present two different variational models. The first one is concerned only

Figure 5.32. Examples of satellite images. By courtesy of the CNES (Centre National d'Etudes Spatiales).

with classification and is based on a level set formulation [290]. The second one is coupled with a restoration process [291] and is inspired by work about phase transition in mechanics. The latter is purely variational and relies on approximation principles via the Γ-convergence theory. For the sake of simplicity, we will make for the two models the following assumptions:

- The discriminating criterion is based on the intensity level of pixels.

- Each class C_i has a Gaussian distribution of intensity $N(\mu_i, \sigma_i)$, where μ_i and σ_i are respectively the mean and the standard deviation of the class C_i.

- The number K of classes and the parameters (μ_i, σ_i) are known (it is a *supervised* classification).

5.4.2 A Level-Set Approach for Image Classification

The classification procedure consists of two steps:

- Defining the classes according to discriminating features. In our case, according to our hypotheses, we choose the parameters of the Gaussian distribution μ_i and σ_i. Of course, other discriminant attributes, such as texture parameter, for example, could be chosen.

- Defining a partitioning process that:
  - Takes into account the first step.
  - Penalizes overlapping regions (pixels with two labels) and the formation of a vacuum.
  - Exhibits regular interfaces between classes, i.e., interfaces with minimal perimeter.

These three properties need to be taken into account in the model. Let us write the precise mathematical formulation. Let Ω be an open bounded domain of R^2 and let $u_0 : \Omega \rightarrow R$ be the observed data function (the gray-level intensity). Let Ω_i be the region defined as[7]:

$$\Omega_i = \{x \in \Omega; \ x \ \text{belongs to the} \ i\text{th class}\}. \tag{5.120}$$

A partitioning of Ω consists in finding a family of sets $\{\Omega_i\}_{i=1,\ldots,K}$ such that

$$\Omega = \bigcup_{i=1}^{K} (\Omega_i \cup \Gamma_i) \quad \text{and} \quad \Omega_i \cap \Omega_j = \emptyset, \quad i \neq j,$$

where $\Gamma_i = \partial\Omega_i \cap \Omega$ is the intersection of the boundary of Ω_i with Ω and $\Gamma_{ij} = \Gamma_{ji} = \Gamma_i \cap \Gamma_j$, $i \neq j$, the interface between Ω_i and Ω_j. Of course, we have $\Gamma_i = \bigcup_{i \neq j} \Gamma_{ij}$ (possibly $\Gamma_{ij} = \emptyset$). We denote by $|\Gamma_i|$ the one-dimensional Hausdorff measure of Γ_i. We have: $|\Gamma_i| = \sum_{i \neq j} |\Gamma_{ij}|$ ($|\emptyset| = 0$).

The classification model we propose for an observed image u_0 consists in searching for a family of sets $\{\Omega_i\}_{i=1,\ldots,K}$ defined by (5.120) and satisfying the following conditions:

(A) $\{\Omega_i\}_{i=1,\ldots,K}$ is a partition of Ω, i.e., $\Omega = \bigcup_{i=1}^{K} \Omega_i \cup \Gamma_i$ and
$\Omega_i \cap \Omega_j = \emptyset$, $i \neq j$.

(B) The partition $\{\Omega_i\}_{i=1,\ldots,K}$ takes into account the Gaussian distribution property of the classes (data term):

$$\Omega_i = \Big\{x \in \Omega; \ \text{the intensity} \ u_0(x) \ \text{has a Gaussian distribution of}$$
$$\text{mean} \ \mu_i \ \text{and of standard deviation} \ \sigma_i\Big\}.$$

(C) The classification is regular, in the sense that the length of each interface Γ_{ij} is minimal.

Conditions (B) and (C) can be expressed in terms of energy minimization:

(B) Minimize, with respect to Ω_i,

$$\sum_i \int_{\Omega_i} \frac{(u_0(x) - \mu_i)^2}{\sigma_i^2} \, dx. \tag{5.121}$$

In fact, in a probabilistic framework, (5.121) means that we want to maximize the conditional probability $\Pr(u_0(x)/x \in \Omega_i)$.

(C) Minimize, with respect to Γ_{ij},

$$\sum_{i,j} \xi_{ij} |\Gamma_{ij}|, \tag{5.122}$$

---

[7] Ω_i can actually be a set of nonconnected regions.

the parameter $\xi_{ij} \in R^+$ being fixed and permitting us to take into account possible information about the length of contours.

The main difficulty in the above formulation comes from the fact that the unknowns are sets and not functions. To overcome this difficulty we propose to use a level set method inspired by the work of Zhao et al. [348] concerning multiphase evolution in fluid dynamics.

Let us suppose that for each $i = 1, \ldots, K$ there exists a Lipschitz function ϕ_i such that

$$\begin{cases} \phi_i(x) > 0 & \text{if } x \in \Omega_i, \\ \phi_i(x) = 0 & \text{if } x \in \Gamma_i, \\ \phi_i(x) < 0 & \text{otherwise}, \end{cases} \qquad (5.123)$$

i.e., the region Ω_i is entirely described by the function ϕ_i. Now let us look at the writing of conditions (A), (B), and (C) in terms of an energy functional involving $\{\phi_i\}_{i=1,\ldots,K}$. This functional will have to contain three terms:

- A term related to condition (A) (partition condition):

$$F^A(\phi_1, \ldots, \phi_K) = \frac{\lambda}{2} \int_\Omega \Big(\sum_{i=1}^K H(\phi_i(x)) - 1 \Big)^2 dx, \quad \lambda \in R^+,$$

  where $H(s)$ is the Heaviside function: $H(s) = 1$ if $s > 0$ and $H(s) = 0$ if $s < 0$. The minimization of F^A with respect to $\{\phi_i\}_{i=1,\ldots,K}$ leads to a solution where the formation of a vacuum (pixels with no labels) and regions overlapping (pixels with more than one label) are penalized.

- A term related to condition (B):

$$F^B(\phi_1, \ldots, \phi_K) = \sum_{i=1}^K e_i \int_\Omega H(\phi_i(x)) \frac{(u_0(x) - \mu_i)^2}{\sigma_i^2} \, dx, \quad \text{with } e_i \in R^+,$$

  where $\{e_i\}_{i=1,\ldots,K}$ are constants that could be useful to take into account, for instance, a bad estimation of the statistics for one of the K classes.

- A third term related to condition (C) (length shortening of interface set):

$$F^C(\phi_1, \ldots, \phi_K) = \sum_{i=1}^K \gamma_i \int_{\phi_i=0} ds, \quad \text{with } \gamma_i \in R^+.$$

Therefore, the complete functional is

$$F(\phi_1, \ldots, \phi_K) = F^A(\phi_1, \ldots, \phi_K) + F^B(\phi_1, \ldots, \phi_K) + F^C(\phi_1, \ldots, \phi_K). \qquad (5.124)$$

Remark In fact, the functional F is closely related to the Mumford and Shah functional (see Section 4.2.2), for which solutions are expected to be piecewise constant.   ∎

Unfortunately, stated as above, the functional F still has some drawbacks from a practical point of view: F is not Gâteaux differentiable, and the length term is not easy to handle numerically. So we have to regularize F. To do this, let δ_α and H_α be respectively the following approximations of the Dirac and Heaviside distributions:

$$\delta_\alpha(s) = \begin{cases} \frac{1}{2\alpha}\left(1 + \cos\left(\frac{\pi s}{\alpha}\right)\right) & \text{if} \quad |s| \leq \alpha, \\ 0 & \text{if} \quad |s| \geq \alpha, \end{cases}$$

$$H_\alpha(s) = \begin{cases} \frac{1}{2}\left(1 + \frac{s}{\alpha} + \frac{1}{\pi}\sin\left(\frac{\pi s}{\alpha}\right)\right) & \text{if} \quad |s| \leq \alpha, \\ 1 & \text{if} \quad s > \alpha, \\ 0 & \text{if} \quad s < -\alpha. \end{cases}$$

Then, we approximate F by

$$F_\alpha(\phi_1, \ldots, \phi_K) = F_\alpha^A(\phi_1, \ldots, \phi_K) + F_\alpha^B(\phi_1, \ldots, \phi_K) + F_\alpha^C(\phi_1, \ldots, \phi_K),$$
$$(5.125)$$

where

$$F_\alpha^A(\phi_1, \ldots, \phi_K) = \frac{\lambda}{2}\int_\Omega \left(\sum_{i=1}^K H_\alpha\left(\phi_i(x)\right) - 1\right)^2 dx,$$

$$F_\alpha^B(\phi_1, \ldots, \phi_K) = \sum_{i=1}^K e_i \int_\Omega H_\alpha\left(\phi_i(x)\right) \frac{(u_0(x) - \mu_i)^2}{\sigma_i^2}\, dx,$$

$$F_\alpha^C(\phi_1, \ldots, \phi_K) = \sum_{i=1}^K \gamma_i \int_{\Omega_i} \delta_\alpha\left(\phi_i(x)\right) |\nabla\phi_i(x)|\, dx.$$

If the definition of the approximated functionals F_α^A and F_α^B comes very naturally, that of F_α^C is less immediate and relies upon the following lemma.

Lemma 5.4.1 *Let* $\phi : R^N \to R$ *be Lipschitz continuous. Then*

$$\lim_{\alpha \to 0} \int_\Omega \delta_\alpha(\phi(x)) |\nabla\phi(x)|\, dx = \int_{\phi=0} ds.$$

Proof Thanks to the coarea formula (Section 2.5.2) we have

$$L_\alpha(\phi) = \int_\Omega \delta_\alpha(\phi(x)) |\nabla\phi(x)|\, dx = \int_R \left[\delta_\alpha(t)\int_{\phi=t} ds\right] dt.$$

By setting $h(t) = \int_{\phi=t} ds$, we get

$$L_\alpha(\phi) = \int_R \delta_\alpha(t)\, h(t)\, dt = \frac{1}{2\alpha} \int_{-\alpha}^{+\alpha} \left(1 + \cos\left(\frac{\pi t}{\alpha}\right)\right) h(t)\, dt.$$

If we take $\theta = t/\alpha$, then

$$L_\alpha(\phi) = \frac{1}{2} \int_{-1}^{+1} (1 + \cos(\pi \theta))\, h(\alpha\theta)\, d\theta,$$

and when $\alpha \to 0$, we obtain

$$\lim_{\alpha \to 0} L_\alpha(\phi) = \frac{1}{2} h(0) \int_{-1}^{+1} (1 + \cos(\pi \theta))\, d\theta = \int_{\phi=0} ds.$$

■

By construction, F_α is Gâteaux differentiable with respect to $\{\phi_i\}_{i=1,\ldots,K}$ and all the integrals are defined over Ω (which is fixed).

Another improvement to avoid oversmoothing of the interfaces between classes (where the gradient of u_0 is high) is to introduce into F_α^C the stopping function

$$g(u_0) = \frac{1}{1 + |\nabla G_\rho * u_0|^2},$$

where G_ρ is the usual Gaussian kernel. This is particularly useful if the data is very noisy, since in this case the parameters γ_i are chosen large enough to ensure a good smoothing inside the classes. So the final minimization problem is

$$\inf_{\phi_i} F_\alpha(\phi_1, \ldots, \phi_K) \tag{5.126}$$

with

$$F_\alpha(\phi_1, \ldots, \phi_K) =$$

$$\frac{\lambda}{2} \int_\Omega \left(\sum_{i=1}^K H_\alpha(\phi_i(x)) - 1 \right)^2 dx + \sum_{i=1}^K e_i \int_\Omega H_\alpha(\phi_i(x)) \frac{(u_0(x) - \mu_i)^2}{\sigma_i^2}\, dx$$

$$+ \sum_{i=1}^K \gamma_i \int_{\Omega_i} g(u_0(x))\, \delta_\alpha(\phi_i(x))\, |\nabla \phi_i(x)|\, dx.$$

The classification problem we are interested in can be formulated as follows: We do not know whether (5.126) has a solution. The question is open and is under investigation. Nevertheless, we may write formally the associated

Euler–Lagrange equations. We get a system of K-coupled PDEs:

$$\delta_\alpha(\phi_i)\left[e_i\frac{(u_0-\mu_i)^2}{\sigma_i^2} - \gamma_i\,g(u_0)\,\mathrm{div}\left(\frac{\nabla\phi_i}{|\nabla\phi_i|}\right) - \gamma_i\frac{\nabla g\cdot\nabla\phi_i}{|\nabla\phi_i|}\right. \tag{5.127}$$

$$\left. + \lambda\left(\sum_{i=1}^{K}H_\alpha(\phi_i)-1\right)\right] = 0$$

for $i=1,\ldots,K$, with Neumann boundary conditions. Observe that unlike the classical active contour equation (see Chapter 4, equation (4.34)), the divergence term is multiplied by $\delta_\alpha(\phi_i)$ and not by $|\nabla\phi_i|$, which clearly gives to this equation a different nature from that of (4.34). If its theoretical study remains an open question, it is an advantage numerically to have the term $\delta_\alpha(\phi_i)$, which delimits a natural narrow band (where the ith is nonzero). To solve (5.127) numerically, we embed it into a dynamical process:

$$\frac{\partial\phi_i}{\partial t} = \delta_\alpha(\phi_i)\left[e_i\frac{(u_0-\mu_i)^2}{\sigma_i^2} - \gamma_i\,g(u_0)\,\mathrm{div}\left(\frac{\nabla\phi_i}{|\nabla\phi_i|}\right) - \frac{\nabla g\cdot\nabla\phi_i}{|\nabla\phi_i|}\right.$$

$$\tag{5.128}$$

$$\left. + \lambda\left(\sum_{i=1}^{K}H_\alpha(\phi_i)-1\right)\right].$$

We discretize (5.128) by using finite difference schemes similar to those for the discretization of (4.34), and we refer to Section A.3.4 for more details.

To illustrate this approach, we present in Figures 5.33 and 5.34 two examples of a synthetic image and the SPOT image.[8]

- In the first example (Figure 5.33) we illustrate how regions interact to avoid overlaping while covering all the space.

- The second example (Figure 5.34) illustrates the algorithm on a real image. Notice in particular the initialization. We use an automatic method for the initialization of the $\{\phi_i\}_{i=1,\ldots,K}$ that we call *seed initialization*. This method consists in cutting the data image u_0 into n windows w_p, $p=1,\ldots,n$, of predefined size. We compute the mean m_p and the standard deviation s_p for the Gaussian distribution of u_0 over each window w_p. Then we select the index k such that $k = argmin_j\; d_B(N(m_p,s_p),N(\mu_j,\sigma_j))$, where d_B is the Bhattacharyya distance,[9] which measures the distance between two Gaussian distributions $N(m_p,s_p)$ and $N(\mu_j,\sigma_j)$. Finally, we initialize

---

[8]Results provided by the INRIA project ARIANA (joint project CNRS–INRIA–UNSA): http://www.inria.fr/ariana/.

[9]$d_B(N(\mu_a,\sigma_a),N(\mu_b,\sigma_b)) = \dfrac{(\mu_a-\mu_b)^2}{4(\sigma_a^2+\sigma_b^2)} + \dfrac{1}{2}\log\dfrac{|\sigma_a^2+\sigma_b^2|}{2\sigma_a\sigma_b}.$

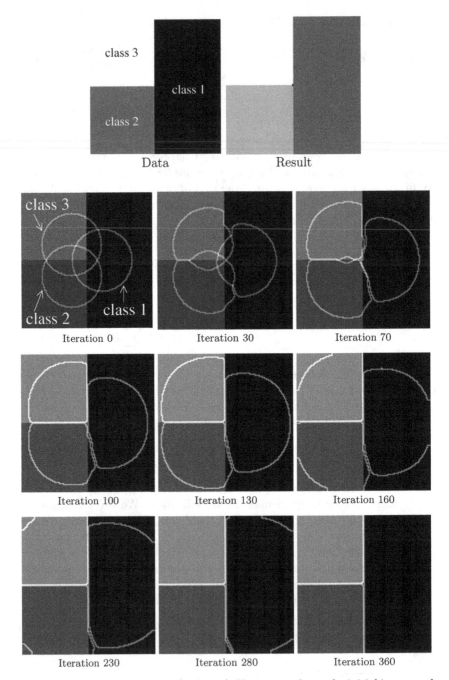

Figure 5.33. Synthetic example (3 classes). Upper row shows the initial image and the result. Iterations of the algorithm are displayed below. This simple example illustrates what is required in the regions: no overlap and full coverage.

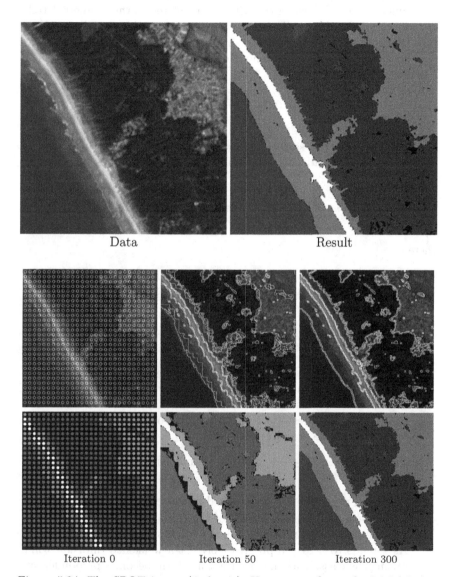

Figure 5.34. The SPOT image (4 classes). Upper row shows the initial image and the result. Iterations of the algorithm are displayed below with two representations: the boundaries of the actual regions, and below, regions are colored according to their class.

the corresponding signed distance functions ϕ_k on each w_p. Windows are not overlapping, and each of them is supporting one and only one function ϕ_k. The size of the windows is related to the smallest details we expect to detect. The major advantages of this initialization are that it is automatic (only the size of the windows has to be fixed), it accelerates the speed of convergence, and it is less sensitive to noise.

5.4.3 A Variational Model for Image Classification and Restoration

The objectives are the same as those described in Section 5.4.2, but in addition we want to add a restoration process. In [291] it is proposed to minimize

$$J_\varepsilon(u) = \int_\Omega (u(x) - u_0(x))^2 \, dx + \lambda^2 \varepsilon \int_\Omega \varphi(|\nabla u(x)|) \, dx + \frac{\eta^2}{\varepsilon} \int_\Omega W(u(x), \mu, \sigma) \, dx,$$

where:

- u_0 is the initial image (the data) and $u(x)$ the image we want to restore and segment into K homogeneous and disjoint classes.

- The function $\varphi(.)$ is a regularization function (with the same role and properties as those defined in Section 3.2 and used several times in this book).

- The last term W is a potential inducing a classification constraint. It takes into account the intensity and the parameters (μ_i, σ_i) of the classes.[10] W has as many minima as the number of classes. It has to attract the values of u toward the label of classes. We will return later to a precise description of W.

- The parameters $\lambda \geq 0$ and $\eta \geq 0$ permit us to adjust the weight of each term.

- $\varepsilon > 0$ is a parameter to be destined to tend to zero. During the first steps of convergence the weight of the third term in J_ε is negligible, and the restoration process (with the two first terms) is predominant. As ε tends to zero, we progressively get a weakened diffusion while raising the classification, since the third term becomes preponderant.

The form of the energy J_ε is borrowed from the van der Waals, Cahn–Hilliard theory of phase transitions in mechanics (see, for example, [4, 80, 155, 239, 310]). To better understand the model, let us recall some results about phase transitions.

---

[10]We recall that each class C_i is characterized by a Gaussian distribution $N(\mu_i, \sigma_i)$

Let us consider a mechanical system made up of two unstable components (or phases). These components may be liquids having different levels of density distribution. The problem is to describe stable configurations and to characterize the interface between the two phases while the system reaches stability. For the sake of clarity, let us consider a single fluid whose energy per unit of volume is a function W of the density distribution u. The function W is supposed to be nonnegative, having two minima in a and b such that $W(a) = W(b) = 0$. Moreover, it is assumed that W is quadratic around a and b and is growing at least linearly at infinity. W is known as a double-well potential. The stable configurations of the system are obtained by solving the following variational problem (see, for instance, [155, 239, 310]):

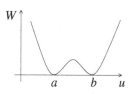

$$\mathcal{P}_\varepsilon \begin{cases} \inf_u E_\varepsilon(u) \text{ with} \\ E_\varepsilon(u) = \varepsilon \int_\Omega |\nabla u(x)|^2 \, dx + \frac{1}{\varepsilon} \int_\Omega W(u(x)) \, dx \\ \text{subject to the constraint } \int_\Omega u(x) \, dx = m, \end{cases}$$

where Ω is a bounded open subset of R^N (the region occupied by the fluid) and m is the total mass of the fluid.

Remark The introduction of the perturbation term permits us to solve the uniqueness problem for

$$\inf_u \left\{ \int_\Omega W(u(x)) \, dx; \int_\Omega u(x) \, dx = m \right\}.$$

∎

The asymptotic behavior of the model as ε tends to zero allows us to characterize stable configurations. This relies on the Γ-convergence theory.

Theorem 5.4.1 *If W satisfies the previously described conditions, then:*

(i) *E_ε Γ-converges to E_0 (for the L^1-strong topology) with*

$$E_0(u) = \begin{cases} K \operatorname{Per}_\Omega(R_1) & \text{if } u(x) \in \{a, b\} \text{ a.e.,} \\ +\infty & \text{otherwise,} \end{cases}$$

with $R_1 = \{x \in \Omega;\ u(x) = a\}$, K defined by

$$K = 2 \inf_{\gamma} \left\{ \int_{-1}^{1} \sqrt{W(\gamma(s))} |\gamma'(s)|\ ds;\ \gamma\ piecewise\ C^1, \right. \tag{5.129}$$
$$\left. \gamma(-1) = a,\ \gamma(1) = b \right\}$$

and $\mathrm{Per}_\Omega(R_1)$ stands for the perimeter of R_1,

$$\mathrm{Per}_\Omega(R_1) = \sup \left\{ \int_\Omega \chi_{R_1}(x)\mathrm{div}(\varphi)\ dx;\ \varphi \in C_0^1(\Omega)^N,\ |\varphi|_{L^\infty(\Omega)} \le 1 \right\},$$

which is the total variation of the characteristic function $\chi_{R_1}(x)$.

(ii) *If u_ε is a sequence of minimizers of E_ε such that u_ε converges to $\overline{u}$ in $L^1(\Omega)$, then $\overline{u}$ is a solution of the problem*

$$\inf_{u \in BV(\Omega)} \left\{ \mathrm{Per}_\Omega(x \in \Omega; u(x) = a);\ W(u(x)) = 0\ a.e.;\ \int_\Omega u(x)\ dx = m \right\}.$$

(iii) *Any sequence (v_ε) such that $E_\varepsilon(v_\varepsilon) \le cte < \infty$, $\forall \varepsilon$, admits a subsequence converging in $L^1(\Omega)$.*

We will give the proof of Theorem 5.4.1 (which is rather technical) at the end of this section.

As a consequence of this result, the action of the term W is quite clear: It forces the solution to take one of the two values a and b. Moreover, since the perimeter of the interface between the two phases $\{u(x) = a\}$ and $\{u(x) = b\}$ is minimal, it follows that this interface is not too irregular.

The transposition of the previous ideas in image analysis is straightforward. Let $u : \Omega \to R$ be a function that represents the intensity of each pixel and let us consider a feature criterion of classification, only based upon the distribution of intensity. Let us assume that the image is compounded of two regions $R_1 = \{x \in \Omega;\ u(x) = a\}$ and $R_2 = \{x \in \Omega;\ u(x) = b\}$. Let u_0 be the observed data corrupted by an additive white Gaussian noise. Following the previous discussion, let us define the energy

$$E_\varepsilon(u) = \int_\Omega (u(x) - u_0(x))^2\ dx + \int_\Omega \left[\varepsilon |\nabla u|^2 + \frac{1}{\varepsilon} W(u(x)) \right]\ dx$$

with W as before. According to Theorem 5.4.1, as $\varepsilon \to 0$ there exists a subsequence of minimizers u_ε of E_ε converging to a smooth segmented image,[11] closed to u_0, and whose

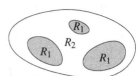

---

[11]The introduction of the additional term $\int_\Omega (u(x) - u_0(x))^2\ dx$ has no effect on the conclusions of Theorem 5.4.1.

pixels belong to R_1 or R_2 a.e. (these two regions being separated by sharp regularized,
i.e., minimal edges). When $\varepsilon > 0$ is fixed (not
too small), the second term in E_ε induces an isotropic smoothing. As ε
decreases, the third term in E_ε forces the solution to choose the two characterizing regions R_1 and R_2. The two values a and b are the labels of R_1
and R_2.

Until now, we have presented the model with two phases (or labels). The
extension to multiple wells (and even vectorial wells) exists, but it is not
quite obvious from a mathematical point of view. The most complete result
is due to Baldo:

Theorem 5.4.2 [33] *Let $\Omega \subset R^N$, $N \geq 2$, be open and bounded with
Lipschitz boundary. Let $W : R_+^n = \{u = (u_1, \ldots, u_n),\ u_i > 0\} \to [0, +\infty[$
be a continuous function such that $W(u) = 0 \iff u = \mu_1, \mu_2, \ldots, \mu_k$,
$\mu_i \in R_+^n$. We also assume the technical hypothesis $\exists \alpha_1, \alpha_2,\ 0 < \alpha_1 < \alpha_2$,
such that $W(u) \geq \sup \{W(v);\ v \in [\alpha_1, \alpha_2]^n\}$ for every $u \notin [\alpha_1, \alpha_2]^n$. For
$u_0 \in L^2(\Omega)^n$, let us define the functional*

$$E_\varepsilon(u) = \int_\Omega |u(x) - u_0(x)|_{R^n}^2\ dx + \int_\Omega \left[\varepsilon |\nabla u|^2 + \frac{1}{\varepsilon} W(u(x))\right]\ dx.$$

*For each ε, let u_ε be a solution of $\inf_{u \in W^{1,2}(\Omega)^n} E_\varepsilon(u)$. Let us suppose that
$u_\varepsilon \to u \in L^2(\Omega)$. Then*

$$u(x) = \sum_{i=1}^k \mu_i\, \chi_{\Omega_i}(x),$$

where $\Omega_1, \Omega_2, \ldots, \Omega_k$ is a partition of Ω that minimizes the energy:

$$\sum_{i=1}^k \int_{\Omega_i} |u(x) - u_0(x)|_{R^n}^2\ dx + \sum_{i=1}^k d(\mu_i, \mu_j) \mathcal{H}^{N-1}(\partial\Omega_i \cap \partial\Omega_j)$$

with

$$d(\mu_i, \mu_j) = \inf_g \left\{\int_0^1 \sqrt{W(g(s))} |g'(s)|;\ g \in C^1(0,1)^n, g(0) = \mu_i, g(1) = \mu_j\right\}.$$

Baldo's result allows us to solve classification problems when the number
of labels is greater than two. The discriminating function $W(u)$ is such
that the label of a class Ω_i is the corresponding mean μ_i, i.e., $\Omega_i = \{x \in
\Omega;\ u(x) = \mu_i\}$, $i = 1, \ldots, k$. So, W necessarily satisfies $W(\mu_i) = 0$, $i =
1, \ldots, k$, and $W(v) > 0$ for $v \neq \mu_i$, $i = 1, \ldots, k$. There exist several ways
of constructing such a potential W. A piecewise quadratic potential can be
used (see [291] for more details).

As far as the regularization is concerned, it has been shown previously that the quadratic smoothing term is too strong (see Section 3.2.2). We can then propose a modified version of the Baldo functional:

$$E_\varepsilon(u) = \int_\Omega (u(x) - u_0(x))^2 \, dx + \int_\Omega \left[\lambda \, \varepsilon \, \varphi(|\nabla u|) + \frac{\eta}{\varepsilon} W(u(x)) \right] dx,$$

with $\varphi(.)$ to be chosen. Let us comment on some experiments:[12]

- To show the interest in changing the smoothing term, the original Baldo functional (i.e., $\varphi(s) = s^2$) has been compared with the following regularization functions: $\varphi(s) = \log \cosh(s)$ and $\varphi(s) = s^2/(1+s^2)$ (see Figure 5.35). Of course, for the two last functions $\varphi(.)$, no mathematical result of convergence exists. Results are purely experimental, and numerical algorithms are based on half-quadratic minimization, as described in Section 3.2.4. It can be observed that more points are misclassified using $\varphi(s) = s^2$, which is due to the oversmoothing of the image.

- The convergence as $\varepsilon \to 0$ is illustrated in the synthetic example from Figure 5.36. For various ε, the minimizer and its associated dual variable are displayed. This example permits us to observe the transition between the restoration process and the classification. As for the decay of ε, it is usually chosen as $\varepsilon^n = \varepsilon_0^n$, where n is the iteration number in ε, and ε_0 may depend on the amount of noise in the sequence. Typical values lie in the interval 0.9 to 0.98 (strong noise). We refer to [289] for more details.

- Finally, we show in Figure 5.37 the results obtained on the SPOT image, already processed using the previous approach (see Figure 5.34). It can be noticed that this approach permits us to retain smaller details.

We close this section by showing how asymptotic results (as ε tends to zero) can be rigorously established via the Γ-convergence theory. We do not prove the general case given in Theorem 5.4.2, but we examine only the two-phases case and the Γ-convergence part (i) of Theorem 5.4.1. Let $W : R \to R$ satisfying the following properties:

$$W \in C^2(R), \ W \geq 0. \tag{5.130}$$

W has exactly two roots, which we label a and b $(a < b)$. We suppose $W'(a) = W'(b) = 0, W''(a) > 0, W''(b) > 0.$ (5.131)

There exist positive constants c_1, c_2, and m and an integer $p \geq 2$ such that $c_1|u|^p \leq W(u) \leq c_2|u|^p$ for $|u| \geq m.$ (5.132)

---

[12]Results provided by the INRIA project ARIANA (joint project CNRS–INRIA–UNSA): http://www.inria.fr/ariana/.

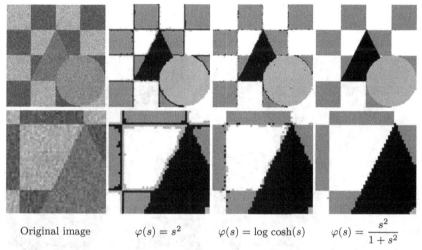

Original image $\varphi(s) = s^2$ $\varphi(s) = \log \cosh(s)$ $\varphi(s) = \dfrac{s^2}{1+s^2}$

Figure 5.35. Influence of the function $\varphi(.)$. The lower line shows a close-up.

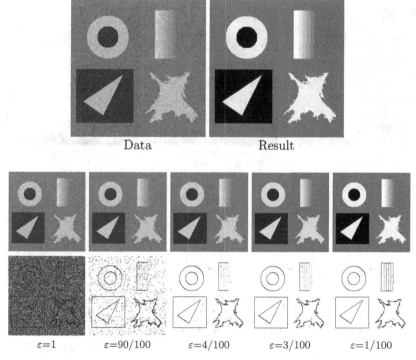

Figure 5.36. Illustration of the convergence as $\varepsilon \to 0$. In this example 6 classes are searched even if a continuous region is present in the image. From $\varepsilon = 1$ to $\varepsilon = 4/100$ the restoration acts, whereas from $\varepsilon = 3/100$ to the end the classification seems predominant (regions appear). This is especially visible on the edge indicator functions (the dual variable) in the lower row.

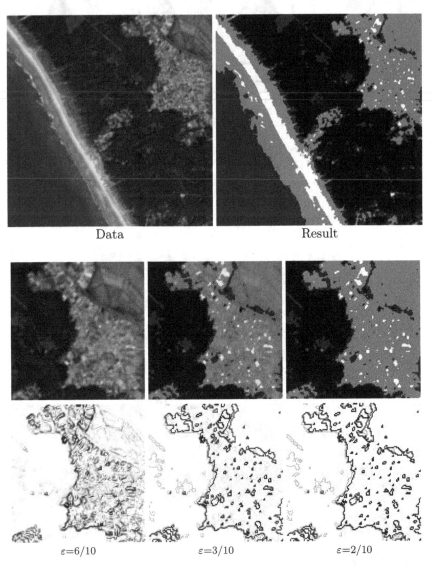

Data   Result

$\varepsilon=6/10$ $\varepsilon=3/10$ $\varepsilon=2/10$

Figure 5.37. The SPOT image (4 classes). Upper row shows the initial image and the result. Iterations of the algorithm as ε tends to zero are displayed below with the associated dual variables. A close-up of the upper left corner is shown. Similar behavior as in Figure 5.36 may be observed.

Now let us define the functionals

$$
E_\varepsilon(u) = \begin{cases} \dfrac{1}{\varepsilon}\displaystyle\int_\Omega W(u)\,dx + \varepsilon\displaystyle\int_\Omega |\nabla u|^2\,dx & \text{if } u \in H^1(\Omega), \\ +\infty & \text{otherwise,} \end{cases}
$$

and

$$
E_0(u) = \begin{cases} K\,\mathrm{Per}_\Omega(\{u = a\}), & u \in BV(\Omega),\ W(u(x)) = 0 \text{ a.e.}, \\ +\infty & \text{otherwise,} \end{cases}
$$

where K is defined by (5.129). To prove that E_ε Γ-converges to E_0, we need to define an auxiliary function $g(.)$, which will play a crucial role afterwards:

$$
g(u) = \inf_{\substack{\gamma(-1)\,=\,a \\ \gamma(1)\,=\,u}} \int_{-1}^{1} \sqrt{W(\gamma(s))}\,|\gamma'(s)|\,ds,
$$

where the infimum is taken over functions $\gamma(.)$ that are Lipschitz continuous. We summarize the properties of $g(.)$ in the following lemma:

Lemma 5.4.2 *For every $u \in R$ there exists a function $\gamma_u : [-1,1] \to R$ such that $\gamma(-1) = a$, $\gamma(1) = u$, and*

$$
g(u) = \int_{-1}^{1} \sqrt{W(\gamma_u(s))}\,|\gamma_u'(s)|\,ds. \tag{5.133}
$$

The function g is Lipschitz continuous and satisfies

$$
|g'(u)| = \sqrt{W(u)} \text{ for a.e. } u. \tag{5.134}
$$

There exists a smooth increasing function $\beta :\,]-\infty,+\infty[\,\to\,]-1,1[$ such that the function $\xi(\tau) = \gamma_b(\beta(\tau))$ satisfies

$$
2g(b) = \int_{-\infty}^{+\infty} \left[W(\xi(\tau)) + |\xi'(\tau)|^2 \right]\,d\tau, \tag{5.135}
$$

$$
\lim_{\tau\to-\infty} \xi(\tau) = a, \quad \lim_{\tau\to+\infty} \xi(\tau) = b, \tag{5.136}
$$

with these limits being attained at an exponential rate.

We refer to Sternberg [309, 310] for the proof.

Proof of (i) of Theorem 5.4.1. E_ε Γ-converges to E_0 for the L^1-strong topology. We follow the proof of Sternberg [309]. According to the definition of Γ-convergence, we have to show the following:

- If $v_\varepsilon \to v_0$ as $\varepsilon \to 0$, then

$$
\lim_{\varepsilon\to0} E_\varepsilon(v_\varepsilon) \geq E_0(v_0). \tag{5.137}
$$

- For any $v_0 \in L^1(\Omega)$, there exists a sequence ρ_ε with $\rho_\varepsilon \to v_0$ in $L^1(\Omega)$ and

$$\overline{\lim_{\varepsilon \to 0}} \, E_\varepsilon(v_\varepsilon) \leq E_0(v_0). \qquad (5.138)$$

Step 1: Proof of (5.137). One need only consider v_0 of the form

$$v_0(x) = \begin{cases} a & \text{if } x \in A, \\ b & \text{if } x \in B, \end{cases} \qquad (5.139)$$

for two disjoint sets A and B with $A \cup B = \Omega$. Otherwise, $v_\varepsilon \to v_0$ in $L^1(\Omega)$ implies $E_\varepsilon(v_\varepsilon) \to +\infty$ (and (5.137) is obvious). So, let us suppose that $v_\varepsilon \to v_0$ in $L^1(\Omega)$ and define $h_\varepsilon(x) = g(v_\varepsilon(x))$. It follows from Lemma 5.4.2 that

$$|\nabla h_\varepsilon(x)| = |\nabla v_\varepsilon(x)| \sqrt{W(v_\varepsilon(x))}. \qquad (5.140)$$

But it is clear that

$$h_\varepsilon \xrightarrow[L^1(\Omega)]{} g(v_0) = \begin{cases} 0 & \text{if } x \in A, \\ g(b) & \text{if } x \in B. \end{cases}$$

Then, from the inequality $a^2 + b^2 \geq 2ab$, (5.140), and the lower semicontinuity of the $BV(\Omega)$-norm for the L^1-strong convergence, we obtain

$$\underline{\lim_{\varepsilon \to 0}} \, E_\varepsilon(v_\varepsilon) \geq \underline{\lim_{\varepsilon \to 0}} \, 2 \int_\Omega \sqrt{W(v_\varepsilon)} |\nabla v_\varepsilon| \, dx = \underline{\lim_{\varepsilon \to 0}} \, 2 \int_\Omega |\nabla h_\varepsilon(x)| \, dx \geq 2 \int_\Omega |Dg(v_0)|.$$

But an easy computation gives

$$\int_\Omega |Dg(v_0)| = g(b) \, \mathrm{Per}_\Omega(\{x; \, v_0(x) = a\}).$$

Thus, we get the desired inequality (5.137):

$$\underline{\lim_{\varepsilon \to 0}} \, E_\varepsilon(v_\varepsilon) \geq K \, \mathrm{Per}_\Omega(\{x; \, v_0(x) = a\})$$

with $K = 2g(b)$.

Step 2: Proof of (5.138). One may assume $v_0 \in BV(\Omega)$ and again take v_0 of the form (5.139), or otherwise the trivial choice $\rho_\varepsilon \equiv v_0$ for each ε is suitable. Let $\Gamma = \partial A \cup \partial B$ and assume $\Gamma \in C^2$ without loss of generality (since one can always approximate a set of finite perimeter by a sequence of sets having smooth boundary [164]). Then, let us define the signed distance function $D : \Omega \to R$ by

$$d(x) = \begin{cases} -\mathrm{dist}(x, \Gamma) & \text{if } x \in A, \\ +\mathrm{dist}(x, \Gamma) & \text{if } x \in B. \end{cases}$$

Near Γ, $d(.)$ is smooth and satisfies

$$|\nabla d| = 1, \quad \lim_{s \to 0} \mathcal{H}^{N-1}\{x;\ d(x) = s\} = \mathcal{H}^{N-1}(\Gamma) = \mathrm{Per}_\Omega A. \qquad (5.141)$$

Finally, let us define the sequence $\rho_\varepsilon(x)$ by

$$\rho_\varepsilon(x) = \begin{cases} \xi\left(\dfrac{-1}{\sqrt{\varepsilon}}\right) & \text{if } d(x) < -\sqrt{\varepsilon}, \\[2ex] \xi\left(\dfrac{d(x)}{\varepsilon}\right) & \text{if } |d(x)| \le \sqrt{\varepsilon}, \\[2ex] \xi\left(\dfrac{1}{\sqrt{\varepsilon}}\right) & \text{if } d(x) \ge \sqrt{\varepsilon}, \end{cases}$$

where the function $\xi(.)$ is the one defined in Lemma 5.4.2. The L^1 convergence of ρ_ε to v_0 follows directly from (5.135). Then, thanks to (5.135)–(5.136), (5.141), and the coarea formula, we get

$$\overline{\lim_{\varepsilon \to 0}}\, E_\varepsilon(\rho_\varepsilon) = \overline{\lim_{\varepsilon \to 0}}\, \frac{1}{\varepsilon} \int\limits_{\{|d(x)| \le \sqrt{\varepsilon}\}} W\left(\xi\left(\frac{d(x)}{\varepsilon}\right)\right) + \left|\xi'\left(\frac{d(x)}{\varepsilon}\right)\right|^2 dx$$

$$= \overline{\lim_{\varepsilon \to 0}}\, \frac{1}{\varepsilon} \int_{-\sqrt{\varepsilon}}^{\sqrt{\varepsilon}} W\left(\xi\left(\frac{s}{\varepsilon}\right)\right) + \left|\xi'\left(\frac{s}{\varepsilon}\right)\right|^2 \mathcal{H}^{N-1}\{x;\ d(x) = s\}\, ds$$

$$= \overline{\lim_{\varepsilon \to}}\, \int_{-1/\sqrt{\varepsilon}}^{1/\sqrt{\varepsilon}} W\left(\xi(\tau)\right) + |\xi'(\tau)|^2 \mathcal{H}^{N-1}\{x;\ d(x) = \varepsilon\tau\}\, d\tau$$

$$\le 2g(b)\left(\overline{\lim_{\varepsilon \to 0}}\, \max_{|s| \le \sqrt{\varepsilon}} \mathcal{H}^{N-1}\{x;\ d(x) = s\}\right) = E_0(v_0),$$

i.e., $\overline{\lim_{\varepsilon \to 0}}\, E_\varepsilon(\rho_\varepsilon) \le E_0(v_0)$, and (5.138) is proven. ∎

Remark Parts (ii) and (iii) of Theorem 5.4.1 are not difficult to prove. We refer to Sternberg [309] for the complete proof. ∎

5.5 Vector-Valued Images

5.5.1 Introduction

In this book, we have mainly considered gray-scale images, i.e., scalar images. This section focuses on vector-valued images. A typical example is color images. A color image can be described by the three components red, green, and blue (RGB). Many other color spaces have been proposed to represent a color. One important motivation is to have a color space with a notion of distance between two colors that reflects human perception. Most of these spaces are built by linear combinations of the RGB components: some examples of color spaces are YUV, YCrCb, XYZ. We refer the reader to [276], which is a classical overview of color spaces.

In fact, most algorithms and results given for scalar images can be extended to vector-valued images. However, extending an algorithm designed for scalar images is not generally as simple as applying the same algorithm to each component independently. For example, for color images, the difficulty in defining a color space reveals that the RGB components are linked and cannot be manipulated independently. So the problem is to take into account the correlation between the image components. Naturally, the answer will strongly depend on the application (restoration, segmentation, inpainting, etc.) and on the type of vector-valued images (2-D vector field, color image, 3-D tensor field, etc).

This section reviews the elegant and unified presentation by Tschumperlé and Deriche [323, 324], which is well adapted for restoration tasks. A vector-valued image will be denoted by

$$u : \Omega \mapsto R^p,$$

$$x \to (u_1, u_2, \ldots, u_p).$$

5.5.2   An Extended Notion of Gradient

For scalar images, the gradient indicates the direction of the most important variation of the intensity. In the vector-valued case it is clearly difficult to establish in which direction the most important variations occur: each component gives different information (see Figure 5.38). So one needs to define a notion to replace the standard gradient.

A possible way to study how a vector-valued image varies is to introduce the structure tensor [322, 336, 347]

$$S(u) = \sum_{i=1}^{p} \nabla u_i \nabla u_i^T, \tag{5.142}$$

where each ∇u_i corresponds to the gradient of the ith canal. The structure tensor S is particularly interesting since its eigenvalues (λ_+, λ_-) define the local min/max vector-valued variations of u in the eigenvectors' directions θ_+ and θ_- respectively.

The spectral elements of S can be then used in a variational approach or a PDE. This is illustrated in Sections 5.5.3 and 5.5.4, on the restoration problem.

5.5.3   The Energy Method

As in for the scalar restoration problem (see Section 3.2), we define a functional to be minimized involving two terms: a data fidelity term and a regularizing term. The energy is

$$E(u) = \int_{\Omega} |u - u_0|^2 \, dx + \mu \int_{\Omega} \phi(\nabla u) \, dx, \tag{5.143}$$

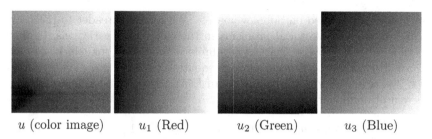

u (color image) u_1 (Red) u_2 (Green) u_3 (Blue)

Figure 5.38. Color image and its three components, namely red, green and blue channels. This figure contains color: it is available on the book web site http://www-sop.inria.fr/books/imath.

where u_0 is the initial image to be reconstructed and μ is a weighting parameter. The real function $\phi(\nabla u)$ is related to the local variations of the image through the eigenvalues (λ_+, λ_-) of the structure tensor S:

$$\phi(\nabla u) = \psi(\lambda_+, \lambda_-).$$

For example, if we choose

$$\psi(\lambda_+, \lambda_-) = \lambda_+ + \lambda_-,$$

we have a term comparable with a Tikhonov regularization [317]. If we want to recover a smoothing analogous to the total variation, we can choose [61, 314]:

$$\psi(\lambda_+, \lambda_-) = \sqrt{\lambda_+ + \lambda_-}. \tag{5.144}$$

Of course, other choices are possible, such as [294, 292]

$$\psi(\lambda_+, \lambda_-) = \sqrt{\lambda_+ - \lambda_-} \text{ or } \psi(\lambda_+, \lambda_-) = \sqrt{\lambda_+}. \tag{5.145}$$

Note that if u is a scalar image, the choices (5.144) and (5.145) coincide with $\phi(\nabla u) = |\nabla u|$.

Another possibility is to consider the regularizing term as a real function of (λ_+, λ_-) of the form

$$\psi(\lambda_+, \lambda_-) = \lambda_- + \lambda_+ g(\lambda_+, \lambda_-),$$

where g can be determined as follows:

- In homogeneous regions, where $\lambda_+ \approx \lambda_-$, we want a large smoothing, so we impose that $\lim_{(\lambda_+ + \lambda_-) \to 0} g(\lambda_+, \lambda_-) = 1$.

- Near edges, one wants to perform a smoothing along the direction θ_-, so we may impose $g(\lambda_+, \lambda_-) \approx 0$ when $\lambda_+ \gg \lambda_-$.

However, it is in general much more difficult to prove the existence of a minimizer in the vector-valued case than in the scalar case. For functionals such as (5.143), the notion of convexity is not necessary in order to get the existence of a minimizer. It is replaced by a weaker notion called quasi-convexity, which implies in many situations the weak lower semicontinuity of functionals defined on Sobolev spaces. We refer the reader to [121] for a deeper study of vectorial problems in the calculus of variations. In this section we will suppose that $E(u)$ admits a minimizer for which we can (formally) derive the Euler–Lagrange equation.

If ψ is regular and u is a minimizer of $E(u)$, then u satisfies the following system of nonlinear PDEs:

$$\frac{\partial u_i}{\partial t} = 2(u_i - u_{0i}) - \mu \, \mathrm{div}(D\nabla u_i), \quad \text{for } i = 1, \ldots, p, \qquad (5.146)$$

where D is a 2×2 diffusion tensor defined as

$$D = \frac{\partial \psi}{\partial \lambda_+} \theta_+ \theta_+^T + \frac{\partial \psi}{\partial \lambda_-} \theta_- \theta_-^T. \qquad (5.147)$$

The formal proof of (5.146) is not difficult but it is rather technical since we have to compute the derivative of (λ_+, λ_-) with respect to u_i (see [323, 324] for a detailed proof). Equation (5.146) is a divergence-based equation, in which the diffusion tensor is defined by the partial derivative of ψ and by the spectral elements of the structure tensor S.

So in this case, PDE (5.146) comes from a variational principle, but of course, one may also propose some PDEs that are derived from an energy. This is the object of the following section.

5.5.4   PDE-Based Methods

With the variational formulation (5.143), we have obtained a diffusion operator of the kind

$$\mathrm{div}(D\nabla u_i), \qquad (5.148)$$

where D was defined by (5.147). We may wonder how to design more general diffusion operators. If we follow the reasoning of the scalar case, it is natural to consider the same kind of diffusion operator (5.148), with another tensor D: intuitively, D should correspond to the local geometry of the diffusion.

Indeed, starting from (5.148) is not the right choice. A striking illustration is that a similar diffusion term (5.148) can be obtained with two qualitatively very different tensors D: if we choose

$$D_1 = \frac{\mathrm{Id}}{|\nabla u|}, \quad D_2 = \frac{\nabla u_i \nabla u_i^T}{|\nabla u|^2},$$

so that D_1 is isotropic and D_2 is purely anisotropic (only one eigenvalue is nonzero), then it is easy to verify that

$$\mathrm{div}(D_1 \nabla u_i) = \mathrm{div}(D_2 \nabla u_i) = \mathrm{div}\left(\frac{\nabla u_i}{|\nabla u_i|}\right).$$

This demonstrates that (5.148) is not the proper choice since it does not take into account the local geometry of the diffusion given by D.

A possible way to try to interpret the diffusion term (5.148) is to decompose it. After some computation, one gets

$$\text{div}(D\nabla u_i) = \text{tr}(DH_i) + \nabla u_i^T \text{Div}(D), \tag{5.149}$$

where $\text{tr}(A)$ denotes the usual sum of the diagonal elements of a matrix A, H_i is the Hessian matrix of the ith component u_i of the vectorial image u, and $\text{Div}(D)$ is defined as a divergence operator acting on matrices and returning vectors. If $D = (d_{i,j})_{i,j=1,2}$ then

$$\text{Div}(D) = \left(\frac{\partial d_{11}}{\partial x_1} + \frac{\partial d_{12}}{\partial x_2}, \frac{\partial d_{21}}{\partial x_1} + \frac{\partial d_{22}}{\partial x_2} \right)^T.$$

Interestingly, the role of the first term in equation (5.149) is clearly related to a smoothing process: it is a weighted sum of second derivatives of u_i. However, the interpretation of the second term in (5.149) is less clear: it involves only first-order derivatives of u_i which resemble transport terms.

For these two reasons, Tschumperlé and Deriche [323, 324] claim that the trace term is the most relevant, and they proposed the following generic vector-valued PDE:

$$\frac{\partial u_i}{\partial t} = \sum_{j=1}^{p} \text{tr}\left(A^{ij} H^i \right) \quad \text{for} \quad i = 1, \ldots, p, \tag{5.150}$$

where the $\left(A^{ij} \right)$ are 2×2 matrices to be specified and the $\left(H^i \right)$ denote the Hessian matrix of u_i. Note that there is no data fidelity term here, since we are focusing only on the regularizing process.

Figure 5.39 shows examples of fixed spatially varying tensor fields T, and the corresponding evolution on a color image: equation (5.150) is applied with $A^{ii} = T$ and $A^{ij} = 0$ for $i \neq j$. One observes the correspondance between local geometry of the diffusion and the diffused image.

Now, for a denoising application, the question is how to define A^{ij} with respect to the structure tensor S. Similarily to the scalar case, some qualitative properties are proposed [324]:

- One would like to avoid mixing diffusion contributions between image components. The coupling should appear only through S. A solution is to consider only one matrix A and define $A^{ij} = \delta_{ij} A$, where A is defined through S (δ_{ij} is the Kronecker symbol).

- In homogeneous regions, where vector variations are small, one should perform an isotropic smoothing. Thus it is imposed that

$$\lim_{\lambda_+ + \lambda_- \to 0} A = \alpha \, \text{Id},$$

where α is a positive constant.

- Near edges one wants to perform a smoothing only along the vector edges θ_-; thus we impose $A = \beta \theta_- \theta_-^T$ when $\lambda_+ \gg \lambda_-$ and λ_+ large. Here β is a decreasing function with respect to $\lambda_+ + \lambda_-$.

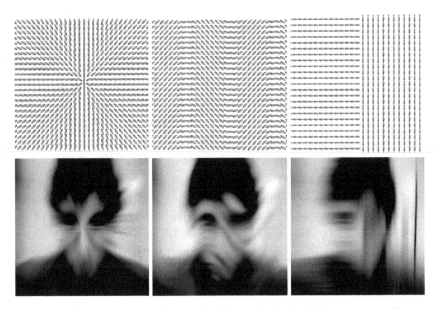

Figure 5.39. Trace-based diffusion (5.150) with fixed diffusion tensors (upper row), from [324]. This figure contains color: please visit the book web site http://www-sop.inria.fr/books/imath for the colored version.

There is a wide range of image processing applications in which the previous model can be applied. Figures 5.40–5.42 illustrate some results of color image restoration, image inpainting, and flow visualization. We refer the reader to [323, 324] for additional applications of this framework.

C++

original image   restored image

Figure 5.40. Color image restoration [324, 320, 321]. This figure contains color: please visit the book web site `http://www-sop.inria.fr/books/imath` for the colored version.

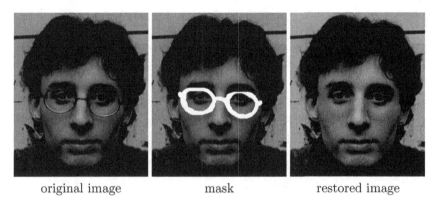

original image   mask   restored image

Figure 5.41. Color image inpainting [324] (see also Section 5.1). This figure contains color: please visit the book web site `http://www-sop.inria.fr/books/imath` for the colored version.

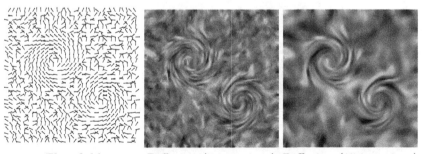

Flow field   Diffusion (5 iterations) Diffusion (15 iterations)

Figure 5.42. Given a flow field (see the left hand side image), the idea is two diffuse a color image initialized by pure noise, following the directions of this flow. Results are shown after five and fifteen iterations [324]. This figure contains color: please visit the book web site `http://www-sop.inria.fr/books/imath` for the colored version.

Appendix A

Introduction to Finite Difference Methods

How to Read This Chapter

This chapter concerns the problem of solving numerically the partial differential equations that we have encountered in this book. Although several kinds of approaches can be considered (like finite elements or spectral methods), the success of finite differences in image analysis is due to the structure of digital images for which we can associate a natural regular grid. This chapter is an introduction to the main notions that are commonly used when one wants to solve a partial differential equation. From Section A.1 to Section A.2 we will consider only the one-dimensional case and focus on the main ideas of finite differences. Section A.3 will be more applied: It will concern the discretization of certain approaches detailed in this book. More precisely:

- Section A.1 introduces the main definitions and theoretical considerations about finite-difference schemes (convergence, consistency, and stability, Lax theorem). Every notion is illustrated by developing explicit calculations for the case of the one-dimensional heat equation. Besides the precise definitions, this will help the reader to understand them in a simple situation.

- Section A.2 concerns hyperbolic equations. We start with the linear case and show that if we do not choose an upwind scheme, then the scheme is always unstable. We then investigate the nonlinear case by focusing on the Burgers equation.

- The purpose of Section A.3 is to show how finite-difference schemes can be used in image analysis. We first introduce in Section A.3.1 the main notation and consider the 2-D heat equation. The remainder of Section A.3 is concerned with the discretization of certain PDEs studied in this book:
  - Restoration by energy minimization (Section A.3.2): We detail the discretization of the divergence term, which can also be found for the Perona and Malik equation.
  - Enhancement by Osher and Rudin's shock filters (Section A.3.3): The main interest is to use a flux limiter called minmod.
  - Curve evolution with level sets and especially segmentation with geodesic active contours (Section A.3.4). For the sake of simplicity, we examine separately each term of the model (mean curvature motion, constant speed motion, advection equation). We essentially write their discretization and give some experimental results.

☞   For general presentations of finite difference methods: [316, 112, 182],

☞   To know more about hyperbolic equations: [166, 212].

A.1 Definitions and Theoretical Considerations Illustrated by the 1-D Parabolic Heat Equation

A.1.1 Getting Started

There are many approaches that are used for discretizing a partial differential equation. Among the most important ones we can mention finite differences, finite elements, and spectral methods.

We focus here on finite differences, which are widely used in image processing. This is due to the structure of a digital image as a set of uniformly distributed pixels (see Section A.3).

To present the main ideas we will consider the following well-posed initial-value problem, written in the one dimensional case:[1]

$$\begin{cases} \mathcal{L}v = F, & t > 0, \ x \in R, \\ v(0, x) = f(x), & x \in R, \end{cases} \tag{A.1}$$

---

[1] This is an initial-value problem, which means that there is no boundary condition. For initial-boundary-value problems, the discussion that follows needs to be slightly adapted, and we refer to [316] for more details.

where v and F are defined on R, and $\mathcal{L}$ is a differential *linear* operator. The function v denotes the exact solution of (A.1).

Example: One of the easiest equations that we may consider is the 1-D heat equation:

$$\frac{\partial v}{\partial t} = \nu \frac{\partial^2 v}{\partial x^2}, \quad t > 0, \quad x \in R, \tag{A.2}$$

where $\nu > 0$ is a constant, which is equivalent to

$$\mathcal{L}v = 0 \quad \text{with} \quad \mathcal{L}v = \frac{\partial v}{\partial t} - \nu \frac{\partial^2 v}{\partial x^2}.$$

The initial condition is $v(0, x) = f(x)$. From now on, we shall use this equation to illustrate the different notions to be defined. ■

Our aim is to solve the PDE (A.1) numerically. We begin by discretizing the spatial domain by placing a grid over the domain. For convenience, we will use a uniform grid, with grid spacing $\triangle x$. Likewise, the temporal domain can be discretized, and we denote by $\triangle t$ the temporal grid spacing. The resulting grid in the time–space domain is illustrated in Figure A.1.

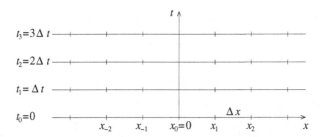

Figure A.1. Grid on the time–space domain.

Solving the problem numerically means finding a discrete function u defined at the points $(n\triangle t, i\triangle x)$ (we will denote by u_i^n the value of u at these points), which is a "good approximation" of v. The function u will be obtained as the solution of a discrete equation, which will be an approximation of (A.1):

$$\begin{cases} L_i^n u_i^n = G_i^n, & i = -\infty, \ldots, +\infty, \\ u_i^0 = f(i\triangle x), \end{cases} \tag{A.3}$$

where L_i^n (respectively G_i^n) corresponds to the discrete approximation of $\mathcal{L}$ (respectively F).[2] Notice that both spatial and temporal derivatives have to be approximated.

Example: Let us show on the 1-D heat equation (A.2) how the discrete equation can

---

[2] Because of discretization, G_i^n is a priori different from F_i^n, which is simply the value of F in $(n\triangle t, i\triangle x)$

be obtained. In fact, the starting point for writing any finite difference scheme is Taylor expansions. For Δt and Δx small, we have

$$v((n+1)\Delta t, i\Delta x) = \left(v + \Delta t \frac{\partial v}{\partial t} + \frac{\Delta t^2}{2} \frac{\partial^2 v}{\partial t^2} \right)(n\Delta t, i\Delta x) + \mathcal{O}(\Delta t^3), \qquad \text{(A.4)}$$

$$v(n\Delta t, (i+1)\Delta x) = \left(v + \Delta x \frac{\partial v}{\partial x} + \frac{\Delta x^2}{2} \frac{\partial^2 v}{\partial x^2} \right)(n\Delta t, i\Delta x) + \mathcal{O}(\Delta x^3), \qquad \text{(A.5)}$$

$$v(n\Delta t, (i-1)\Delta x) = \left(v - \Delta x \frac{\partial v}{\partial x} + \frac{\Delta x^2}{2} \frac{\partial^2 v}{\partial x^2} \right)(n\Delta t, i\Delta x) + \mathcal{O}(\Delta x^3). \qquad \text{(A.6)}$$

We recall that $g = \mathcal{O}(\phi(s))$ for $s \in S$ if there exists a constant C such that $|f(s)| \leq C|\phi(s)|$ for all $s \in S$. We say that $g(s)$ is "big $\mathcal{O}$" of $\phi(s)$ or that $g(s)$ is of order $\phi(s)$. In the previous Taylor expansions, notice that the constant C naturally depends on the high-order derivatives of v.

By Equation (A.4), we have

$$\frac{\partial v}{\partial t}(n\Delta t, i\Delta x) = \frac{v_i^{n+1} - v_i^n}{\Delta t} + \mathcal{O}(\Delta t),$$

where we have set $v_i^n = v(n\Delta t, i\Delta x)$. Similarly, by using previous Taylor expansions, we may propose an approximation of the second spatial derivative. By adding (A.5) and (A.6), we have

$$\frac{\partial^2 v}{\partial x^2}(n\Delta t, i\Delta x) = \frac{v_{i+1}^n - 2v_i^n + v_{i-1}^n}{\Delta x^2} + \mathcal{O}(\Delta x^2). \qquad \text{(A.7)}$$

Consequently, if we consider the differential operator $\mathcal{L}$ from (A.2), we have

$$\frac{\partial v}{\partial t}(n\Delta t, i\Delta x) - \nu \frac{\partial^2 v}{\partial x^2}(n\Delta t, i\Delta x) = \frac{v_i^{n+1} - v_i^n}{\Delta t} - \nu \frac{v_{i+1}^n - 2v_i^n + v_{i-1}^n}{\Delta x^2} + \mathcal{O}(\Delta t) + \mathcal{O}(\Delta x^2). \qquad \text{(A.8)}$$

So a reasonable approximation of equation (A.2) is

$$L_i^n u = 0 \quad \text{with} \quad L_i^n = \frac{u_i^{n+1} - u_i^n}{\Delta t} - \nu \frac{u_{i+1}^n - 2u_i^n + u_{i-1}^n}{\Delta x^2}. \qquad \text{(A.9)}$$

This difference equation (A.9) can also be rewritten in the following form:

$$u_i^{n+1} = (1 - 2r)u_i^n + r\left(u_{i+1}^n + u_{i-1}^n \right), \quad \text{where} \quad r = \nu \Delta t / \Delta x^2. \qquad \text{(A.10)}$$

This shows clearly that this scheme is explicit, which means that the values at time $(n+1)\Delta t$ are obtained only from the values at time $n\Delta t$. We will mention how to write implicit schemes at the end of this section. $\blacksquare$

☛ *It is important to realize that the discretized equation replaces the original equation by a new one, and that an exact solution of the discretized problem will lead to an approximate solution of the original PDE, since we introduce a discretization error (the error of replacing a continuous equation by a discrete one).*

For a given approximation (A.3), we would like to know precisely the relations between the discrete equation with the PDE and their respective solutions. In other words, what does it mean that u is an approximation of v, and can we quantify it? Are there any conditions on the grid size (Δt, Δx) to yield a "good" approximation? To answer these questions we define precisely the notions of convergence, consistency, and stability in the next sections.

A.1.2 Convergence

The first notion that is essential is to understand what it means that the discrete solution u of (A.3) is an approximation of, or converges to, the solution v of (A.1). To be more precise, we can define pointwise convergence as follows:

Definition A.1.1 (pointwise convergent scheme) *The scheme* (A.3) *approximating the partial differential equation* (A.1) *is pointwise convergent if for any x and t, as $((n+1)\triangle t, i\triangle x)$ converges to (t, x), then u_i^n converges to $v(t, x)$ as $\triangle x$ and $\triangle t$ converge to 0.*

Example: Let us show that the solution of the difference scheme (A.10),

$$\begin{cases} u_i^{n+1} = (1 - 2r)u_i^n + r\left(u_{i+1}^n + u_{i-1}^n\right), & x \in R, \\ u_i^0 = f(i\triangle x), \end{cases} \tag{A.11}$$

where $r = \nu\triangle t/\triangle x^2$, converges pointwise to the solution of the initial-value problem (A.2):

$$\begin{cases} \dfrac{\partial v}{\partial t} = \nu\dfrac{\partial^2 v}{\partial x^2}, & x \in R, \\ v(0, x) = f(x). \end{cases} \tag{A.12}$$

We will assume that $0 \le r \le \frac{1}{2}$ in order to have all the coefficients positive in the difference equation. We need to estimate

$$z_i^n = u_i^n - v(n\triangle t, i\triangle x),$$

where v is the exact solution of the initial-value problem (A.12). Equation (A.8) becomes

$$v_i^{n+1} = (1 - 2r)v_i^n + r\left(v_{i+1}^n + v_{i-1}^n\right) + \mathcal{O}(\triangle t^2) + \mathcal{O}(\triangle t\triangle x^2). \tag{A.13}$$

Then by subtracting equation (A.13) from (A.11), we have:

$$z_i^{n+1} = (1 - 2r)z_i^n + r\left(z_{i+1}^n + z_{i-1}^n\right) + \mathcal{O}(\triangle t^2) + \mathcal{O}(\triangle t\triangle x^2), \tag{A.14}$$

and then (since we assumed $0 \le r \le \frac{1}{2}$)

$$|z_i^{n+1}| \le (1 - 2r)|z_i^n| + r|z_{i+1}^n| + r|z_{i-1}^n| + C(\triangle t^2 + \triangle t\triangle x^2), \tag{A.15}$$

where C is a constant associated with the "big $\mathcal{O}$" terms and depends on the assumed bounds of the higher-order derivatives of v, in space and time. In fact, we will assume that the derivatives v_{tt} and v_{xxxx} (which would appear in the subsequent terms of the Taylor expansion of (A.8)) are uniformly bounded on $[0, t] \times R$. So, by taking the supremum with respect to i in (A.15) we obtain

$$Z^{n+1} \le Z^n + C(\triangle t^2 + \triangle t\triangle x^2) \quad \text{with} \quad Z^n = |z^n|_{\ell\infty} \equiv \sup_{i \in Z} \{|z_i^n|\}. \tag{A.16}$$

Applying (A.16) repeatedly yields

$$Z^{n+1} \le Z^n + C(\triangle t^2 + \triangle t\triangle x^2) \le Z^{n-1} + 2C(\triangle t^2 + \triangle t\triangle x^2)$$
$$\le \cdots \le Z^0 + (n + 1)C(\triangle t^2 + \triangle t\triangle x^2).$$

Since $Z^0 = 0$, the previous inequality implies

$$\left| u_i^{n+1} - v((n+1)\triangle t, i\triangle x) \right| \le (n+1)\triangle t \, C(\triangle t + \triangle x^2). \tag{A.17}$$

Thus we see that the right-hand side of (A.17) goes to zero as $(n+1)\triangle t \to t$, and $\triangle t$, $\triangle x \to 0$, which means that u converges to v pointwise. Notice that in fact, we have just proven a stronger result than the pointwise convergence:

$$Z^{n+1} = \left| z^{n+1} \right|_{\ell^\infty} \to 0 \tag{A.18}$$

as $(n+1)\triangle t \to t$, and $\triangle t$, $\triangle x \to 0$.   ∎

The pointwise convergence is in general difficult to prove. So we shall instead use a definition of convergence in terms of an l^p-norm $(p < \infty)$ of a difference between the solution of the PDE and the solution of the difference equation. In the following definition, we will use the notation

$$u^{n+1} = (\ldots, u_{-1}^n, u_0^n, u_1^n, \ldots),$$
$$v^{n+1} = (\ldots, v_{-1}^n, v_0^n, v_1^n, \ldots).$$

Definition A.1.2 (convergent scheme) *The scheme* (A.3) *approximating the partial differential equation* (A.1) *is a convergent scheme at time t if as $(n+1)\triangle t \to t$,*

$$\left| u^{n+1} - v^{n+1} \right|_* \to 0 \tag{A.19}$$

as $\triangle x \to 0$ and $\triangle t \to 0$, and where $|\,.\,|_*$ is a norm to be specified.

This definition shows that whenever convergence is being discussed, the norm that is used must be specified. Its choice depends on the problem to be solved. For $z = (\ldots, z_{-1}, z_0, z_1, \ldots)$, typical examples include

$$|z|_{\ell^\infty} = \sup_{i \in Z} \{ |z_i| \}, \quad |z|_{\ell^2} = \sqrt{\sum_{i=-\infty}^{i=+\infty} |z_i|^2}, \quad \text{and} \quad |z|_{\ell^2, \triangle x} = \sqrt{\sum_{i=-\infty}^{i=+\infty} |z_i|^2 \triangle x}. \tag{A.20}$$

Another important piece of information that we may be interested in is the rate of convergence, i.e., how fast the solution of the difference equation converges to the solution of the PDE. This order of convergence is defined as follows:

Definition A.1.3 (order of convergence) *A difference scheme* (A.3) *approximating the partial differential equation* (A.1) *is a convergent scheme of order (p, q) if for any t, as $(n+1)\triangle t$ converges to t,*

$$\left| u^{n+1} - v^{n+1} \right|_* = \mathcal{O}(\triangle x^p) + \mathcal{O}(\triangle t^q) \tag{A.21}$$

as $\triangle x$ and $\triangle t$ converge to 0.

Example: For the approximation (A.11) of the heat equation, we have in fact proven its convergence for the ℓ^∞ norm (A.18). Moreover, we can verify that this scheme is of order $(2, 1)$.   ■

The convergence is usually something difficult to prove. Most of the time, its proof is based on the Lax theorem, which we present in the next section.

A.1.3   The Lax Theorem

This theorem gives a sufficient condition for a two-level difference[3] scheme to be convergent:

Theorem A.1.1 (Lax) *A consistent two-level difference scheme for a well-posed linear initial value problem is convergent if and only if it is stable.*

In this theorem we have introduced two new notions:

- *Consistency:* This concerns the error introduced by the discretization of the equation. This error should tend to zero as $\triangle t$ and $\triangle x$ go to zero.

- *Stability:* The intuitive idea is that small errors in the initial condition should cause small errors in the solution. This is similar to the definition of well-posedness of a PDE.

Most of the schemes that are used are consistent. The major problem will be to prove their stability.

The two next sections define precisely these two notions and give the main ideas to ensure that they are satisfied.

A.1.4   Consistency

As in the case of convergence, we can first define the property of a scheme to be *pointwise* consistent with the PDE:

Definition A.1.4 (pointwise consistent) *The scheme (A.3) approximating the partial differential equation (A.1) is pointwise consistent at point (t, x) if for any smooth function $\phi = \phi(t, x)$,*

$$(\mathcal{L}\phi - F)|_i^n - [L_i^n \phi(n\triangle t, i\triangle x) - G_i^n] \to 0 \qquad (A.22)$$

as $\triangle x$, $\triangle t \to 0$ and $((n+1)\triangle t, i\triangle x) \to (t, x)$.

Example: Notice that from equality (A.8), we have in fact just proven that the scheme (A.10) is pointwise consistent with the PDE (A.2).   ■

---

[3] A two-level difference scheme is a scheme where only two different levels of time are present in the difference equation, typically $H(u^{n+1}, u^n) = 0$.

As in the case of convergence, it is usually more interesting to have a definition in terms of norms and not only pointwise. If we write the two-level
scheme as

$$u^{n+1} = Qu^n + \Delta t\, G^n, \tag{A.23}$$

where $u^n = (\ldots, u^n_{-1}, u^n_0, u^n_1, \ldots)$, $G^{n+1} = (\ldots, G^n_{-1}, G^n_0, G^n_1, \ldots)$, and Q is
an operator acting on the appropriate space, then a stronger definition of
consistency can be given as follows:

Definition A.1.5 (consistent) *The scheme* (A.3) *is consistent with the
partial differential equation* (A.1) *in a norm* $|\,.\,|_*$ *if the solution v of the
partial differential equation satisfies:*

$$v^{n+1} = Qv^n + \Delta t G^n + \Delta t \tau^n,$$

where τ^n is such that

$$|\tau^n|_* \to 0$$

as Δx, $\Delta t \to 0$.

The term τ^n is called the truncature term. We may be more precise and
define also the order in which τ^n goes to 0.

Definition A.1.6 (truncature error, order of accuracy) *The difference scheme* (A.3) *is said to be accurate of order* (p, q) *if*

$$|\tau^n|_* = \mathcal{O}(\Delta x^p) + \mathcal{O}(\Delta t^q).$$

Remark It is easy to see that if a scheme is of order (p, q), $p, q \geq 1$, then
it is a consistent scheme. Also, it can be verified that if a scheme is either
consistent or accurate of order (p, q), the scheme is pointwise consistent. ∎

Example: Let us discuss the consistency of the scheme

$$\frac{u^{n+1}_i - u^n_i}{\Delta t} = \nu \frac{u^n_{i+1} - 2u^n_i + u^n_{i-1}}{\Delta x}$$

with the PDE $\dfrac{\partial v}{\partial t} = \nu \dfrac{\partial^2 v}{\partial x^2}$, $x \in R$, $t > 0$. If we denote by v the solution of the PDE,
then equation (A.8) becomes

$$\frac{v^{n+1}_i - v^n_i}{\Delta t} - \nu \frac{v^n_{i+1} - 2v^n_i + v^n_{i-1}}{\Delta x^2} = \mathcal{O}(\Delta t) + \mathcal{O}(\Delta x^2).$$

As we can see, we need to be more precise to apply Definitions A.1.6 and A.1.5. In
particular, we need to know exactly the terms in $\mathcal{O}(\Delta t) + \mathcal{O}(\Delta x^2)$. In fact, similar
calculations have to be done but using Taylor expansions with remainder instead of
standard Taylor expansions.
    After rewriting the difference scheme in the form of (A.23),

$$u^{n+1}_i = (1 - 2r)u^n_i + r\left(u^n_{i+1} + u^n_{i-1}\right), \quad \text{where } r = \nu \Delta t / \Delta x^2,$$

we can define the truncature error by

$$\Delta t \tau_i^n = v_i^{n+1} - \left\{ (1 - 2r)v_i^n + r\left(v_{i+1}^n + v_{i-1}^n\right) \right\}, \tag{A.24}$$

where v is a solution of the PDE. Then we need to develop the right-hand term of (A.24) by using Taylor expansions with remainder. After some calculation, there exist $t_1 \in]n\Delta t, (n+1)\Delta t[$, $x_1 \in](i-1)\Delta x, i\Delta x[$ and $x_2 \in]i\Delta x, (i+1)\Delta x[$ such that

$$\Delta t \tau_i^n = \frac{\partial^2 v}{\partial t^2}(t_1, i\Delta x) \frac{\Delta t}{2} - \nu \left(\frac{\partial^4 v}{\partial x^4}(n\Delta t, x_1) + \frac{\partial^4 v}{\partial x^4}(n\Delta t, x_2) \right) \frac{\Delta x^2}{24}. \tag{A.25}$$

Notice that as we have mentioned, when we write simply $\mathcal{O}(\Delta t) + \mathcal{O}(\Delta x^2)$, we have to be aware that the coefficients involved are not constants but depend on certain derivatives of the solution. This also means that as soon as we will talk about consistency, we will need to make some smoothness assumptions.

To apply Definition A.1.5, we need to choose a norm. If we assume that

$$\frac{\partial^2 v}{\partial t^2} \quad \text{and} \quad \frac{\partial^4 v}{\partial x^4} \quad \text{are uniformly bounded on } [0, T] \times R \text{ for some } T,$$

then we can then choose the sup-norm to get that this scheme is accurate of order $(2, 1)$ with respect to this norm. Otherwise, if we assume

$$\sum_{i=-\infty}^{i=+\infty} \left| \left(\frac{\partial^2 v}{\partial t^2} \right)_i^n \right|^2 \Delta x < A < \infty \quad \text{and} \quad \sum_{i=-\infty}^{i=+\infty} \left| \left(\frac{\partial^4 v}{\partial x^4} \right)_i^n \right|^2 \Delta x < B < \infty,$$

for any Δx and Δt, then the difference scheme is accurate of order $(2, 1)$ with respect to the $\ell^{2, \Delta x}$ norm.   ∎

One important remark that comes out of the previous example is that as soon as one considers the problem of consistency, one needs to choose a norm. It is also important to note that this choice is, in fact, related to some smoothness assumptions on the solution.

Finally, we would like to mention that proving consistency can be very difficult, especially for implicit schemes. We refer the interested reader to [316] for more details.

A.1.5   Stability

To conclude this section, we need to discuss the problem of stability, which is necessary for applying the Lax theorem. Though stability is much easier to establish than convergence, it is still often difficult to prove that a given scheme is stable. Many definitions of stability can be found in the literature, and we present below one that is commonly used:

Definition A.1.7 (stable scheme) *The two-level difference scheme*

$$\begin{cases} u^{n+1} = Qu^n, n \geq 0, \\ u^0 \quad given, \end{cases} \tag{A.26}$$

where $u^n = (\ldots, u_{-1}^n, u_0^n, u_1^n, \ldots)$*, is said to be stable with respect to the norm* $| \cdot |_*$ *if there exist positive constants* Δx_0 *and* Δt_0*, and nonnegative* *constants K and β such that*

$$\left|u^{n+1}\right|_* \le K e^{\beta t} \left|u^0\right|_* \qquad (A.27)$$

for $0 \le t = (n+1)\triangle t$, $0 < \triangle x \le \triangle x_0$ and $0 < \triangle t \le \triangle t_0$.

Remarks From Definition A.1.7 we may observe the following:

- This definition has been established for homogeneous schemes (A.26). If we have a nonhomogeneous scheme, it can be proved that the stability of the associated homogeneous scheme, along with the convergence, is enough to prove its convergence.

- As for convergence and consistency, we will need to define which norm is used.

- This definition of stability does allow the solution to grow with time.

■

As we already mentioned, there are other definitions for stability. In particular, another common definition is one that does not allow for exponential growth. The inequality (A.27) then becomes

$$\left|u^{n+1}\right|_* \le K \left|u^0\right|_* , \qquad (A.28)$$

which clearly implies (A.27). The interest of (A.27) is that it permits us to include more general situations.

Example: Let us show that the scheme

$$u_i^{n+1} = (1 - 2r)u_i^n + r\left(u_{i+1}^n + u_{i-1}^n\right) \qquad (A.29)$$

is stable for the sup-norm. If we assume that $r \le \frac{1}{2}$, (A.29) yields

$$|u_i^{n+1}| \le (1 - 2r)|u_i^n| + r|u_{i+1}^n| + r|u_{i-1}^n| \le |u^n|_{\ell\infty} .$$

If we take the supremum over the right-hand side, we get

$$|u^{n+1}|_{\ell\infty} \le |u^n|_{\ell\infty} .$$

Hence inequality (A.27) is satisfied with $K = 1$ and $\beta = 0$. Notice that in order to prove the stability, we have assumed that $r = \nu\triangle t/\triangle x^2 \le \frac{1}{2}$. In this case we say that the scheme is *conditionally stable*. In the case where there is no restriction on $\triangle x$ and $\triangle t$, we say that the scheme is *unconditionally stable*. ■

The previous example was a simple case where we were able to prove directly stability, i.e., inequality (A.27) or (A.28). In fact, there are several tools that can be used to prove it. The one that is probably the most commonly used is Fourier analysis, which is used for linear difference schemes with constant coefficients. We recall in Table A.1 the definitions of the Fourier transform and the inverse Fourier transform, for the continuous and discrete settings (i.e., for a vector $u^n = (\ldots, u_{-1}^n, u_0^n, u_1^n, \ldots) \in \ell^2$).

We also recall an important property of the Fourier transform, which is Parseval's identity (see [154] for more details).

	Continuous setting	Discrete setting
Fourier transform	$\hat{v}(t,\omega) = \dfrac{1}{\sqrt{2\pi}} \displaystyle\int\limits_{-\infty}^{+\infty} e^{-i\omega x} v(t,x)dx$	$\hat{u}(\xi) = \dfrac{1}{\sqrt{2\pi}} \displaystyle\sum_{m=-\infty}^{m=+\infty} e^{-im\xi} u_m$
inverse Fourier transform	$v(t,x) = \dfrac{1}{\sqrt{2\pi}} \displaystyle\int\limits_{-\infty}^{+\infty} e^{i\omega x} \hat{v}(t,\omega)d\omega$	$u_m = \dfrac{1}{\sqrt{2\pi}} \displaystyle\int\limits_{-\pi}^{+\pi} e^{im\xi} \hat{u}(\xi)d\xi$
Parseval's identity	$|v|_{L^2(R)} = |\hat{v}|_{L^2(R)}$	$|u|_{\ell^2} = |\hat{u}|_{L^2(-\pi,\pi)}$

Table A.1. Some recalls about the Fourier transform ($\underline{i}^2 = 1$).

Interestingly, the discrete transform has similar properties to those of the continuous one. In particular, to prove the stability of a difference scheme, we will use two main ideas:

- The first is that taking the Fourier transform of a PDE turns it into an ODE. Spatial derivatives are turned into products. For example, we can easily verify that

$$\widehat{v_{xx}}(t,\omega) = -\,\omega^2\,\,\hat{v}(t,\omega).$$

An analogous idea is valid in the discrete case. Let us consider the "standard" approximation of the second-order derivative that we have been using until now (with $\Delta x = 1$ just to simplify notation):

$$u_{xx}|_k = u_{k+1} - 2u_k + u_{k-1}. \tag{A.30}$$

Then the Fourier transform of $\{u_{xx}\}$ in the discrete setting is

$$\widehat{(u_{xx})} = \frac{1}{\sqrt{2\pi}} \sum_{k=-\infty}^{k=+\infty} e^{-ik\xi} u_{xx}|_k \tag{A.31}$$

$$\underset{(A.30)}{=} \frac{1}{\sqrt{2\pi}} \sum_{k=-\infty}^{k=+\infty} e^{-ik\xi} u_{k+1} - 2\underbrace{\frac{1}{\sqrt{2\pi}} \sum_{k=-\infty}^{k=+\infty} e^{-ik\xi} u_k}_{\hat{u}(\xi)} + \frac{1}{\sqrt{2\pi}} \sum_{k=-\infty}^{k=+\infty} e^{-ik\xi} u_{k-1}.$$

By suitable changes of variable in the previous expression for the first sum $(m = k + 1)$ and the third one $(m = k - 1)$, we have

$$\widehat{(u_{xx})} = \frac{1}{\sqrt{2\pi}} \sum_{m=-\infty}^{m=+\infty} e^{-i(m-1)\xi} u_m - 2\hat{u}(\xi) + \frac{1}{\sqrt{2\pi}} \sum_{m=-\infty}^{m=+\infty} e^{-i(m+1)\xi} u_m$$

$$= (e^{-i\xi} - 2 + e^{+i\xi})\hat{u}(\xi) = \boxed{-4\sin^2\left(\frac{\xi}{2}\right)} \hat{u}(\xi). \tag{A.32}$$

- The second concerns Parseval's identity. The main interest of this identity is that it is equivalent to prove the inequality (A.27) in the transform space or in the solution space. As a matter of fact, in Definition A.1.7 of stability, the inequality that was required in terms of the energy norm was of the form

$$\left|u^{n+1}\right|_{\ell^2} \le Ke^{\beta(n+1)\Delta t} \left|u^0\right|_{\ell^2}. \tag{A.33}$$

But since $|u|_{\ell^2, \Delta x} = \sqrt{\Delta x}\, |u|_{\ell^2} = \sqrt{\Delta x}\, |\hat{u}|_{\ell^2}$, if we can find a K and a β such that

$$\left|\hat{u}^{n+1}\right|_{\ell^2} \le Ke^{\beta(n+1)\Delta t} \left|\hat{u}^0\right|_{\ell^2}, \tag{A.34}$$

then the same K and β will also satisfy (A.33). So the sequence $\{u^n\}$ will be stable if and only if the sequence $\{\hat{u}^n\}$ is stable in $L^2(-\pi, \pi)$.

These ideas are applied in the following example, where we show how to prove the stability of the discrete scheme associated with the 1-D heat equation.

Example: Let us prove the stability of the difference scheme

$$u_k^{n+1} = ru_{k+1}^n + (1 - 2r)u_k^n + ru_{k-1}^n, \tag{A.35}$$

where $r = \nu\Delta t/\Delta x^2 \le \frac{1}{2}$. By doing similar computations as in (A.31)–(A.32), taking the Fourier transform of u^{n+1} with (A.35) leads to

$$\hat{u}^{n+1}(\xi) = r\frac{1}{\sqrt{2\pi}} \sum_{k=-\infty}^{k=+\infty} e^{-ik\xi} u_{k+1}^n + (1 - 2r)\hat{u}(\xi) + r\frac{1}{\sqrt{2\pi}} \sum_{k=-\infty}^{k=+\infty} e^{-ik\xi} u_{k-1}^n$$

$$= \left(1 - 4r\sin^2\left(\frac{\xi}{2}\right)\right) \hat{u}^n(\xi). \tag{A.36}$$

The coefficient of $\hat{u}^n$ on the right-hand side of (A.36) is called the symbol of the difference scheme (A.35). We denote it by $\rho(\xi)$. Then if we apply the result of (A.36) $n + 1$ times, we get

$$\hat{u}^{n+1}(\xi) = (\rho(\xi))^{n+1}\hat{u}^0(\xi).$$

So the condition (A.34) will be satisfied with $K = 1$ and $\beta = 0$ as soon as $|\rho(\xi)| \le 1$ for all $\xi \in [-\pi, \pi]$, i.e.,

$$\left|1 - 4r\sin^2\left(\frac{\xi}{2}\right)\right| \le 1. \tag{A.37}$$

This condition is satisfied if $r \le \frac{1}{2}$. Thus $r \le \frac{1}{2}$ is a sufficient condition for stability (and along with the consistency, for convergence). It is also necessary. If $r > \frac{1}{2}$, then at least for some ξ, $|\rho(\xi)| > 1$, and then $|\rho(\xi)|^{n+1}$ will be greater than $Ke^{\beta(n+1)\triangle t}$ for any K and β.[4] ∎

Another approach that is often used is to consider a discrete Fourier mode for the problem

$$u_k^n = \xi^n e^{ijk\pi\triangle x}, \tag{A.38}$$

where $0 \le j \le M$ and the superscript on the ξ term is a multiplicative exponent. The idea is then to insert this general Fourier mode into the difference scheme and find the expression for ξ. A necessary condition for stability is obtained by restricting $\triangle x$ and $\triangle t$ so that $|\xi| \le 1$ (the ξ^n term will not grow without bound). This method is usually referred to as the discrete von Neumann criterion for stability (see [316] for more details).

Example: Let us apply the discrete von Neumann criterion for stability for the difference scheme

$$u_k^{n+1} = ru_{k+1}^n + (1 - 2r)u_k^n + ru_{k-1}^n. \tag{A.39}$$

By inserting the general Fourier mode

$$u_k^n = \xi^n e^{ijk\pi\triangle x}$$

in the difference scheme, we easily obtain

$$\xi^{n+1}e^{ijk\pi\triangle x} = \xi^n e^{ijk\pi\triangle x}\left(re^{-ij\pi\triangle x} + (1 - 2r) + re^{+ij\pi\triangle x}\right).$$

Thus if we divide both sides of the above equation by $\xi^n e^{ijk\pi\triangle x}$, we get

$$\xi = re^{-ij\pi\triangle x} + (1 - 2r) + re^{+ij\pi\triangle x} = 1 - 4r\sin^2\left(\frac{j\pi\triangle x}{2}\right).$$

By saying that $|\xi| \le 1$, we recover the previous result (A.37). ∎

Example: As we already mentioned, the discretization (A.9) initially proposed for the one-dimensional heat equation was explicit. The values of u at time $(n+1)\triangle t$ were fully determined by the values of u at time $n\triangle t$. One may investigate more general schemes such as

$$u_i^{n+1} = u_i^n + \frac{\nu\triangle t}{h^2}\left(\lambda(u_{i+1}^{n+1} - 2u_i^{n+1} + u_{i-1}^{n+1}) + (1 - \lambda)(u_{i+1}^n - 2u_i^n + u_{i-1}^n)\right).$$

If $\lambda \ne 0$, the scheme is now implicit: One needs to solve a linear system in order to know the solution at time $(n+1)\triangle t$. For $\lambda = 1$ the scheme is fully implicit, and for $\lambda = 0.5$ the so-called Crank–Nicholson scheme is obtained. This difference between explicit and implicit schemes can be simply represented as in Figure A.2.
Depending on the value of λ, we have the following:

- For $\lambda = 0$, the scheme is explicit, of order $\mathcal{O}(\triangle t, h^2)$, and stable under the condition $\triangle t \le \frac{h^2}{2\nu}$. This condition implies that the time step has to be chosen small enough, which will naturally slow down the resolution of the equation.

---

[4]This is true, since for any sequence of $\triangle t$ (chosen so that $(n+1)\triangle t \to t$) and choice of $\triangle x$ (so that r remains constant) the expression $|\rho(\xi)|^{n+1}$ becomes unbounded, while for sufficiently large values of n, $Ke^{\beta(n+1)\triangle t}$ will be bounded by $Ke^{\beta(t_0+1)}$ for some $t_0 > t$, t_0 near t.

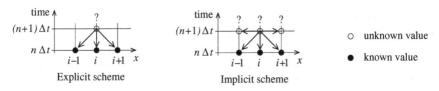

Figure A.2. Relations of dependency: The estimation of u_i^{n+1} depends on the neighbors indicated by an arrow.

- For $\lambda = \frac{1}{2}$, the scheme is implicit, of order $\mathcal{O}(\Delta t^2, h^2)$, and unconditionally stable.

- For $\lambda > \frac{1}{2}$, the scheme is implicit, of order $\mathcal{O}(\Delta t, h^2)$, and unconditionally stable.

We leave it as an exercise to the reader to verify these results, that is, proving consistency.[5] and stability.   ∎

A.2   Hyperbolic Equations

Let us consider the one-dimensional linear advection equation (also called the transport or wave equation):

$$
\begin{cases}
\dfrac{\partial v}{\partial t}(t, x) + a\dfrac{\partial v}{\partial x}(t, x) = 0, \quad t > 0, \quad x \in R \\
v(0, x) = v_0(x),
\end{cases}
\tag{A.40}
$$

where a is a constant. It can be easily verified that the solution is

$$
v(t, x) = v_0(x - at).
\tag{A.41}
$$

Consequently, $v(t, x)$ is constant on lines of slope a, which are called characteristics. This means that the information is propagated in the direction of the sign of a, for example from left to right if a is positive.

In order to solve (A.40) numerically we have to approximate the temporal and spatial derivatives of u. As for the case of the heat equation (see Section A.1), the method is based on the Taylor expansions of v, which we recall

---

[5]In the example considered throughout this section we have expanded the functions about the index point $(n\Delta t, i\Delta x)$, and it was reasonably obvious that this was the correct point about which to expand. However, in some situations, the consistency cannot be proved if the point about which to expand is not adapted. The decision about which point to expand must be made by carefully considering how we expect the difference scheme to approximate the PDE. Typically, to prove the consistency of the Crank–Nicholson scheme, it is logical to consider the consistency of the scheme at the point $\left((n + \frac{1}{2})\Delta t, i\Delta x\right)$

here:

$$v((n+1)\Delta t, i\Delta x) = \left(v + \Delta t \frac{\partial v}{\partial t} + \frac{\Delta t^2}{2} \frac{\partial^2 v}{\partial t^2}\right)(n\Delta t, i\Delta x) + \mathcal{O}(\Delta t^3),$$

$$\text{(A.42)}$$

$$v(n\Delta t, (i+1)\Delta x) = \left(v + \Delta x \frac{\partial v}{\partial x} + \frac{\Delta x^2}{2} \frac{\partial^2 v}{\partial x^2}\right)(n\Delta t, i\Delta x) + \mathcal{O}(\Delta x^3),$$

$$\text{(A.43)}$$

$$v(n\Delta t, (i-1)\Delta x) = \left(v - \Delta x \frac{\partial v}{\partial x} + \frac{\Delta x^2}{2} \frac{\partial^2 v}{\partial x^2}\right)(n\Delta t, i\Delta x) + \mathcal{O}(\Delta x^3).$$

$$\text{(A.44)}$$

The temporal derivative can be approximated using (A.42) by

$$\frac{\partial v}{\partial t}(n\Delta t, i\Delta x) = \frac{v_i^{n+1} - v_i^n}{\Delta t} + \mathcal{O}(\Delta t).$$

As far as the spatial derivative is concerned, there are several possibilities:

- From (A.43) we have $\frac{\partial v}{\partial x}(n\Delta t, i\Delta x) = \frac{v_{i+1}^n - v_i^n}{\Delta x} + \mathcal{O}(\Delta x)$ (forward difference).

- From (A.44) we have $\frac{\partial v}{\partial x}(n\Delta t, i\Delta x) = \frac{v_i^n - v_{i-1}^n}{\Delta x} + \mathcal{O}(\Delta x)$ (backward difference).

- By subtracting (A.44) from (A.43) we have
  $\frac{\partial v}{\partial x}(n\Delta t, i\Delta x) = \frac{v_{i+1}^n - v_{i-1}^n}{2\Delta x} + \mathcal{O}(\Delta x^2)$ (centered difference).

Consequently, there are three different possibilities for the discrete scheme of (A.40):

$$u_i^{n+1} = u_i^n + a\Delta t \begin{cases} \delta_x^+ u_i^n & \left(\equiv \dfrac{u_{i+1}^n - u_i^n}{\Delta x}\right) & \text{forward scheme,} \\[2mm] \delta_x u_i^n & \left(\equiv \dfrac{u_{i+1}^n - u_{i-1}^n}{2\Delta x}\right) & \text{centered scheme,} \\[2mm] \delta_x^- u_i^n & \left(\equiv \dfrac{u_i^n - u_{i-1}^n}{\Delta x}\right) & \text{backward scheme.} \end{cases}$$

Let us first consider the centered approximation, which is of order $(2,1)$:

$$u_n^{n+1} = u_i^n - a\Delta t \delta_x u_i^n = u_i^n - \frac{a\Delta t}{2\Delta x}(u_{i+1}^n - u_{i-1}^n). \qquad \text{(A.45)}$$

Let us examine its stability. As explained in Section A.1.5, a common approach is to use the discrete von Neumann criterion, which consists in inserting the general Fourier mode

$$u_k^n = \xi^n e^{ijk\pi\Delta x} \quad (0 \le j \le M) \qquad \text{(A.46)}$$

into the difference scheme. We recall that the superscript on the ξ term is a multiplicative exponent. By doing analogous calculations as for the difference scheme (A.39), we find that

$$\xi = 1 - i\frac{a\Delta t}{\Delta x}\sin(j\pi\Delta x) \text{ and then } |\xi|^2 = 1 + \left(\frac{a\Delta t}{\Delta x}\right)^2 \sin^2(j\pi\Delta x) \geq 1 \ \forall j,$$

which means that this scheme is always unstable.

☛ *The reason for this is that we did not take into account the nature of the equation. As there is a propagation, that is, the information is propagated in a certain direction, the numerical scheme should take it into account.*

To take this observation into account, one may propose the following scheme:

$$u_i^{n+1} = u_i^n - \begin{cases} a\,\Delta t\,\delta_x^- u_i^n & \text{if } a > 0, \\ a\,\Delta t\,\delta_x^+ u_i^n & \text{if } a < 0. \end{cases}$$

This can be rewritten as

$$u_i^{n+1} = u_i^n - \Delta t[\max(0,a)\delta_x^- u_i^n + \min(0,a)\delta_x^+ u_i^n]. \tag{A.47}$$

We call (A.47) an upwind scheme, because it uses values in the direction of information propagation. Let us see again the stability.

- Case $a > 0$: By replacing (A.46) in (A.47), we let the reader see for himself that we get

$$|\xi| = 1 - 2C(1-C)(1 - \cos(j\pi\Delta x)) \quad \text{with} \quad C = \frac{a\Delta t}{\Delta x} > 0.$$

  It will be less than or equal to 1 if and only if $C \leq 1$.

- Case $a < 0$: If we set $C = \frac{a\Delta t}{\Delta x} < 0$, similar calculus yield $-C \leq 1$.

To summarize, the stability condition is

$$|a|\frac{\Delta t}{\Delta x} \leq 1. \tag{A.48}$$

It is usually called CFL, in reference to the authors Courant–Friedrichs–Lewy (in 1928).

This condition may be interpreted in terms of domain of dependence. In the continuous case, as has been mentioned, the information is propagated along the characteristics, and their equation is $\frac{dt}{dx} = \frac{1}{a}$. In the discrete case, if $a > 0$, then (A.47) becomes

$$u_i^{n+1} = \left(1 - \frac{a\Delta t}{\Delta x}\right)u_i^n + \frac{a\Delta t}{\Delta x}u_{i-1}^n.$$

This allows us to define the discrete domain of dependence of u_i^{n+1}: u_i^{n+1} depends on u_i^n and u_{i-1}^n, u_i^n depends on u_i^{n-1} and u_{i-1}^{n-1}, etc. (see Figure

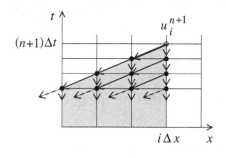

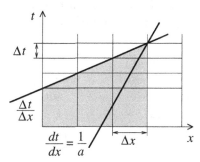

(a) Discrete domain of dependence of u_i^{n+1} for (A.47) with $a > 0$

(b) Interpretation of the CFL condition

Figure A.3. Definition and interpretation of the discrete domain of dependence.

A.3). Moreover, the CFL condition means

$$\frac{\Delta t}{\Delta x} \leq \frac{1}{a} = \frac{dt}{dx},$$

which signifies that the characteristic line has to be included in the discrete domain of dependence.

☞ *The discrete domain of dependence must contain the exact continuous domain of dependence.*

In the nonlinear case the study is of course more complicated. For example, let us examine the nonlinear Burgers equation

$$\frac{\partial v}{\partial t} + v \frac{\partial v}{\partial x} = 0 \tag{A.49}$$

with the initial condition

$$v(0, x) = v_0(x) = \begin{cases} 1 & \text{if } x \leq 0, \\ 1 - x & \text{if } 0 < x < 1, \\ 0 & \text{if } x \geq 1. \end{cases} \tag{A.50}$$

Here the propagation speed depends on the value of u itself. We may try as in the linear case to get an explicit solution of (A.49)–(A.50). As it is classical for hyperbolic equations, the method of characteristics can be used. Let us suppose that u is a smooth solution of (A.49)–(A.50) and let $x(t)$ be an integral curve of the differential equation

$$\frac{dx}{dt}(t) = v(t, x(t)), \quad x(0) = x_0.$$

We claim that u is constant along the characteristic curve $x(t)$. Indeed, since u is a solution of (A.49), we have

$$\frac{d}{dt}(v(t, x(t))) = \frac{\partial v}{\partial t}(t, x(t)) + \frac{dx}{dt}(t)\frac{\partial v}{\partial x}(t, x(t))$$

$$= \frac{\partial v}{\partial t}(t, x(t)) + v(t, x(t))\frac{\partial v}{\partial x}(t, x(t)) = 0.$$

Therefore

$$\frac{dx}{dt}(t) = v(0, x(0)) = v_0(x_0)$$

and

$$x(t) = x_0 + t\, v_0(x_0).$$

According to the definition of u_0 we deduce that

$$x(t) = \begin{cases} x_0 + t & \text{if } x_0 \leq 0, \\ x_0 + (1 - x_0)\, t & \text{if } 0 \leq x_0 \leq 1, \\ x_0 & \text{if } x_0 \geq 1. \end{cases}$$

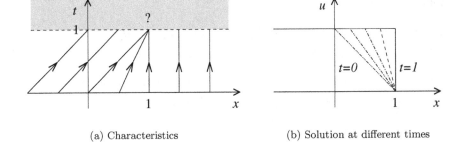

(a) Characteristics   (b) Solution at different times

Figure A.4. Behavior of the Burgers equation (A.49). The left-hand figure represents the characteristics while the right-hand one shows some solution at different times. As we can observe, characteristics intersect at $t = 1$, and after this time, it is not clear how to define the solution.

For $t < 1$ the characteristics do not intersect. Hence, given a point (t, x), $t < 1$, we draw the characteristic passing through this point, and we determine the corresponding point x_0:

$$x_0 = \begin{cases} x - t & \text{if } x \leq t, \\ \dfrac{x - t}{1 - t} & \text{if } t \leq x \leq 1, \\ x & \text{if } x \geq 1, \end{cases}$$

and we get the following continuous solution for $t < 1$:

$$u(t,x) = \begin{cases} 1 & \text{if } x \leq t, \\ \dfrac{1-x}{1-t} & \text{if } t \leq x \leq 1, \\ 0 & \text{if } x \geq 1. \end{cases}$$

It consists of a front moving from left to right for $t < 1$. At $t = 1$, the characteristics collide, and beyond this collision time it is not clear how to define the solution uniquely. This discontinuity phenomenon is known as shock. What about for $t \geq 1$? Is it possible to define a unique solution? We must devise some way to interpret a less regular notion of solution. Let $\varphi : [0, +\infty[\times R \to R$ be smooth with compact support. We call φ a test function. We first observe that (A.49) can be written in a conservative form:

$$\frac{\partial v}{\partial t} + \frac{1}{2}\frac{\partial}{\partial x}(v^2) = 0.$$

Then multiplying the above equality by φ, we deduce

$$\int_0^\infty \int_{-\infty}^\infty \left(\frac{\partial v}{\partial t} + \frac{1}{2}\frac{\partial}{\partial x}(v^2) \right) \varphi \, dx \, dt = 0,$$

and by integrating by parts this last equation

$$\int_0^\infty \int_{-\infty}^\infty v(t,x)\frac{\partial \varphi}{\partial t}(t,x) \, dx \, dt + \int_{-\infty}^\infty v_0(x)\varphi(0,x) \, dx \qquad (A.51)$$

$$+ \int_0^\infty \int_{-\infty}^\infty \frac{1}{2}v^2(t,x)\frac{\partial \varphi}{\partial x}(t,x) \, dx \, dt = 0.$$

We derive (A.51) supposing v to be smooth, but it is still valid if v is only bounded.

Definition A.2.1 *We say that $v \in L^\infty((0,\infty) \times R)$ is a weak solution of (A.49) if equality (A.51) holds for each test function φ.*

So according to this definition we are now going to search for discontinuous solutions. But before doing that, what can be deduced from (A.51)? Let us suppose in some open domain $\Omega \subset (0,\infty) \times R$ that v is smooth on either side of a smooth curve $x = \xi(t)$. Let us denote by Ω_L (respectively Ω_R) the part of Ω on the left (respectively on the right) of $x = \xi(t)$. We assume that v has limits v_- and v_+ on each side of $x = \xi(t)$: $v_\pm = \lim_{\varepsilon \to 0} u((t,\xi(t)) \pm \varepsilon N)$, where N is the normal vector to $x = \xi(t)$ given by $N = (-\xi'(t), 1)^T$.

Now by choosing a test function φ with compact support in Ω but that does not vanish along $x = \xi(t)$ we get

$$\iint\limits_{\Omega_L} \left(v(t,x)\frac{\partial \varphi}{\partial t}(t,x) + \frac{1}{2}v^2(t,x)\frac{\partial \varphi}{\partial x}(t,x) \right) \, dx\, dt$$

$$+ \iint\limits_{\Omega_R} \left(v(t,x)\frac{\partial \varphi}{\partial t}(t,x) + \frac{1}{2}v^2(t,x)\frac{\partial \varphi}{\partial x}(t,x) \right) \, dx\, dt = 0.$$

But since φ has compact support within Ω, we have

$$\iint\limits_{\Omega_L} \left(v(t,x)\frac{\partial \varphi}{\partial t}(t,x) + \frac{1}{2}v^2(t,x)\frac{\partial \varphi}{\partial x}(t,x) \right) \, dx\, dt$$

$$= -\iint\limits_{\Omega_L} \left(\frac{\partial v}{\partial t} + \frac{1}{2}\frac{\partial}{\partial x}(v^2) \right) \varphi(t,x) \, dx\, dt + \int\limits_{x=\xi(t)} \left(-\xi'(t)v_- + \frac{v_-^2}{2} \right) \varphi(t,x)dl$$

$$= \int\limits_{x=\xi(t)} \left(-\xi'(t)v_- + \frac{v_-^2}{2} \right) \varphi \, dl,$$

since v is a smooth solution satisfying (A.49) in Ω_L. Similarly, we have

$$\iint\limits_{\Omega_R} \left(v(t,x)\frac{\partial \varphi}{\partial t}(t,x) + \frac{1}{2}v^2(t,x)\frac{\partial \varphi}{\partial x}(t,x) \right) \, dx\, dt$$

$$= -\int\limits_{x=\xi(t)} \left(-\xi'(t)v_+ + \frac{v_+^2}{2} \right) \varphi(t,x)dl.$$

Adding these two last identities, we obtain

$$\int\limits_{x=\xi(t)} \left(-\xi'(t)v_- + \frac{v_-^2}{2} \right) \varphi(t,x)dl \ - \int\limits_{x=\xi(t)} \left(-\xi'(t)v_+ + \frac{v_+^2}{2} \right) \varphi(t,x)dl = 0. \tag{A.52}$$

Since φ is arbitrary, we easily deduce from (A.52)

$$\xi'(t)\,(v_+ - v_-) = \frac{1}{2}\left(v_+^2 - v_-^2 \right). \tag{A.53}$$

For a general equation of the form $\frac{\partial v}{\partial t} + \frac{\partial}{\partial x}(f(v)) = 0$, we should have obtained

$$\xi'(t)\,(v_+ - v_-) = (f(v_+) - f(v_-)). \tag{A.54}$$

Identity (A.54) is known as the *Rankine–Hugoniot* condition. It may be read as

$$\text{speed of discontinuity} \times \text{jump of } v = \text{jump of } f(v).$$

Unfortunately, (A.53) and (A.54) are necessary conditions for the existence of discontinuous solutions but they are not sufficient to ensure uniqueness. For example, if we consider again (A.49) with the initial condition

$$v_0(x) = \begin{cases} 0 & x < 0, \\ 1 & x \geq 0, \end{cases}$$

then it is easy to check that

$$v(t, x) = \begin{cases} 0 & \text{if } x < t/2, \\ 1 & \text{if } x > t/2, \end{cases}$$

is a weak solution of (A.49) satisfying the Rankine–Hugoniot condition. However, we can find another such solution

$$\hat{u}(t, x) = \begin{cases} 1 & \text{if } x > t, \\ x/t & \text{if } 0 < x < t, \\ 0 & \text{if } x < 0. \end{cases}$$

☛ *Thus, in general, weak solutions are not unique, and we have to find a further criterion that ensures uniqueness. Such a condition exists; it is called entropy.*

We do not continue the investigation of the theoretical difficulties of hyperbolic equations of conservation laws, since the general theory is complex, and it is far beyond the scope of this Appendix to review it. We refer the interested reader to [166, 212] for the complete theory. Of course, these difficulties are still present when we try to discretize these equations. For example, let us consider again Burgers equation

$$\begin{cases} \dfrac{\partial v}{\partial t} + \dfrac{1}{2}\dfrac{\partial}{\partial x}(v^2) = 0, \\ v(0, x) = \begin{cases} 1 & \text{if } x < 0, \\ 0 & \text{otherwise.} \end{cases} \end{cases} \tag{A.55}$$

If we rewrite (A.55) in the quasilinear form

$$\frac{\partial v}{\partial t} + v\frac{\partial v}{\partial x} = 0, \tag{A.56}$$

then a natural finite-difference scheme inspired from the upwind method for (A.40) and assuming that $v \geq 0$ is:

$$\begin{cases} u_i^{n+1} = u_i^n - \dfrac{\Delta t}{\Delta x}u_i^n(u_i^n - u_{i-1}^n), \\ u_i^0 = \begin{cases} 1 & \text{if } j < 0, \\ 0 & \text{otherwise.} \end{cases} \end{cases} \tag{A.57}$$

Then it is easy to verify that $u_i^n = u_i^0$ for all i and n regardless of the step sizes Δt and Δx. Therefore, as Δt and Δx tend to zero, the numerical solution converges to the function $\overline{v}(t, x) = v_0(x)$. Unfortunately, $\overline{v}(t, x)$ is not a weak solution. The reason is that discretizing the Burgers equation written in the form (A.56) is not equivalent for nonsmooth solutions. For nonsmooth solutions, the product $v v_x$ does not necessarily have a meaning (even weakly). So, studying the Burgers equation written in a conservative form is the right approach. But in this case we have to define the numerical schemes that agree with this form. These schemes exist and are called *conservative schemes.*

Let us consider a general hyperbolic equation of conservation laws:

$$
\begin{cases}
\dfrac{\partial v}{\partial t} + \dfrac{1}{2}\dfrac{\partial}{\partial x}(f(v)) = 0, \\
v(0, x) = v_0(x).
\end{cases}
\tag{A.58}
$$

We say that the numerical scheme is in conservation form if it can be written as

$$
u_i^{n+1} = u_i^n - \frac{\Delta t}{\Delta x}\left[F(u_{i-p}^n, u_{i-p+1}^n, \ldots, u_{i+q}^n) - F(u_{i-p-1}^n, u_{i-p}^n, \ldots, u_{i+q-1}^n)\right]
\tag{A.59}
$$

for some function F of $(p+q+1)$ arguments. F is called the numerical flux function. Of course, some consistency relations between F and f have to be satisfied. For example, if $p = 0$ and $q = 1$, then (A.59) becomes

$$
u_i^{n+1} = u_i^n - \frac{\Delta t}{\Delta x}\left[F(u_i^n, u_{i+1}^n) - F(u_{i-1}^n, u_i^n)\right].
\tag{A.60}
$$

In fact, for hyperbolic equations it is often preferable to view u_i^n as an approximation of an average of $v(n\Delta t, x)$ defined by

$$
u_i^n = \frac{1}{\Delta x}\int_{x_{i-\frac{1}{2}}}^{x_{i+\frac{1}{2}}} v(n\Delta t, x)\, dx,
\tag{A.61}
$$

where $x_{i\pm\frac{1}{2}} = \left(i \pm \frac{1}{2}\right)\Delta x$. From the definition of a weak solution of (A.58) and by choosing a particular test function φ, we can show that if u is a weak solution, then

$$
\int_{x_{i-\frac{1}{2}}}^{x_{i+\frac{1}{2}}} v((n+1)\Delta t, x)\, dx
$$

$$
= \int_{x_{i-\frac{1}{2}}}^{x_{i+\frac{1}{2}}} v(n\Delta t, x)\, dx - \int_{n\Delta t}^{(n+1)\Delta t} \left[f(v(t, x_{i+\frac{1}{2}})) - f(v(t, x_{i-\frac{1}{2}})).\right] dt
$$

Then dividing by Δx, we get from (A.61)

$$
u_i^{n+1} = u_i^n - \frac{1}{\Delta x}\int_{n\Delta t}^{(n+1)\Delta t} \left[f(v(t, x_{i+\frac{1}{2}})) - f(v(t, x_{i-\frac{1}{2}}))\right] dt.
\tag{A.62}
$$

So, comparing (A.60) and (A.62) it is natural to choose

$$F(u_i^n, u_{i+1}^n) = \frac{1}{\triangle t} \int_{n\triangle t}^{(n+1)\triangle t} f(v(t, x_{i+\frac{1}{2}})) \, dt, \qquad (A.63)$$

and then the scheme defined by (A.60) will be consistent with the original conservation law if F reduces to f for the case of a constant solution; i.e., if $v(t, x) \equiv c$, then necessarily

$$F(c, c) = f(c) \quad \forall c \in R. \qquad (A.64)$$

This is the definition of a consistent scheme. This notion is very important, since according to the Lax–Wendroff theorem [166, 212], if the numerical scheme is consistent and in a conservative form, and if the resulting sequence of approximated solutions converges, then necessarily the limiting function is a weak solution of the conservation law.

Unfortunately, consistency and conservative form are not sufficient in general to capture the correct discontinuous solution. For example, schemes might develop undesirable oscillations. These conditions are related to the entropy condition mentioned above. Since we have not developed at all this notion, we will say no more about the numerical approximation of hyperbolic equations, and we refer to [166, 212] for more development.

We summarize all the numerical concerns by saying that a monotone (which means that the numerical flux function F is a monotone increasing function of each of its arguments), consistent and conservative scheme always captures the solution we would like to get (the unique entropic weak solution).

A.3   Difference Schemes in Image Analysis

A.3.1   Getting Started

In this section we would like to show how certain PDEs studied in this book can be discretized. The generic form of these PDEs is

$$\begin{cases} \mathcal{L}v = F & (t, x, y) \in R^+ \times \Omega, \\ \dfrac{\partial v}{\partial N}(t, x, y) = 0 & \text{on } R^+ \times \partial\Omega \quad \text{(Neumann boundary condition)}, \\ v(0, x, y) = f(x, y) & \text{(initial condition)}, \end{cases} \qquad (A.65)$$

where Ω is the image domain and N is the normal to the boundary of Ω, denoted by $\partial\Omega$. $\mathcal{L}$ is generally a second-order differential operator such as (see, for example, Section 3.3)

$$\frac{\partial v}{\partial t}(t, x, y) + H(x, y, v(t, x, y), \nabla v(t, x, y), \nabla^2 v(t, x, y)) = 0.$$

Example: One of the simplest PDEs presented in image analysis is the heat equation (see Section 3.3.1, where it is analyzed):

$$\begin{cases} \dfrac{\partial v}{\partial t} = \nu \Delta v = \nu \left(\dfrac{\partial^2 v}{\partial x^2} + \dfrac{\partial^2 v}{\partial y^2} \right) & (t, x, y) \in R^+ \times \Omega \\ \dfrac{\partial v}{\partial N}(t, x, y) = 0 \ \text{ on } \ R^+ \times \partial\Omega, \\ v(0, x, y) = f(x, y), \end{cases} \qquad (A.66)$$

where ν is a positive constant. ∎

As already mentioned, finite differences are widely used in image processing, which is due to the digital structure of an image as a set of pixels uniformly distributed (see Section 1.2). It is then very easy and natural to associate with an image a uniform grid, as presented in Figure A.5. Since there is no reason to choose it differently, the grid spacing in the x and y directions is usually equal:

$$\triangle x = \triangle y = h.$$

Notice that in many articles from the computer vision literature, it is even chosen as $h = 1$, which means that the pixel size is chosen as the unit of reference. We will call the positions (ih, jh) vertices, nodes, or pixels equivalently. We will denote by $v_{i,j}^n$ (respectively $u_{i,j}^n$) the value of the exact solution (respectively the discrete solution) at location (ih, jh) and time $n\triangle t$.

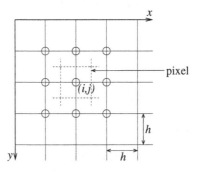

Figure A.5. Grid on the space domain. The circles indicate the vertices that belong to the 3×3 neighborhood of the vertex (i, j).

Example: The PDE (A.66) is an initial-boundary value problem. To discretize it, we need to consider the following:

- The equation. To find the difference scheme associated with the heat equation (A.66), we can proceed as in the one-dimensional case (see Section A.1.1), that is, by using Taylor expansions expanded about the index point $(n\triangle t, ih, jh)$. Naturally, the simplest method is to consider separately the discretization of each second-order derivative in x and y, which is equivalent to using the one-dimensional approximation. By doing so, we obtain

$$\frac{\partial v}{\partial t} - \nu \Delta v \bigg|_{i,j}^{n} = \frac{v_{i,j}^{n+1} - v_{i,j}^{n}}{\Delta t}$$

$$- \nu \frac{v_{i+1,j}^{n} + v_{i-1,j}^{n} + v_{i,j+1}^{n} + v_{i,j-1}^{n} - 4v_{i,j}^{n}}{h^2} + \mathcal{O}(\Delta t) + \mathcal{O}(h^2).$$

Then the difference scheme that we can propose is

$$u_{i,j}^{n+1} = u_{i,j}^{n} + \frac{\nu \Delta t}{h^2} \left(u_{i+1,j}^{n} + u_{i-1,j}^{n} + u_{i,j+1}^{n} + u_{i,j-1}^{n} - 4u_{i,j}^{n} \right). \qquad (A.67)$$

It is of order $(2, 1)$.

- The boundary condition. The Neumann boundary condition can be taken into account by a symmetry procedure. If the value of a pixel (vertex) that is outside the domain is needed, we use the value of the pixel that is symmetric with respect to the boundaries.

- The initial condition is simply: $u_{i,j}^{0} = g_{i,j}$, where g is the discretization of f.

To illustrate this algorithm, we show in Figure A.6 some iterations as applied to a very simple image.

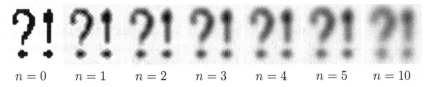

$$n = 0 \qquad n = 1 \qquad n = 2 \qquad n = 3 \qquad n = 4 \qquad n = 5 \qquad n = 10$$

Figure A.6. Example of results with the scheme (A.67) at different times (iterations), as applied to a simple and small image (32×32).

Notice that this example shows clearly the propagation of the information as the number of iterations increases.   ■

Remark The scheme (A.67) was obtained by discretizing the Laplacian as a sum of the second-order derivatives in the x and y directions (see Figure A.8):

$$\Delta v|_{i,j} \approx \frac{v_{i+1,j}^{n} + v_{i-1,j}^{n} + v_{i,j+1}^{n} + v_{i,j-1}^{n} - 4v_{i,j}^{n}}{h^2}. \qquad (A.68)$$

Clearly, this discretization does not take into account the 2-D nature and properties of this operator. To illustrate what we mean by "2-D nature and properties," we can remark that the Laplacian operator is rotationally invariant. If we apply a rotation of center (x, y) to the image v (with any angle $\theta \in [0, 2\pi[$), then $\Delta v(x, y)$ remains constant for all θ, as it should be for the discretization. Naturally, as we consider a discrete domain, we may ask only that $\Delta v|_{i,j}$ keep constant under rotations of $\pi/4$, as depicted in Figure A.7. This is not the case for discretization (A.68), since we obtain 1 or 2 (for $h = 1$) depending on the situation.

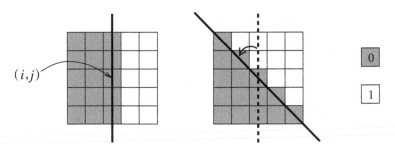

Figure A.7. Example of a binary image representing a vertical edge and the same image after a rotation of $\pi/4$ radians. Rotationally invariant operators estimated at the point in the middle should yield the same value in both situations.

To overcome this difficulty, we need to use the complete 3×3 neighborhood. We may propose the following approximation:

$$\Delta v|_{i,j} \approx \lambda \frac{v_{i+1,j} + v_{i-1,j} + v_{i,j+1} + v_{i,j-1} - 4v_{i,j}}{h^2} \tag{A.69}$$
$$+ (1 - \lambda) \frac{v_{i+1,j+1} + v_{i-1,j+1} + v_{i+1,j-1} + v_{i-1,j-1} - 4v_{i,j}}{2h^2},$$

where $\lambda \in [0, 1]$ is a constant to be chosen. We can verify that this approximation is consistent. Applying this operator (A.69) in the two situations from Figure A.7 and saying that both results should be equal yields $\lambda = \frac{1}{3}$, hence the approximation.

$\Delta v|_{i,j}$

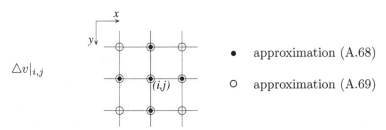

$\bullet$ approximation (A.68)

$\circ$ approximation (A.69)

Figure A.8. Representation of the vertices involved in the finite-difference schemes.

Similarly, we can propose a discretization for the first-order derivatives in x and y that is consistent with the fact that the norm of the gradient is invariant under rotation. As we will see further, a second-order centered approximation of the first derivative in x is

$$\frac{\partial v}{\partial x}\bigg|_{i,j} \approx \delta_x v_{i,j} = \frac{v_{i+1,j} - v_{i-1,j}}{2h}, \tag{A.70}$$

which can also be written in the y direction. The vertices involved in the estimation (A.70) are represented in Figure A.9. As for the case of the Laplacian, these approximations are in fact "one-dimensional" and do not really take advantage of the fact that the data is of dimension 2. This is visible if we consider the value of the norm of the gradient of u in the two situations described in Figure A.7: We obtain either $\frac{1}{2}$ or $\frac{1}{\sqrt{2}}$. The solution is to use more pixels in the estimation of the derivatives. In particular, we may suggest the following approximation:

$$\left.\frac{\partial v}{\partial x}\right|_{i,j} \approx \lambda \frac{v_{i+1,j} - v_{i-1,j}}{2h} \tag{A.71}$$
$$+ \frac{(1-\lambda)}{2}\left(\frac{v_{i+1,j+1} - v_{i-1,j+1}}{2h} + \frac{v_{i+1,j-1} - v_{i-1,j-1}}{2h}\right),$$

where λ is a parameter to be chosen. Applying the operator (A.71) in the two situations of Figure A.9 and by saying that both results should be equal yields $\lambda = \sqrt{2} - 1$, hence the approximation.

$$\left.\frac{\partial v}{\partial x}\right|_{i,j}$$

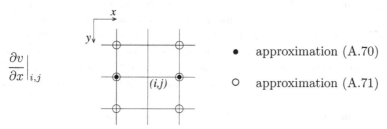

•   approximation (A.70)

○   approximation (A.71)

Figure A.9. Representation of the vertices involved in the finite-difference schemes.

Finally, we would like to mention that not only is using more points in the approximations good for rotation-invariance properties, but, practically, the result is also less sensitive to noise. The reason is that it is equivalent to perform a smoothing of the data before the estimation. ∎

A.3.2 Image Restoration by Energy Minimization

We first consider the image restoration problem as presented in Section 3.2. By introducing a dual variable b, the problem becomes to minimize with respect to v and b the functional

$$J_\varepsilon(v,\, b) = \frac{1}{2}\int_\Omega |Rv - v_0|^2\, dx + \lambda \int_\Omega (b\,|\nabla v|^2 + \psi_\varepsilon(b))dx.$$

The so-called half-quadratic minimization algorithm consists in minimizing $\boxed{\text{C++}}$ successively J_ε with respect to each variable. The algorithm is (see Section 3.2.4 for more detail) as follows:

_____For (v^0, b^0) given_____

- $v^{n+1} = \underset{v}{argmin}\, J_\varepsilon(v, b^n)$ i.e., $\begin{cases} R^* R v^{n+1} - \operatorname{div}(b^n \nabla v^{n+1}) = 0 \text{ in } \Omega, \\ b^n \dfrac{\partial v^{n+1}}{\partial N} = 0 \text{ on } \partial\Omega. \end{cases}$

- $b^{n+1} = \underset{b}{argmin}\, J_\varepsilon(v^{n+1}, b)$ i.e., $b^{n+1} = \dfrac{\phi'\left(|\nabla v^{n+1}|\right)}{2|\nabla v^{n+1}|}$.

- Go back to the first step until convergence.

_____The limit (v^∞, b^∞) is the solution_____

As far as discretization is concerned, the only term that may be difficult to approximate is the divergence operator. So for $b \geq 0$ and v given at nodes (i, j) the problem is to find an approximation at the node (i, j) for $\operatorname{div}(b\nabla v)$. This kind of term is naturally present as soon as there is a regularization with a ϕ function (see, for instance, optical flow, Section 5.3.2, or sequence segmentation, Section 5.3.3). The diffusion operator used in the Perona and Malik model is also of the same kind (see Sections 3.3.1 and 3.3.2).

Since this divergence operator may be rewritten as:

$$\operatorname{div}(b\nabla v) = \frac{\partial}{\partial x}\left(b\frac{\partial v}{\partial x}\right) + \frac{\partial}{\partial y}\left(b\frac{\partial v}{\partial y}\right),$$

we can use the previous one-dimensional approximation and combine them. For example, if we use the central finite-difference approximation (A.70), we have

$$\operatorname{div}(b\nabla v)|_{i,j} \approx \delta_x(b_{i,j}\delta_x v_{i,j}) + \delta_y(b_{i,j}\delta_y v_{i,j}) \tag{A.72}$$

$$= \frac{1}{4h^2}\Big(b_{i+1,j}v_{i+2,j} + b_{i-1,j}v_{i-2,j} + b_{i,j+1}v_{i,j+2} + b_{i,j-1}v_{i,j-2}$$

$$- (b_{i+1,j} + b_{i-1,j} + b_{i,j+1} + b_{i,j-1})v_{i,j}\Big).$$

The main drawback of this representation is that it involves only the points $((i \pm 2)h, (j \pm 2)h)$, and none of the 3×3 neighborhood (see also Figure A.10). This may be nonrobust for noisy data or when there is considerable variation in this region. Another possibility is to combine forward and backward differences (see Section A.2):

$$\operatorname{div}(b\nabla v)|_{i,j} \approx \delta_x^+(b_{i,j}\delta_x^- v_{i,j}) + \delta_y^+(b_{i,j}\delta_y^- v_{i,j})$$

$$= \frac{1}{h^2}\Big(b_{i+1,j}v_{i+1,j} + b_{i,j}v_{i-1,j} + b_{i,j+1}v_{i,j+1} + b_{i,j}v_{i,j-1}$$

$$- (b_{i+1,j} + b_{i,j+1} + 2b_{i,j})v_{i,j}\Big).$$

This approximation now involves the 3×3 neighborhood, but it introduces an asymmetry: The values of b at $((i-1)h, jh)$ and $(ih, (j-1)h)$ are not used. A solution is to use the following approximation for the derivatives:

$$\delta_x^* v_{i,j} = \frac{v_{i+\frac{1}{2},j} - v_{i-\frac{1}{2},j}}{h} \quad \text{and} \quad \delta_y^* v_{i,j} = \frac{v_{i,j+\frac{1}{2}} - v_{i,j-\frac{1}{2}}}{h},$$

where $v_{i\pm\frac{1}{2},j\pm\frac{1}{2}}$ is the value of v at location $((i\pm\frac{1}{2})h, (j\pm\frac{1}{2})h)$, which can be obtained by interpolation. As for (A.70), it is a second-order approximation. Then we have

$$\text{div}(b\nabla v)|_{i,j} \approx \delta_x^*(b_{i,j}\delta_x^* v_{i,j}) + \delta_y^*(b_{i,j}\delta_y^* v_{i,j}) \tag{A.73}$$

$$= \frac{1}{h^2}\Big(b_{+0}v_{i+1,j} + b_{-0}v_{i-1,j} + b_{0+}v_{i,j+1} + b_{0-}v_{i,j-1}$$

$$- (b_{+0} + b_{-0} + b_{0+} + b_{0-})v_{i,j} \Big),$$

where $b_{\pm 0} = b_{i\pm\frac{1}{2},j}$ and $b_{0\pm} = b_{i,j\pm\frac{1}{2}}$. Notice that since we applied the operators δ_x^* and δ_y^* twice, this approximation uses the values of v only at $((i\pm 1)h, (j\pm 1)h)$ (see Figure A.10). However, interpolation is needed for b.

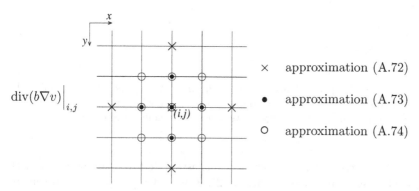

$\times$	approximation (A.72)
$\bullet$	approximation (A.73)
$\circ$	approximation (A.74)

$\text{div}(b\nabla v)\big|_{i,j}$

Figure A.10. Vertices involved in the approximation of the divergence term for the different schemes.

As mentioned previously for the estimation of the Laplacian, it would also be interesting to take into account the diagonal values. Then, we can look for an approximation such that

$$\text{div}(b\nabla v)|_{i,j} \approx \tag{A.74}$$

$$\frac{\lambda_p}{h^2}\left(b_{+0}v_{i+1,j} + b_{-0}v_{i-1,j} + b_{0+}v_{i,j+1} + b_{0-}v_{i,j-1} - \beta_p v_{i,j}\right)$$

$$+ \frac{\lambda_d}{h^2}\left(b_{++}v_{i+1,j+1} + b_{--}v_{i-1,j-1} + b_{-+}v_{i-1,j+1} + b_{+-}v_{i+1,j-1} - \beta_d v_{i,j}\right)$$

$$\tag{A.75}$$

with $b_{\pm\pm} = b_{i\pm\frac{1}{2}, j\pm\frac{1}{2}}$,

$$\begin{cases} \beta_p = b_{0+} + b_{0-} + b_{+0} + b_{-0}, \\ \beta_d = b_{++} + b_{--} + b_{+-} + b_{-+}, \end{cases}$$

and where λ_p and λ_d are two weights to be chosen. The first condition is that the scheme must be consistent, and it can be verified that this implies

$$\lambda_p + 2\lambda_d = 1. \tag{A.76}$$

Now there remains one degree of freedom. Two possibilities can been considered:

- The first is to choose (λ_p, λ_d) constant, and for instance equal to $(\frac{1}{2}, \frac{1}{4})$, giving a privilege to the principal directions.

- The second is to choose (λ_p, λ_d) by taking into account the orientation of the gradient of v, as described in Figure A.11.

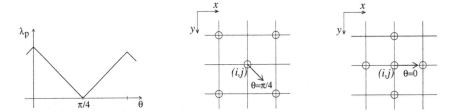

Figure A.11. Adaptive choice of the coefficients (λ_p, λ_d) as a function of θ, the orientation of ∇v. The two right-hand figures show which points will be used in the discretization of the divergence term in two specific situations.

We tested these different discretizations as applied to a simple image with geometric structures (see Figure A.12). From left to right, an improvement in the results can be perceived (by observing the restoration of the horizontal and vertical edges). It is the adaptive choice that gives the best result.

A.3.3 Image Enhancement by the Osher and Rudin Shock Filters

This section concerns the shock-filter equation discussed in Section 3.3.3 and proposed by Osher and Rudin [261]:

$$\frac{\partial v}{\partial t} = -|\nabla v| \, F(L(v)), \tag{A.77}$$

where:

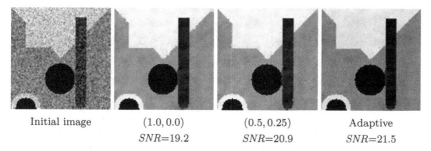

Initial image	(1.0, 0.0)	(0.5, 0.25)	Adaptive
	$SNR{=}19.2$	$SNR{=}20.9$	$SNR{=}21.5$

Figure A.12. Numerical tests for the different discretizations of the divergence term (the choice of (λ_p, λ_d) is indicated below the images).

- F is a Lipschitz function satisfying $F(0) = 0$, $\text{sign}(s)F(s) > 0$ $(s \neq 0)$, for example $F(s) = \text{sign}(s)$.

- L is a second-order edge detector, for example

$$L(v) = \Delta v = v_{xx} + v_{yy} \quad \text{or} \quad L(v) = \frac{1}{|\nabla v|^2} \left(v_x^2\, v_{xx} + 2 v_x v_y v_{xy} + v_y^2 v_{yy} \right),$$

which corresponds to the second derivative of v in the direction of the normal to the isophotes.

Equation (A.77) involves two kinds of terms: a first-order term $|\nabla v|$ and a second-order term $F(L(v))$:

- L is discretized with central finite differences.

- $|\nabla v|$ has to be approximated with more care. v_x and v_y are approximated using the minmod operator $m(\alpha, \beta)$. For instance,

$$v_x|_i = m(\delta_x^- v_i, \delta_x^+ v_i),$$

where

$$m(\alpha, \beta) = \begin{cases} \text{sign}(\alpha)\, \min(|\alpha|, |\beta|) & \text{if } \alpha\beta > 0, \\ 0 & \text{if } \alpha\beta \leq 0. \end{cases}$$

This function is usually called a flux limiter. As shown in Figure A.13, it permits us to choose the lowest slope, or zero in case of a local extremum (this prevents instabilities due to noise).

To summarize, the approximation of (A.77) is then given by

$$u_{i,j}^{n+1} = u_{i,j}^n - \frac{\Delta t}{h} \sqrt{(m(\delta_x^+ u_{i,j}^n,\ \delta_x^- u_{i,j}^n))^2 + (m(\delta_y^+ u_{i,j}^n,\ \delta_y^- u_{i,j}^n))^2}\ F_{i,j}(L(u^n)),$$

where $F_{i,j}(L(u)) = F(L_{i,j}(u))$. We show an example in Figure A.14 and refer to Section 3.3.3 for more detail.

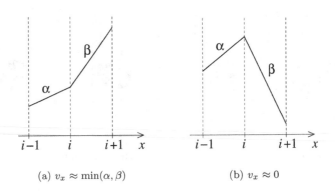

(a) $v_x \approx \min(\alpha, \beta)$ (b) $v_x \approx 0$

Figure A.13. Approximation of the first derivative using the minmod function.

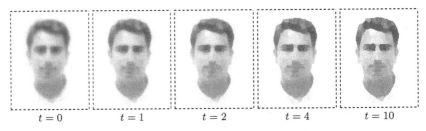

$t = 0$ $t = 1$ $t = 2$ $t = 4$ $t = 10$

Figure A.14. Example of the shock filter on a blurred image of a face. It shows clearly that this filter reconstructs a piecewise constant image that is not satisfying from a perceptual point of view (the result does not look like a real image).

A.3.4   Curve Evolution with the Level-Set Method

In this section we briefly discuss the discretization of the PDEs governing curve evolution. We examine only the case where these curves are identified as level sets of the same function $u(t, x)$ (Eulerian formulation). Of course, we have in mind the geodesic active contours model given by (see Section 4.3.3)

$$\frac{\partial v}{\partial t} = g(|\nabla I|)\, |\nabla v|\, \operatorname{div}\left(\frac{\nabla v}{|\nabla v|}\right) + \alpha\, g(|\nabla I|)\, |\nabla v| + \nabla g \cdot \nabla v. \quad (A.78)$$

As mentioned before, (A.78) involves two kinds of terms: a parabolic term (the first one) and hyperbolic terms (the last two). One can easily imagine that the discretization of each term needs an appropriate treatment, according to its nature (parabolic or hyperbolic). The main idea is that parabolic terms can be discretized by central finite differences, while hyperbolic terms need to be approximated by nonoscillatory upwind schemes. For the sake of clarity we start by examining the evolution driven by each of these terms.

For a detailed description of the above schemes and other numerical questions not developed here, we refer the reader to [300, 262].

MEAN CURVATURE MOTION

Let us consider

$$
\begin{cases}
\dfrac{\partial v}{\partial t} = |\nabla v| \; \mathrm{div}\left(\dfrac{\nabla v}{|\nabla v|}\right), \\
v(0, x, y) = v_0(x, y).
\end{cases}
\tag{A.79}
$$

Equation (A.79) is a parabolic equation and has diffusive effects (like the heat equation). So, the use of upwind schemes is inappropriate, and classical central differences are used,

$$
u_{i,j}^{n+1} = u_{i,j}^n + \Delta t \sqrt{(\delta_x u_{i,j}^n)^2 + (\delta_y u_{i,j}^n)^2} \; K_{i,j}^n,
$$

where $K_{i,j}^n$ is the central finite-difference approximation of the curvature:

$$
K = \mathrm{div}\left(\dfrac{\nabla v}{|\nabla v|}\right) = \dfrac{v_{xx} v_y^2 + v_{yy} v_x^2 - 2 v_x v_y v_{xy}}{(v_x^2 + v_y^2)^{3/2}}.
$$

We let the reader write the expression of $K_{i,j}^n$. Unfortunately, the discretization of (A.79) is not as easy as it may appear. At some points (t, x), ∇v can be undefined, or $|\nabla v| = 0$, or $|\nabla v| = +\infty$. This situation can occur even in very simple cases [273]. For example, let us consider the shrinking of a unit circle in two dimensions which corresponds to

$$
v(0, x, y) = \sqrt{x^2 + y^2} - 1,
\tag{A.80}
$$

i.e., v_0 is the signed distance to the unit circle. Equation (A.79) is rotationally invariant, and if we search for the solution of the form

$$
v(t, x, y) = \phi(t, \sqrt{x^2 + y^2}),
$$

we easily get $v(t, x, y) = \sqrt{x^2 + y^2 + 2t} - 1$, from which we deduce

$$
\nabla v = \dfrac{1}{\sqrt{x^2 + y^2 + 2t}} \begin{pmatrix} x \\ y \end{pmatrix}, \qquad |\nabla v| = \dfrac{\sqrt{x^2 + y^2}}{\sqrt{x^2 + y^2 + 2t}},
$$

$$
\dfrac{\nabla v}{|\nabla v|} = \dfrac{1}{\sqrt{x^2 + y^2}} \begin{pmatrix} x \\ y \end{pmatrix}, \qquad \mathrm{div}\left(\dfrac{\nabla v}{|\nabla v|}\right) = \dfrac{1}{\sqrt{x^2 + y^2}},
$$

so the two last quantities are not defined at the origin, and effectively a spike occurs at the origin (see [273]). Moreover, the interface $\Gamma(t) = \{(x, y); \; v(t, x, y) = 0\}$ is the circle $x^2 + y^2 = 1 - 2t$, and on $\Gamma(t)$ we have $|\nabla v|(t) = \sqrt{1 - 2t}$. Therefore $v(t, x, y)$ becomes more and more flat as the interface evolves and disappears at $t = \frac{1}{2}$. To circumvent this type of problem, we have to find a numerical trick that prevents the gradient norm from vanishing (or blowing up). This can be realized by reinitializing the function v from time to time to a signed distance function.

More precisely, we run (A.79) until some step n; then we solve the   C++
auxiliary PDE

$$\begin{cases} \dfrac{\partial \phi}{\partial t} + \text{sign}(v)(|\nabla \phi| - 1) = 0, \\ \phi(0, x, y) = v(n\triangle t, x, y). \end{cases} \qquad (\text{A.81})$$

The resulting solution (as t tends to infinity), denoted by ϕ^∞, is a signed
distance function whose zero-level set is the same as that of the function
$v(n\triangle t, x, y)$. We refer to Section 4.3.4 for more details about this equation.
Then we can run again (A.79) with the initial data $v(0, x, y) = \phi^\infty(x, y)$.
Practically, this reinitialization has to be done every $n = 20$ iterations of
the curve evolution equation, and it is usually performed about every 5 to
10 iterations of (A.81).[6]

Remark There exists another way to avoid doing the reinitialization step.
It consists in considering a modified equation (A.79) that has the property
of maintaining the norm of the gradient of the solution equal to one. For
further details see [168, 273].   ■

An example of mean curvature motion is shown in Figure A.15. Notice
that if we let the evolution run until convergence, any curve transforms
into a circle and then collapses.

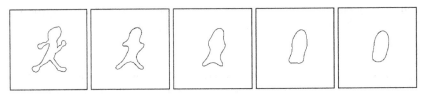

Figure A.15. Example of mean curvature motion.

CONSTANT SPEED EVOLUTION

The second example is given by

$$\begin{cases} \dfrac{\partial v}{\partial t} = c\, |\nabla v|, \\ v(0, x, y) = v_0(x, y), \end{cases} \qquad (\text{A.82})$$

where c is a constant. This equation describes a motion in the direction
normal to the front (the corresponding Lagrangian formulation of (A.82)
is $\frac{\partial \Gamma}{\partial t}(t, p) = c\, N(t, p)$, where N is the normal to $\Gamma(t)$). For $c = 1$, it
is also referred to as *grass fire*, since it simulates a grass fire wave-front
propagation.

---

[6]These numbers are just an indication and naturally depend on the kind of equation
to be solved and on the time steps.

Equation (A.82) is approximated by a nonoscillatory upwind scheme:

$$u_{i,j}^{n+1} = u_{i,j}^n + \Delta t \nabla^+ u_{i,j}^n,$$

where

$$\nabla^+ u_{i,j}^n = \left[\max(\delta_x^- u_{i,j}^n, 0)^2 + \min(\delta_x^+ u_{i,j}^n, 0)^2 \right.$$
$$\left. + \max(\delta_y^- u_{i,j}^n, 0)^2 + \min(\delta_y^+ u_{i,j}^n, 0)^2 \right]^{1/2}.$$

We show in Figures A.16 and A.17 two examples of constant speed motions.

Figure A.16. Example of constant speed motion ($c = 1$, grass fire).

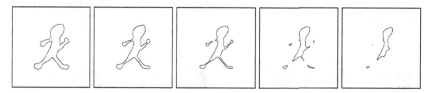

Figure A.17. Example of constant speed motion ($c = -1$).

Remark Motions like equation (A.82) and more generally with a monotone speed have the following property: Every point is crossed once and only once by the curve during its evolution. Notice that this is not the case for mean curvature motion. This property can be used to derive an efficient numerical approach called the *fast marching algorithm* [325, 299]. It is beyond the scope of this Appendix to explain this method, and we refer to the original articles and to [300] for more detail. ∎

THE PURE ADVECTION EQUATION

We consider here the equation

$$\begin{cases} \dfrac{\partial v}{\partial t} = A(x,y) \cdot \nabla v, \\ v(0,x,y) = v_0(x,y), \end{cases} \tag{A.83}$$

where $A(x,y) = (A_1(x,y), A_2(x,y))$. For (A.83) we use a simple upwind scheme, i.e., we check the sign of each component of A and construct a one-side upwind difference in the appropriate direction:

$$u_{i,j}^{n+1} = u_{i,j}^n + \Delta t \Big[\max((A_1)_{i,j}^n, 0)\delta_x^- u_{i,j}^n + \min((A_1)_{i,j}^n, 0)\delta_x^+ u_{i,j}^n$$
$$\max((A_2)_{i,j}^n, 0)\delta_y^- u_{i,j}^n + \min((A_2)_{i,j}^n, 0)\delta_y^+ u_{i,j}^n \Big].$$

IMAGE SEGMENTATION BY THE GEODESIC ACTIVE CONTOUR MODEL

C++

Now we can consider the geodesic active contour model (A.78), which can be seen as the sum of the previous discretization. So the discrete scheme is

$$u_{i,j}^{n+1} = u_{i,j}^n + \Delta t \Big[g_{i,j}\, K_{i,j}^n [(\delta_x u_{i,j}^n)^2 + (\delta_y u_{i,j}^n)^2]^{\frac{1}{2}}$$
$$+ \alpha[\max(g_{i,j}, 0)\nabla^+ + \min(g_{i,j}, 0)\nabla^-]u_{i,j}^n$$
$$+ \max(g_{x\,i,j}, 0)\delta_x^- u_{i,j}^n + \min(g_{x\,i,j}, 0)\,\delta_x^+ u_{i,j}^n$$
$$+ \max(g_{y\,i,j}, 0)\,\delta_y^- u_{i,j}^n + \min(g_{y\,i,j}, 0)\,\delta_y^+ u_{i,j}^n \Big].$$

where $\nabla^- u_{i,j}^n$ is obtained from $\nabla^+ u_{i,j}^n$ by inverting the signs plus and minus.

Figure A.18 shows a typical example (see Figure 4.13 for the complete evolution).

Figure A.18. Example of segmentation. Different iterations are displayed.

Remark From a numerical point of view, all the equations presented in this section involve local operations. Since we are interested only in the curve, it is enough to update the values in a band around the current position of the curve, also called a *narrow band*. Naturally, this region (band) has to be updated as the curve evolves. See, for instance, [300] for more details. ∎

Appendix B
Experiment Yourself!

How to Read This Chapter

The objective of this book was to explain the underlying mathematics of the PDE-based and variational approaches used in image processing. In Appendix A we explain the main ideas of finite difference methods, which allow one to discretize continuous equations.

As a further step, we wish to provide programming tools so that readers can implement the approaches and test them on their own data.

A web site is associated with this second edition. This web site contains some related links, complementary of information, and also source code that allow the reader to test easily some variational and PDE-based approaches.

http://www-sop.inria.fr/books/imath

Section B.1 justifies the technical choices for the software developments, which is essentially the chosen image processing library. Section B.2 gives a nonexhaustive list of the C++ codes available online and shows an example of CImg code.

B.1   The CImg Library

The chosen programming language is the object-oriented language C++, which is freeware and a very efficient language. Bjarne Stroustrup [311] is the designer and original implementor of C++. The interested reader will also find a wide variety of books and online tutorials on this language.

Many image processing libraries are proposed online. We chose the CImg library, which stands for "Cool Image," developed by David Tschumperlé in 2000. The CImg library is simple to use and efficient:

- The CImg Library is an open source C++ toolkit for image processing. It provides simple classes and functions to load, save, process, and display images in your own C++ code.

- It is highly portable and is fully functional on Unix/X11, Windows, MacOS X, and FreeBSD operating systems. It should compile on other systems as well (possibly without display capabilities).

- It consists of only a single header file CImg.h, which must be included in your C++ program source.

- It contains useful image processing algorithms for image loading/saving, displaying, resizing/rotating, filtering, object drawing (text, lines, faces, curves, ellipses,...), etc.

- Images are instanced by a class able to represent images up to 4 dimensions wide (x, y, z, v) (from 1-D scalar signals to 3-D volumes of vector-valued pixels), with template pixel types.

- It depends on a minimal number of libraries: you can compile it with only standard C libraries. No need for exotic libraries and complex dependencies.

- Additional features appear with the use of ImageMagick, libpng, or libjpeg: install the ImageMagick package or link your code with libpng and libjpeg to be able to load and save compressed image formats (GIF, BMP, TIF, JPG, PNG, ...). Available for any platform.

The CImg package includes full documentation and many examples to help the developer in his first steps.

`http://cimg.sourceforge.net`

B.2   What Is Available Online?

Table B.1 gives a sample of approaches whose CImg codes are available from the book web site. The proposed codes correspond in general to approaches explained in the book. The symbol $\boxed{\texttt{C++}}$ written in the margin will indicate to the reader that some code is available. The list in Table B.1 is not exhaustive, and we invite the reader to consult regularly the book web site, which will contain updated information.

Tables B.2 and B.3 give an idea of the CImg code necessary to run a heat equation. Table B.2 is the main code, which calls the function `get_isotropic2d` defined in Table B.3. We refer the reader to the book web site for more information.

External contributors are encouraged to submit their own C++ source codes. The procedure is described at the book web site. We hope that such an initiative will enable readers to experiment and compare different approaches without too much effort. We thank in advance new contributors, who will help us to develop this free source code database for PDE-based approaches.

Image restoration		
Description	Ref.	Sections
Half Quadratic Minimization	[108]	3.2.4 and A.3.2
Heat equation	[43, 5, 198]	3.3.1
Perona-Malik equation	[275]	3.3.1 and 3.3.2
Coherence enhancing diffusion	[336]	3.3.1
Shock filters	[261]	3.3.3 and A.3.3
Level-sets and segmentation		
Description	Ref.	Sections
Mean curvature motion	[147]	4.3.2 and A.3.4
Geodesic active contours	[85, 223, 222]	4.3.5 and A.3.4
Distance function initialization	[223, 312]	4.3.4 and A.3.4
Vector–valued regularization	[324]	5.5
Some applications		
Description	Ref.	Sections
Inpainting	[320, 321]	5.1
Optical flow	[184]	5.3.2

Table B.1. Example of approaches proposed as CImg source code. Please consult the book web site for an updated list. Contributors are welcome.

```
#define cimg_plugin "pde_plugin.h"
#include "CImg.h"
using namespace cimg_library;

int main(int argc,char **argv) {

  CImg<float> img("my_image.gif");
  CImgDisplay disp(img,"heat flow");

  img.get_isotropic2d(1000,-5,&disp);                                    10

  while (!disp.closed) disp.wait();

  return 0;
}
```

Table B.2. CImg main code for the heat equation.

```
//! 2D isotropic smoothing with the classical heat flow PDE
/**
   Return an image that has been smoothed by the classical isotropic heat flow PDE
   \param nb_iter = Number of PDE iterations
   \param dt = PDE time step. If dt<0, -dt represents the maximum PDE velocity
               that will be applied on image pixels (adaptative time step)
   \param disp = Display used to show the PDE evolution. If disp==NULL,
                 no display is performed.
**/
CImg<T> get_isotropic2d(const int nb_iter=100, const float dt=0,                10
                        CImgDisplay *disp=NULL) const {

    CImg<float> img(*this), veloc(*this,false);

    // Iterative process
    for (unsigned int iter=0; iter<nb_iter; iter++) {

        // Estimation of the Laplacian
        CImg_3x3(I,float);
        cimg_mapV(img,k) cimg_map3x3(img,x,y,0,k,I) {                            20
            const float
                ixx = Inc + Ipc − 2*Icc,
                iyy = Icn + Icp − 2*Icc;
            veloc(x,y,k) = ixx + iyy;
        }

        // Estimation of the optimal time step
        float xdt = dt;
        if (dt<0) {
            CImgStats stats(veloc,false);                                       30
            xdt = −dt/cimg::abs(cimg::max(stats.min,stats.max));
        }

        // Update the image
        img += xdt*veloc;

        // Display resulting image
        if (disp) {
            if (disp−>resized) disp−>resize();
            img.display(*disp);                                                 40
        }
    }

    // Returns final result
    return img;
}
```

Table B.3. CImg code for heat equation.

References

[1] D. Adalsteinsson and J.A. Sethian. The fast construction of extension velocities in level set methods. *Journal of Computational Physics*, 148(1):2–22,
1999.

[2] R. Adams. *Sobolev spaces*, volume 65 of *Pure and Applied Mathematics,
Series of Monographs and Textbooks*. Academic Press, Inc., New York, San
Francisco, London, 1975.

[3] A.A. Alatan and L. Onural. Gibbs random field model based 3-D motion estimation by weakened rigidity. In *Proceedings of the International
Conference on Image Processing*, volume II, pages 790–794, 1994.

[4] S. Allen and J. Cahn. A microscopic theory for antiphase boundary motion and its application to antiphase domain coarsening. *Acta Metallurgica*,
27:1085–1095, 1979.

[5] L. Alvarez, F. Guichard, P.L. Lions, and J.M. Morel. Axiomatisation et
nouveaux opérateurs de la morphologie mathématique. *Comptes Rendus
de l'Académie des Sciences*, pages 265–268, 1992. t. 315, Série I.

[6] L. Alvarez, F. Guichard, P.L. Lions, and J.M. Morel. Axioms and fundamental equations of image processing. *Archive for Rational Mechanics and
Analysis*, 123(3):199–257, 1993.

[7] L. Alvarez, P.L. Lions, and J.M. Morel. Image selective smoothing and edge
detection by nonlinear diffusion (II). *SIAM Journal of Numerical Analysis*,
29:845–866, 1992.

[8] L. Alvarez and L. Mazorra. Signal and image restoration using shock filters
and anisotropic diffusion. *SIAM Journal of Numerical Analysis*, 31(2):590–
605, 1994.

[9] L. Alvarez, J. Weickert, and J. Sànchez. A scale-space approach to nonlocal optical flow calculations. In Mads Nielsen, P. Johansen, O.F. Olsen, and J. Weickert, editors, *Scale-Space Theories in Computer Vision*, volume 1682 of *Lecture Notes in Computer Science*, pages 235–246. Springer–Verlag, 1999.

[10] O. Alvarez, E.N. Barron, and H. Ishii. Hopf-lax formulas for semicontinuous data. *Indiana Univ. Math. J.*, 48:993–1035, 1999.

[11] L. Ambrosio. A compactness theorem for a new class of functions of bounded variation. *Bolletino della Unione Matematica Italiana*, VII(4):857–881, 1989.

[12] L. Ambrosio. Variational problems in *SBV* and image segmentation. *Acta Applicandae Mathematicae*, 17:1–40, 1989.

[13] L. Ambrosio. Existence theory for a new class of variational problems. *Archive for Rational Mechanics and Analysis*, 111:291–322, 1990.

[14] L. Ambrosio and G. Dal Maso. A general chain rule for distributional derivatives. *Proceedings of the American Mathematical Society*, 108(3):691–702, 1990.

[15] L. Ambrosio, N. Fusco, and D. Pallara. *Functions of Bounded Variation and Free Discontinuity Problems*. Oxford Mathematical Monographs. Oxford University Press, March 2000.

[16] L. Ambrosio and S. Masnou. A direct variational approach to a problem arising in image reconstruction. *Interfaces and Free Boundaries*, 5:63–81, 2003.

[17] L. Ambrosio and V.M. Tortorelli. Approximation of functionals depending on jumps by elliptic functionals via Γ-convergence. *Communications on Pure and Applied Mathematics*, XLIII:999–1036, 1990.

[18] F. Andreu-Vaillo, V. Caselles, and J.M. Maz. *Parabolic quasilinear equations minimizing linear growth functionals*. Birkhser Verlag, Basel ; Boston, 2004.

[19] G. Anzellotti. Pairings between measures and bounded functions and compensated compactness. *Ann. Mat. Pura Appl.*, IV(135):293–318, 1983.

[20] M. Attouch. Epi-convergence and duality. Convergence of sequences of marginal and Lagrangian functions. Applications to homogenization problems in mechanics. *Lecture Notes in Mathematics*, 1190:21–56, 1986.

[21] G. Aubert and J.F. Aujol. Modeling very oscillating signals. application to image processing. *Applied Mathematics and Optimization*, 51(2), March 2005.

[22] G. Aubert and L. Blanc-Féraud. Some remarks on the equivalence between 2D and 3D classical snakes and geodesic active contours. *The International Journal of Computer Vision*, 34(1):19–28, 1999.

[23] G. Aubert, L. Blanc-Féraud, M. Barlaud, and P. Charbonnier. A deterministic algorithm for edge-preserving computed imaging using Legendre transform. In *Proceedings of the International Conference on Pattern Recognition*, volume III, pages 188–191, Jerusalem, Israel, October 1994. Computer Society Press.

[24] G. Aubert, R. Deriche, and P. Kornprobst. Computing optical flow via variational techniques. *SIAM Journal of Applied Mathematics*, 60(1):156–182, 1999.

[25] G. Aubert and P. Kornprobst. A mathematical study of the relaxed optical flow problem in the space BV. *SIAM Journal on Mathematical Analysis*, 30(6):1282–1308, 1999.

[26] G. Aubert and P. Kornprobst. Mathematics of image processing. In J.-P. Françoise, G.L. Naber, and S.T. Tsou, editors, *Encyclopedia of Mathematical Physics*, volume 3, pages 1–9, Oxford, 2006. Elsevier.

[27] G. Aubert and R. Tahraoui. Résultats d'existence en optimisation non convexe. *Applicable Analysis*, 18:75–100, 1984.

[28] G. Aubert and L. Vese. A variational method in image recovery. *SIAM Journal of Numerical Analysis*, 34(5):1948–1979, 1997.

[29] J.-F. Aujol and G. Aubert. Signed distance functions and viscosity solutions of discontinuous hamilton-jacobi equations. Technical Report 4507, INRIA, July 2002.

[30] J.-F. Aujol and A. Chambolle. Dual norms and image decomposition models. *The International Journal of Computer Vision*, 63(1):85–104, 2005.

[31] J.-F. Aujol and B. Matei. Simultaneous structure and texture compact representation. In *ACIVS*, 2004.

[32] J.F. Aujol, G. Aubert, L. Blanc-Féraud, and A. Chambolle. Image decomposition into a bounded variation component and an oscillating component. *Journal of Mathematical Imaging and Vision*, 22(1), January 2005.

[33] S. Baldo. Minimal interface criterion for phase transitions in mixtures of Cahn–Hilliard fluids. *Annales de l'Institut Henri Poincaré, Analyse Non Linéaire*, 7(2):67–90, 1990.

[34] B. Ballester, V. Caselles, J. Verdera, M. Bertalmio, and G. Sapiro. A variational model for filling-in gray level and color images. In *Proceedings of the 8th International Conference on Computer Vision*, volume 1, pages 10–16, Vancouver, Canada, 2001. IEEE Computer Society, IEEE Computer Society Press.

[35] C. Ballester, M. Bertalmio, V. Caselles, G. Sapiro, and J. Verdera. Filling-in by joint interpolation of vector fields and gray levels. *IEEE Transactions on Image Processing*, 10(8):1200–1211, 2001.

[36] C. Ballester, V. Caselles, and J. Verdera. Disocclusion by joint interpolation of vector fields and gray levels. *Multiscale Modeling and Simulation*, 2:80–123, 2003.

[37] G.I. Barenblatt, M. Bertsch, R. Dal Passo, and M. Ughi. A degenerate pseudoparabolic regularization of a nonlinear forward–backward heat equation arising in the theory of heat and mass exchange in stably stratified turbulent shear flow. *SIAM Journal on Mathematical Analysis*, 24:1414–1439, 1993.

[38] G. Barles. Discontinuous Viscosity solutions of first order Hamilton–Jacobi equations: a guided visit. *Nonlinear Analysis TMA*, 20(9):1123–1134, 1993.

[39] G. Barles. *Solutions de Viscosité des Equations de Hamilton–Jacobi*. Springer–Verlag, 1994.

[40] E.N. Barron and R. Jensen. Optimal control and semicontinuous viscosity solutions. *Proceedings of the American Mathematical Society*, 113:49–79, 1991.

[41] J.L. Barron, D.J. Fleet, and S.S. Beauchemin. Performance of optical flow techniques. *The International Journal of Computer Vision*, 12(1):43–77, 1994.

[42] A.C. Barroso, G. Bouchitté, G. Buttazzo, and I. Fonseca. Relaxation of bulk and interfacial energies. *Archive for Rational Mechanics and Analysis*, 135:107–173, 1996.

[43] Bart M. ter Haar Romeny. *Geometry-driven diffusion in computer vision*. Computational imaging and vision. Kluwer Academic Publishers, 1994.

[44] J. Bect, L. Blanc-Féraud, G. Aubert, and A. Chambolle. A l^1-unified variational framework for image restoration. In T. Pajdla and J. Matas, editors, *Proceedings of the 8th European Conference on Computer Vision*, volume IV, Prague, Czech Republic, 2004. Springer–Verlag.

[45] G. Bellettini and A. Coscia. Approximation of a functional depending on jumps and corners. *Bolletino della Unione Matematica Italiana*, 8(1):151–181, 1994.

[46] G. Bellettini and A. Coscia. Discrete approximation of a free discontinuity problem. *Numerical Functional Analysis and Optimization*, 15(3–4):201–224, 1994.

[47] D. Béréziat, I. Herlin, and L. Younes. A generalized optical flow constraint and its physical interpretation. In *Proceedings of the International Conference on Computer Vision and Pattern Recognition*, volume 2, pages 487–492, Hilton Head Island, South Carolina, June 2000. IEEE Computer Society.

[48] M. Bertalmio, A. Bertozzi, and G. Sapiro. Navier-stokes, fluid-dynamics and image and video inpainting. In *Proceedings of CVPR'01*, pages 355–362, 2001.

[49] M. Bertalmio, V. Caselles, G. Haro, and G. Sapiro. *PDE-based image and surface inpainting*, chapter 3, pages 33–62. In Paragios et al. [266], 2005.

[50] M. Bertalmio, G. Sapiro, V. Caselles, and C. Ballester. Image inpainting. In Kurt Akeley, editor, *Proceedings of the SIGGRAPH*, pages 417–424. ACM Press, ACM SIGGRAPH, Addison Wesley Longman, 2000.

[51] M. Bertalmio, L. Vese, G. Sapiro, and S. Osher. Simultaneous structure and texture image inpainting. *IEEE Transactions on Pattern Analysis and Machine Intelligence*, 12(8):882–889, 2003.

[52] M. Bertero and P. Boccacci. *Introduction to Inverse Problems in Imaging*. Bristol: IoP, Institute of Physics Publishing, 1998.

[53] M. Berthod, Z. Kato, S. Yu, and J. Zerubia. Bayesian image classification using Markov Random Fields. *Image and Vision Computing*, 14(4):285–293, 1996.

[54] M.J. Black. *Robust incremental optical flow*. PhD thesis, Yale University, Department of Computer Science, 1992.

[55] M.J. Black and P. Anandan. The robust estimation of multiple motions: Parametric and piecewise-smooth flow fields. *CVGIP: Image Understanding*, 63(1):75–104, 1996.

[56] M.J. Black, D.J. Fleet, and Y. Yacoob. Robustly estimating changes in image appearance. *Computer Vision and Image Understanding*, 78:8–31, 2000.

[57] M.J. Black and P. Rangarajan. On the unification of line processes, outlier rejection, and robust statistics with applications in early vision. *The International Journal of Computer Vision*, 19(1):57–91, 1996.

[58] A. Blake and M. Isard. *Active Contours*. Springer–Verlag, 1998.

[59] A. Blake and A. Zisserman. *Visual Reconstruction*. MIT Press, 1987.

[60] L. Blanc-Féraud, M. Barlaud, and T. Gaidon. Motion estimation involving discontinuities in a multiresolution scheme. *Optical Engineering*, 32(7):1475–1482, July 1993.

[61] P. Blomgren and T.F. Chan. Color tv: Total variation methods for restoration of vector-valued images. *IEEE Trans. Imag. Proc.*, 7(3):304–309, 1998. Special Issue on Partial Differential Equations and Geometry-Driven Diffusion in Image Processing and Analysis.

[62] A. Bonnet. Caractérisation des minima globaux de la fonctionnelle de Mumford–Shah en segmentation d'images. *Comptes Rendus de l'Académie des Sciences*, 321(I):1121–1126, 1995.

[63] A. Bonnet. Sur la régularité des bords de minima de la fonctionnelle de Mumford–Shah. *Comptes Rendus de l'Académie des Sciences*, 321(I):1275–1279, 1995.

[64] A. Bonnet. On the regularity of the edge set of Mumford–Shah minimizers. *Progress in Nonlinear Differential Equations*, 25:93–103, 1996.

[65] R. Bornard. *Approches Probabilistes Appliquées a la Restauration d'Archives Télévisées*. PhD thesis, Ecole Centrale des Arts et Manufactures "Ecole Centrale Paris" (In English), November 2002.

[66] G. Bouchitté, A. Braides, and G. Buttazzo. Relaxation results for some free discontinuity problems. *J. Reine Angew. Math.*, 458:1–18, 1995.

[67] G. Bouchitté, I. Fonseca, and L. Mascarenhas. A global method for relaxation. Technical report, Université de Toulon et du Var, May 1997.

[68] C. Bouman and K. Sauer. A generalized Gaussian model for edge-preserving MAP estimation. *IEEE Transactions on Image Processing*, 2(3):296–310, 1993.

[69] B. Bourdin. Image segmentation with a finite element method. *M2AN, Math. Model. Numer. Anal.*, 33(2):229–244, 1999.

[70] J. Bourgain and H. Brezis. On the equation div$y = f$ and application to control of phases. *Journal of the American Mathematical Society*, 16(2):293–426, 2002.

[71] A. Braides. *Approximation of free-discontinuity problems*, volume 1694 of *Lecture Notes in Mathematics*. Springer–Verlag, 1998.

[72] A. Braides and G. Dal Maso. Non-local approximation of the Mumford–Shah functional. *Calc. Var. Partial Differ. Equ.*, 5(4):293–322, 1997.

[73] H. Brezis. *Opérateurs Maximaux Monotones et Semi-Groupes de Contractions dans les Espaces de Hilbert.* North-Holland Publishing Comp, Amsterdam-London, 1973.

[74] H. Brezis. *Analyse fonctionnelle. Théorie et applications.* Masson, 1983.

[75] R. W. Brockett and P. Maragos. Evolution equations for continuous-scale morphological filtering. *IEEE Transactions on Image Processing*, 42(12):3377–3386, December 1994.

[76] T. Brox, A. Bruhn, N. Papenberg, and J. Weickert. High accuracy optical flow estimation based on a theory for warping. In T. Pajdla and J. Matas, editors, *Proceedings of the 8th European Conference on Computer Vision*, volume 4, pages 25–36, Prague, Czech Republic, 2004. Springer–Verlag.

[77] A. Buades, B. Coll, and J.M. Morel. On image denoising method. Technical report, CMLA, 2004.

[78] A. Buades, B. Coll, and J.M. Morel. Neighborhood filters and PDE's. Technical Report 04, CMLA, 2005.

[79] G. Buttazzo. *Semicontinuity, Relaxation and Integral Representation in the Calculus of Variations.* Number 207 in Pitman Research Notes in Mathematics Series. Longman Scientific & Technical, 1989.

[80] J. Cahn and J.E. Hilliard. Free energy of a nonuniform system. I. interfacial free energy. *Journal of Chemical Physics*, 28(1):258–267, 1958.

[81] J. F. Canny. A computational approach to edge detection. *IEEE Transactions on Pattern Analysis and Machine Intelligence*, 8(6):769–798, November 1986.

[82] J.F. Canny. Finding edges and lines in images. Technical Report AI-TR-720, Massachusetts Institute of Technology, Artificial Intelligence Laboratory, June 1983.

[83] F. Cao. Partial differential equations and mathematical morphology. *J. Math. Pures Appl., IX.*, 77(9):909–941, 1998.

[84] M. Carriero, A. Leaci, and F. Tomarelli. A second order model in image segmentation: Blake & Zisserman functional. *Nonlinear Differ. Equ. Appl.*, 25:57–72, 1996.

[85] V. Caselles, F. Catte, T. Coll, and F. Dibos. A geometric model for active contours. *Nümerische Mathematik*, 66:1–31, 1993.

[86] V. Caselles, R. Kimmel, and G. Sapiro. Geodesic active contours. In *Proceedings of the 5th International Conference on Computer Vision*, pages 694–699, Boston, MA, June 1995. IEEE Computer Society Press.

[87] V. Caselles, R. Kimmel, and G. Sapiro. Geodesic active contours. *The International Journal of Computer Vision*, 22(1):61–79, 1997.

[88] F. Catte, P.L. Lions, J.M. Morel, and T. Coll. Image selective smoothing and edge detection by nonlinear diffusion. *SIAM Journal of Numerical Analysis*, 29(1):182–193, February 1992.

[89] T. Cazenave and A. Haraux. *Introduction aux Problèmes d'Evolution Semi-Linéaires. (Introduction to Semilinear Evolution Problems).* Mathématiques & Applications. Ellipses, 1990.

[90] A. Chambolle. Image segmentation by variational methods: Mumford and Shah functional and the discrete approximation. *SIAM Journal of Applied Mathematics*, 55(3):827–863, 1995.

[91] A. Chambolle. An algorithm for total variation minimization and applications. *Journal of Mathematical Imaging and Vision*, 20(1-2):89–97, March 2004.

[92] A. Chambolle and G. Dal Maso. Discrete approximations of the Mumford–Shah functional in dimension two. *M2AN*, 33(4):651–672, 1999. (also available as Technical report 9820 from Université Paris Dauphine, Ceremade).

[93] A. Chambolle, R.A. DeVore, N.Y. Lee, and B.J. Lucier. Nonlinear wavelet image processing: variational problems, compression, and noise removal through wavelet shrinkage. *IEEE Transactions on Image Processing*, 7(3):319–334, 1998.

[94] A. Chambolle and P.L. Lions. Image recovery via total variation minimization and related problems. *Nümerische Mathematik*, 76(2):167–188, 1997.

[95] T. Chan, S.H. Kang, and J. Shen. Euler's elastica and curvature based inpainting. *SIAM J. Appl. Math.*, 2002.

[96] T. Chan, B.Y. Sandberg, and L. Vese. Active contours without edges for vector-valued images. *Journal of Visual Communication and Image Representation*, 11:130–141, 2000.

[97] T. Chan and J. Shen. Mathematical models for local deterministic inpaintings. Technical Report 00-11, Department of Mathematics, UCLA, Los Angeles, March 2000.

[98] T. Chan and J. Shen. Mathematical models for local non-texture inpaintings. *SIAM Journal of Applied Mathematics*, 62:1019–1043, 2002.

[99] T. Chan and L. Vese. Active contours without edges. Technical Report 98-53, UCLA CAM Report, 1999.

[100] T. Chan and L. Vese. Image segmentation using level sets and the piecewise-constant Mumford–Shah model. Technical Report 00-14, UCLA CAM Report, 2000.

[101] T. Chan and L. Vese. Active contours without edges. *IEEE Transactions on Image Processing*, 10(2):266–277, February 2001.

[102] T. Chan and L. Vese. A level set algorithm for minimizing the Mumford–Shah functional in image processing. *1st IEEE Workshop on Variational and Level Set Methods in Computer Vision*, pages 161–168, 2001.

[103] T.F. Chan and J. Shen. Non-texture inpainting by curvature-driven diffusions (cdd). *J. Visual Comm. Image Rep.*, 12(4):436–449, 2001.

[104] T.F. Chan and J. Shen. *Image Processing and Analysis - Variational, PDE, wavelet, and stochastic methods.* SIAM, Philadelphia, 2005.

[105] Laurent Chanas. *Méthodes Variationnelles pour la Restauration de Séquences d'Images Fortemenent Dégradées. Application aux Images Infrarouges Eblouies par Laser.* PhD thesis, Université de Cergy-Pontoise, October 2001.

[106] P. Charbonnier. *Reconstruction d'image: régularisation avec prise en compte des discontinuités.* PhD thesis, Université de Nice-Sophia Antipolis, 1994.

[107] P. Charbonnier, G. Aubert, M. Blanc-Féraud, and M. Barlaud. Two deterministic half-quadratic regularization algorithms for computed imaging. In *Proceedings of the International Conference on Image Processing*, volume II, pages 168–172, 1994.

[108] P. Charbonnier, L. Blanc-Féraud, G. Aubert, and M. Barlaud. Deterministic edge-preserving regularization in computed imaging. *IEEE Transactions on Image Processing*, 6(2):298–311, 1997.

[109] R. Chellapa and A. Jain. *Markov Random Fields: Theory and Application.* Academic Press, Boston, 1993.

[110] Y.G. Chen, Y. Giga, and S. Goto. Uniqueness and existence of viscosity solutions of generalized mean curvature flow equations. *Journal on Differential Geometry*, 33:749–786, 1991.

[111] M. Chipot, R. March, M. Rosati, and G. Vergara Caffarelli. Analysis of a nonconvex problem related to signal selective smoothing. *Mathematical Models and Methods in Applied Science*, 7(3):313–328, 1997.

[112] P.G. Ciarlet and J.L. Lions, editors. *Handbook of Numerical Analysis. Volume I. Finite Difference Methods. Solution of Equations in R^n.* North-Holland, 1990.

[113] J.P. Cocquerez, L. Chanas, and J. Blanc-Talon. Simultaneous inpainting and motion estimation of highly degraded video-sequences. In *Scandinavian Conference on Image Analysis*, pages 523–530. Springer-Verlag, 2003. LNCS, 2749.

[114] I. Cohen. Nonlinear variational method for optical flow computation. In *Scandinavian Conference on Image Analysis*, volume 1, pages 523–530, 1993.

[115] T. Corpetti, E. Mémin, and P. Pérez. Estimating fluid optical flow. In *Proceedings of the International Conference on Pattern Recognition*, volume 3, pages 1045–1048, Barcelona, Spain, September 2000. Computer Society Press.

[116] M. G. Crandall and H. Ishii. The maximum principle for semicontinuous functions. *Differential and Integral Equations*, 3(6):1001–1014, 1990.

[117] M.G. Crandall. Viscosity solutions of Hamilton–Jacobi equations. In *Nonlinear Problems: Present and Future, Proc. 1st Los Alamos Conf., 1981*, volume 61, pages 117–125. North-Holland Math. Stud., 1982.

[118] M.G. Crandall, H. Ishii, and P.-L. Lions. User's guide to viscosity solutions of second order partial differential equations. *Bull. Amer. Soc.*, 27:1–67, 1992.

[119] M.G. Crandall and P.L. Lions. Condition d'unicité pour les solutions généralisées des équations de Hamilton–Jacobi du premier ordre. *Comptes Rendus de l'Académie des Sciences*, (292):183–186, 1981.

[120] A. Criminisi, P. Pérez, and K. Toyama. Region filling and object removal by examplar-based image inpainting. *IEEE Transactions on Image Processing*, 13(9), 2004.

[121] B. Dacorogna. *Direct Methods in the Calculus of Variations.* Number 78 in Applied Mathematical Sciences. Springer–Verlag, 1989.

[122] G. Dal Maso. *An Introduction to Γ-Convergence.* Progress in Nonlinear Differential Equations and their Applications. Birkhäuser, 1993.

[123] R. Dautray and J.-L. Lions. *Mathematical Analysis and Numerical Methods for Science and Technology.* Springer, 2000. 6 volumes.

[124] G. David. *Singular sets of minimizers for the Mumford-Shah functional.* Birkhäuser, 2005.

[125] E. De Giorgi. Convergence problems for functionals and operators. *Recent methods in non-linear analysis, Proc. Int. Meet., Rome 1978*, pages 131–188, 1979.

[126] E. De Giorgi. Free discontinuity problems in calculus of variations. *Frontiers in pure and applied mathematics, Coll. Pap. Ded. J.L. Lions Occas. 60th Birthday*, pages 55–62, 1991.

[127] E. De Giorgi and L. Ambrosio. Un nuovo tipo di funzionale del calcolo delle variazioni. *Att. Accad. Naz. Lincei, Rend. Cl Sci. Fis. Mat. Nat.*, 82:199–210, 1988.

[128] E. De Giorgi, M. Carriero, and A. Leaci. Existence theorem for a maximum problem with a free discontinuity set. *Archive for Rational Mechanics and Analysis*, 108:195–218, 1989.

[129] E. Decencière Ferrandière. *Restauration Automatique de Films Anciens.* PhD thesis, Ecole Nationale Superieure des Mines de Paris, December 1997.

[130] F. Demengel and R. Temam. Convex functions of a measure and applications. *Indiana University Mathematics Journal*, 33:673–709, 1984.

[131] G. Demoment. Image reconstruction and restoration: overview of common estimation structures and problems. *IEEE Transactions on Acoustics, Speech and Signal Processing*, 37(12):2024–2036, 1989.

[132] R. Deriche. Using Canny's criteria to derive a recursively implemented optimal edge detector. *The International Journal of Computer Vision*, 1(2):167–187, May 1987.

[133] R. Deriche, P. Kornprobst, and G. Aubert. Optical flow estimation while preserving its discontinuities: A variational approach. In *Proceedings of the 2nd Asian Conference on Computer Vision*, volume 2, pages 71–80, Singapore, December 1995.

[134] A. Dervieux and F. Thomasset. A finite element method for the simulation of Rayleigh-Taylor instability. *Lecture Notes in Mathematics*, 771:145–159, 1979.

[135] B. Despas and F. Helt, editors. *La restauration numérique des films cinématographiques.* Commission Supérieure et Technique de l'Image et du Son, 1997.

[136] V. Devlaminck and J.P. Dubus. Estimation of compressible or incompressible deformable motions for density images. In *Proceedings of the International Conference on Image Processing*, volume I, pages 125–128, 1996.

358   References

[137] R.A. DeVore, B. Jawerth, and B. Lucier. Image compression through wavelet transform coding. *IEEE Transactions on Information Theory, Special Issue Wavelet Transforms Multires. Anal.*, 38:719–746, 1992.

[138] R.A. DeVore and P. Popov. Interpolation of Besov spaces. *Proceedings of the American Mathematical Society*, 305:397–414, 1988.

[139] E. DiBenedetto. *Degenerate Parabolic Equations*. Springer–Verlag, 1993.

[140] D. Donoho. De-noising by soft-thresholding. *IEEE Transactions on Information Theory*, 41:613–627, 1995.

[141] D. Donoho. Nonlinear solution of linear inverse problems by wavelet-vaguelet decomposition. *Applied and Computational Harmonic Analysis*, 2:101–126, 1995.

[142] D. Donoho and I. Johnstone. Adapting to unknown smoothness via wavelet shrinkage. *Journal of American Statistical Association*, 90:1200–1224, 1995.

[143] A.A. Efros and T.K. Leung. Texture synthesis by non-parametric sampling. In *Proceedings of the 7th International Conference on Computer Vision*, volume 2, pages 1033–1038, Kerkyra, Greece, 1999. IEEE Computer Society, IEEE Computer Society Press.

[144] I. Ekeland and R. Temam. *Convex analysis and variational problems. Translated by Minerva Translations, Ltd., London.*, volume 1 of *Studies in Mathematics and its Applications*. Amsterdam, Oxford: North-Holland Publishing Company; New York: American Elsevier Publishing Company, 1976.

[145] A.I. El-Fallah and G.E. Ford. On mean curvature diffusion in nonlinear image filtering. *Pattern Recognition Letters*, 19:433–437, 1998.

[146] W. Enkelmann. Investigation of multigrid algorithms for the estimation of optical flow fields in image sequences. *Computer Vision, Graphics, and Image Processing*, 43:150–177, 1988.

[147] C.L. Epstein and M. Gage. The curve shortening flow. In A.J. Chorin, A.J. Majda, and P.D. Lax, editors, *Wave Motion: Theory, Modelling and Computation*. Springer–Verlag, 1987.

[148] L.C. Evans. *Partial Differential Equations*, volume 19 of *Graduate Studies in Mathematics*. Proceedings of the American Mathematical Society, 1998.

[149] L.C. Evans and R.F. Gariepy. *Measure Theory and Fine Properties of Functions*. CRC Press, 1992.

[150] C. Eveland, K. Konolige, and R.C. Bolles. Background modeling for segmentation of video-rate stereo sequences. *International Conference on Computer Vision and Pattern Recognition*, pages 266–272, 1998.

[151] R. Fattal, D. Lischinski, and M. Werman. Gradient domain high dynamic range compression. In Kurt Akeley, editor, *Proceedings of the SIGGRAPH*, pages 249–256. ACM Press, ACM SIGGRAPH, Addison Wesley Longman, 2002.

[152] J.M. Fitzpatrick. The existence of geometrical density-image transformations corresponding to object motion. *Computer Vision, Graphics, and Image Processing*, 44(2):155–174, 1988.

[153] G.B. Folland. *Real Analysis. Modern Techniques and their Applications.* Pure and Applied Mathematics. A Wiley-Interscience Publication, 1984.

[154] G.B. Folland. *Fourier Analysis and its Applications.* The Wadsworth & Brooks/Cole Mathematics Series. Brooks/Cole Advanced Books & Software, 1992.

[155] I. Fonseca and L. Tartar. The gradient theory of phase transitions for systems with two potential wells. *Proc. R. Soc. Edinb., Sect.*, 111(A)(1/2):89–102, 1989.

[156] E. François and P. Bouthemy. The derivation of qualitative information in motion analysis. In O.D. Faugeras, editor, *Proceedings of the 1st European Conference on Computer Vision*, pages 226–230, Antibes, France, April 1990. Springer, Berlin, Heidelberg.

[157] B. Galvin, B. McCane, K. Novins, D. Mason, and S. Mills. Recovering motion fields: an evaluation of eight optical flow algorithms. *British Machine Vision Conference*, pages 195–204, 1998.

[158] D. Geman. *Random fields and inverse problems in imaging*, volume 1427 of *Lecture Notes in Mathematics, "Ecole d'été de Saint Flour"*. Springer–Verlag, 1992.

[159] D. Geman and G. Reynolds. Constrained restoration and the recovery of discontinuities. *IEEE Transactions on Pattern Analysis and Machine Intelligence*, 14(3):367–383, 1993.

[160] S. Geman and D. Geman. Stochastic relaxation, Gibbs distributions, and the Bayesian restoration of images. *IEEE Transactions on Pattern Analysis and Machine Intelligence*, 6(6):721–741, 1984.

[161] D. Gilbarg and N.S. Trudinger. *Elliptic Partial Differential Equations of Second Order. 2nd ed.*, volume XIII of *Grundlehren der Mathematischen Wissenschaften*. Springer–Verlag, 1983.

[162] G.L. Gimel'farb. *Image Textures and Gibbs Random Fields.* Kluwer Academic Publishers, 1999.

[163] E. De Giorgi and T. Franzoni. Su un tipo di convergenza variazionale. *Atti Accad. Naz. Lincei Rend. Cl. Sci. Fis. Mat. Natur.*, 68:842–850, 1975.

[164] E. Giusti. *Minimal Surfaces and Functions of Bounded Variation.* Birkhäuser, 1984.

[165] M. Gobbino. Finite difference approximation of the Mumford–Shah functional. *Communications on Pure Applied Mathematics*, 51(2):197–228, 1998.

[166] E. Godlewski and P.A. Raviart. *Hyperbolic Systems of Conservation Laws*, volume 3/4 of *Mathématiques et Applications*. Ellipses, 1991.

[167] C. Goffman and J. Serrin. Sublinear functions of measures and variational integrals. *Duke Mathematical Journal*, 31:159–178, 1964.

[168] J. Gomes and O. Faugeras. Reconciling distance functions and level sets. *Journal of Visual Communication and Image Representation*, 11(2):209–223, 2000.

[169] J. Gomes and O. Faugeras. Representing and evolving smooth manifolds of arbitrary dimension embedded in R^n as the intersection of n hypersurfaces: the vector distance functions. Technical Report 4012, INRIA, 2000.

[170] José Gomes and Olivier Faugeras. The vector distance function. *The International Journal of Computer Vision*, 52(2/3):161–187, 2003.

[171] Y. Gousseau and J.M. Morel. Are natural images of Bounded Variations? *SIAM J. Math. Anal.*, 33(3):634–648, 2001.

[172] W.E.L. Grimson, C. Stauffer, R. Romano, and L. Lee. Using adaptive tracking to classify and monitor activities in a site. In *Proceedings of the International Conference on Computer Vision and Pattern Recognition*, pages 22–31, Santa Barbara, California, June 1998. IEEE Computer Society.

[173] P. Grisvard. *Elliptic Problems in Nonsmooth Domains*, volume XIV of *Monographs and Studies in Mathematics*. Pitman Advanced Publishing Program. Boston-London-Melbourne: Pitman Publishing Inc., 1985.

[174] F. Guichard. Multiscale analysis of movies. *Proceedings of the eighth workshop on image and multidimensional signal processing, IEEE*, pages 236–237, 1993.

[175] F. Guichard. A morphological, affine, and Galilean invariant scale-space for movies. *IEEE Transactions on Image Processing*, 7(3):444–456, 1998.

[176] F. Guichard and J.M. Morel. Image iterative smoothing and P.D.E.'s. Questions Mathématiques en Traitement du Signal et de l'Image. Technical Report 12, Institut Henri Poincaré, 1998.

[177] F. Guichard and L. Rudin. Accurate estimation of discontinuous optical flow by minimizing divergence related functionals. In *Proceedings of the International Conference on Image Processing*, volume I, pages 497–500, 1996.

[178] S.N. Gupta and J.L. Prince. On div-curl regularization for motion estimation in 3-d volumetric imaging. In *Proceedings of the International Conference on Image Processing*, pages 929–932, 1996.

[179] G.D. Hager and P.N. Belhumeur. Efficient region tracking with parametric models of geometry and illumination. *IEEE Transactions on Pattern Analysis and Machine Intelligence*, 27(10):1025–1039, 1998.

[180] R. Haralick. Digital step edges from zero crossing of second directional derivatives. *IEEE Transactions on Pattern Analysis and Machine Intelligence*, 6(1):58–68, January 1984.

[181] H.W. Haussecker and D.J. Fleet. Computing optical flow with physical models of brightness variations. In *Proceedings of the International Conference on Computer Vision and Pattern Recognition*, volume 2, pages 760–767, Hilton Head Island, South Carolina, June 2000. IEEE Computer Society.

[182] C. Hirsch. *Numerical computation of internal and external flows. Volume 1: Fundamentals of numerical discretization*. Wiley Series in Numerical Methods in Engineering; Wiley-Interscience Publication, 1988.

[183] B.K. Horn. *Robot Vision*. MIT Press, 1986.

[184] B.K. Horn and B.G. Schunck. Determining Optical Flow. *Artificial Intelligence*, 17:185–203, 1981.

[185] M. Irani and S. Peleg. Motion analysis for image enhancement: resolution, occlusion, and transparency. *Journal on Visual Communications and Image Representation*, 4(4):324–335, 1993.

[186] H. Ishii. Existence and uniqueness of solutions of Hamilton Jacobi equations. *funkcialaj ekvacioj*, 29:167–188, 1986.

[187] H. Ishii and M. Ramaswamy. Uniqueness results for a class of Hamilton-Jacobi equations with singular coefficients. *Comm. Par. Diff. Eq.*, 20:2187–2213, 1995.

[188] Anil K. Jain. *Fundamentals of Digital Image Processing*. Prentice-Hall International Editions, 1989.

[189] G. Kanizsa. *Grammatica Del Vedere*. Bologna: Il Mulino, 1980.

[190] M. Kass, A. Witkin, and D. Terzopoulos. Snakes: Active contour models. In *First International Conference on Computer Vision*, pages 259–268, London, June 1987.

[191] S. Kichenassamy. The Perona–Malik paradox. *SIAM Journal of Applied Mathematics*, 57(5):1328–1342, 1997.

[192] S. Kichenassamy, A. Kumar, P. Olver, A. Tannenbaum, and A. Yezzi. Gradient flows and geometric active contour models. In *Proceedings of the 5th International Conference on Computer Vision*, pages 810–815, Boston, MA, June 1995. IEEE Computer Society Press.

[193] S. Kichenassamy, A. Kumar, P. Olver, A. Tannenbaum, and A. Yezzi. Conformal curvature flows: from phase transitions to active vision. *Archive for Rational Mechanics and Analysis*, 134:275–301, 1996.

[194] R. Kimmel. *Numerical Geometry of Images: Theory, Algorithms, and Applications*. Springer, November 2004.

[195] R. Kimmel, R. Malladi, and N. Sochen. Images as embedded maps and minimal surfaces: movies, color, texture, and volumetric medical images. *International Journal of Computer Vision*, 39(2):111–129, September 2000.

[196] D. Kinderlehrer and G. Stampacchia. *An Introduction to Variational Inequalities and Their Applications*. Academic Press, 1980.

[197] A. Kirsch. *An Introduction to the Mathematical Theory of Inverse Problems*, volume 120 of *Applied Mathematical Sciences*. Springer, 1996.

[198] J.J. Koenderink. The structure of images. *Biological Cybernetics*, 50:363–370, 1984.

[199] A. Kokaram. *Motion Picture Restoration*. PhD thesis, Cambridge University, England, May 1993.

[200] A. Kokaram. *Motion Picture Restoration: Digital Algorithms for Artefact Suppression in Degraded Motion Picture Film and Video*. Springer, 1998.

[201] A. Kokaram. On Missing Data Treatment for Degraded Video and Film Archives: A Survey and a New Bayesian Approach. *IEEE Transactions on Image Processing*, 13(3):397–415, March 2004.

[202] A. Kokaram, R. Bornard, A. Rareş, D. Sidorov, J-H. Chenot, L. Laborelli, and J. Biemond. Robust and Automatic Digital Restoration Systems: Coping with Reality. In *International Broadcasting Convention (IBC)*, pages 405–411, Amsterdam, The Netherlands, September 2002.

[203] P. Kornprobst. *Contributions à la Restauration d'Images et à l'Analyse de Séquences: Approches Variationnelles et Solutions de Viscosité.* PhD thesis, Université de Nice-Sophia Antipolis, 1998.

[204] P. Kornprobst, R. Deriche, and G. Aubert. Image coupling, restoration and enhancement via PDE's. In *Proceedings of the International Conference on Image Processing*, volume 4, pages 458–461, Santa Barbara, California, October 1997.

[205] P. Kornprobst, R. Deriche, and G. Aubert. Nonlinear operators in image restoration. In *Proceedings of the International Conference on Computer Vision and Pattern Recognition*, pages 325–331, Puerto Rico, June 1997. IEEE Computer Society, IEEE.

[206] P. Kornprobst, R. Deriche, and G. Aubert. Image sequence analysis via partial differential equations. *Journal of Mathematical Imaging and Vision*, 11(1):5–26, October 1999.

[207] A. Kumar, A. Tannenbaum, and G. Balas. Optical flow: a curve evolution approach. *IEEE Transactions on Image Processing*, 5:598–611, 1996.

[208] O.A. Ladyzhenskaya, V.A. Solonnikov, and N.N. Ural'ceva. *Linear and Quasilinear Equations of Parabolic Type.* Proceedings of the American Mathematical Society, Providence, RI, 1968.

[209] F. Lauze. *Computational Methods for Motion Recovery, Motion Compensated Inpainting and Applications.* PhD thesis, The IT University of Copenhagen, October 2004.

[210] F. Lauze and M. Nielsen. A variational algorithm for motion compensated inpainting. In S. Barman A. Hoppe and T. Ellis, editors, *British Machine Vision Conference*, volume 2, pages 777–787. BMVA, 2004.

[211] F. Lauze and M. Nielsen. From inpainting to active contours. In *IEEE Workshop on Variational and Level Set Methods*, pages 97–108, 2005.

[212] R.J. LeVeque. *Numerical Methods for Conservation Laws.* Birkhäuser, Basel, 1992.

[213] M. Levoy, K. Pulli, B. Curless, S. Rusinkiewicz, D. Koller, L. Pereira, M. Ginzton, S. Anderson, J. Davis, J. Ginsberg, J. Shade, and D. Fulk. The digital Michelangelo projetc: 3D scanning of large statues. In Kurt Akeley, editor, *Proceedings of the SIGGRAPH*, pages 269–276. ACM Press, ACM SIGGRAPH, Addison Wesley Longman, 2000.

[214] H. Lewis. *Principle and Applications of Imaging Radar, Manual of Remote Sensing, 3rd edition*, volume 2. J. Wiley and Sons, 1998.

[215] S.Z. Li. *Markov Random Field Modeling in Computer Vision.* Springer–Verlag, 1995.

[216] J.L. Lions and E. Magenes. *Problèmes aux Limites Non Homogènes et Applications.* Paris, Dunod, 1968. Volumes 1, 2, 3.

[217] P.-L. Lions. *Generalized Solutions of Hamilton–Jacobi Equations.* Number 69 in Research Notes in Mathematics. Pitman Advanced Publishing Program, 1982.

[218] M.M. Lipschutz. *Differential Geometry, Theory and Problems.* Schaum's Outline Series. McGraw-Hill, 1969.

[219] W. Lohmiller and J.J. Slotine. Global convergence rates of nonlinear diffusion for time-varying images. In Mads Nielsen, P. Johansen, O.F. Olsen, and J. Weickert, editors, *Scale-Space Theories in Computer Vision*, volume 1682 of *Lecture Notes in Computer Science*, pages 525–529. Springer–Verlag, 1999.

[220] B. Lucas. *Generalized image matching by method of differences*. PhD thesis, Carnegie-Mellon University, 1984.

[221] B. Lucas and T. Kanade. An iterative image registration technique with an application to stereo vision. In *International Joint Conference on Artificial Intelligence*, pages 674–679, 1981.

[222] R. Malladi, J.A. Sethian, and B.C. Vemuri. Evolutionary fronts for topology-independent shape modeling and recovery. In J-O. Eklundh, editor, *Proceedings of the 3rd European Conference on Computer Vision*, volume 800 of *Lecture Notes in Computer Science*, pages 3–13, Stockholm, Sweden, May 1994. Springer–Verlag.

[223] R. Malladi, J.A. Sethian, and B.C. Vemuri. Shape modeling with front propagation: A level set approach. *IEEE Transactions on Pattern Analysis and Machine Intelligence*, 17(2):158–175, February 1995.

[224] S. Mallat. *A Wavelet Tour of Signal Processing*. Academic Press, 1998.

[225] P. Maragos. A representation theory for morphological image and signal processing. *IEEE Transactions on Pattern Analysis and Machine Intelligence*, 11(6):586–599, June 1989.

[226] P. Maragos. *Partial Differential Equations for Morphological Scale-Spaces and Eikonal Applications*, pages 587–612. Elsevier Acad. Press, 2005.

[227] R. March. Visual reconstructions with discontinuities using variational methods. *Image and Vision Computing*, 10:30–38, 1992.

[228] D. Marr. *Vision*. W.H. Freeman and Co., 1982.

[229] D. Marr and E. Hildreth. Theory of edge detection. *Proceedings of the Royal Society London, B*, 207:187–217, 1980.

[230] S. Masnou and J.M. Morel. Level lines based disocclusion. *International Conference on Image Processing*, III:259–263, 1998.

[231] U. Massari and I. Tamanini. Regularity properties of optimal segmentations. *J. Reine Angew. Math.*, 420:61–84, 1991.

[232] G. Matheron. *Random Sets and Integral Geometry*. John Wiley & Sons, 1975.

[233] M. Mattavelli and A. Nicoulin. Motion estimation relaxing the constant brightness constraint. In *Proceedings of the International Conference on Image Processing*, volume II, pages 770–774, 1994.

[234] E. Mémin and P. Pérez. A multigrid approach for hierarchical motion estimation. In *Proceedings of the 6th International Conference on Computer Vision*, pages 933–938. IEEE Computer Society Press, Bombay, India, January 1998.

[235] Y. Meyer. *Oscillating patterns in image processing and nonlinear evolution equations*, volume 22 of *University Lecture Series*. American Mathematical Society, Providence, RI, 2001.

[236] A.A. Miljutin. A priori estimates for solutions of second order linear elliptic equations. *Mat. Sb.*, 51(93):459–474, 1960.

[237] C. Miranda. *Partial Differential Equations of Elliptic Type.* Springer-Verlag, 1970.

[238] A. Mitiche and P. Bouthemy. Computation and analysis of image motion: a synopsis of current problems and methods. *The International Journal of Computer Vision*, 19(1):29–55, July 1996.

[239] L. Modica. The gradient theory of phase transitions and the minimal interface criterion. *Archive for Rational Mechanics and Analysis*, 98:123–142, 1987.

[240] L. Moisan. Analyse multiéchelle de films pour la reconstruction du relief. *Comptes Rendus de l'Académie des Sciences*, 320-I:279–284, 1995.

[241] L. Moisan. Multiscale analysis of movies for depth recovery. In *Proceedings of the International Conference on Image Processing*, volume 3, pages 25–28, Washington, USA, 1995.

[242] L. Moisan. Perspective invariant multiscale analysis of movies. In Leonid I. Rudin and Simon K. Bramble, editors, *Investigative and Trial Image Processing*, volume 2567 of *Proceedings of the Proceedings of the International Society for Optical Engineering*, pages 84–94, July 1995.

[243] J.M. Morel and S. Solimini. Segmentation of images by variational methods: A constructive approach. *Rev. Math. Univ. Complut. Madrid*, 1:169–182, 1988.

[244] J.M. Morel and S. Solimini. Segmentation d'images par methode variationnelle: Une preuve constructive d'existence. *Comptes Rendus de l'Académie des Sciences*, 308(15):465–470, 1989.

[245] J.M. Morel and S. Solimini. *Variational Methods in Image Segmentation.* Progress in Nonlinear Differential Equations and Their Applications. Birkhäuser, Basel, 1995.

[246] D. Mumford and J. Shah. Boundary detection by minimizing functionals. In *Proceedings of the International Conference on Computer Vision and Pattern Recognition*, pages 22–26, San Francisco, CA, June 1985. IEEE.

[247] D. Mumford and J. Shah. Optimal approximations by piecewise smooth functions and associated variational problems. *Communications on Pure and Applied Mathematics*, 42:577–684, 1989.

[248] H.H. Nagel. Constraints for the estimation of displacement vector fields from image sequences. In *International Joint Conference on Artificial Intelligence*, pages 156–160, 1983.

[249] H.H. Nagel. Displacement vectors derived from second order intensity variations in image sequences. *Computer Vision, Graphics, and Image Processing*, 21:85–117, 1983.

[250] S.K. Nayar and S.G. Narasimhan. Vision in bad weather. In *Proceedings of the 6th International Conference on Computer Vision*, pages 820–827, Bombay, India, January 1998. IEEE Computer Society, IEEE Computer Society Press.

[251] S. Negahdaripour. Revised definition of optical flow: integration of radiometric and geometric cues for dynamic scene analysis. *IEEE Transactions on Pattern Analysis and Machine Intelligence*, 20(9):961–979, 1998.

[252] S. Negahdaripour and C.H. Yu. A generalized brightness change model for computing optical flow. In *Proceedings of the 4th International Conference on Computer Vision*, pages 2–11, Berlin, Germany, May 1993. IEEE Computer Society Press.

[253] P. Nési. Variational approach to optical flow estimation managing discontinuities. *Image and Vision Computing*, 11(7):419–439, September 1993.

[254] M. Nikolova. Local strong homogeneity of a regularized estimator. *SIAM Journal of Applied Mathematics*, 61(2):633–658, 2000.

[255] M. Nitzberg, D. Mumford, and T. Shiota. *Filtering, Segmentation and Depth*, volume 662 of *Lecture Notes in Computer Science*. Springer, 1993.

[256] M. Nitzberg and T. Shiota. Nonlinear image filtering with edge and corner enhancement. *IEEE Transactions on Pattern Analysis and Machine Intelligence*, 14:826–833, 1992.

[257] J.A. Noble. *Description of image surfaces*. PhD thesis, Robotics research group, Department of Engineering Science, Oxford University, 1989.

[258] J.M. Odobez and P. Bouthemy. Robust multiresolution estimation of parametric motion models. *Journal of Visual Communication and Image Representation*, 6(4):348–365, December 1995.

[259] M. Orkisz and P. Clarysse. Estimation du flot optique en présence de discontinuités: une revue. *Traitement du Signal*, 13(5):489–513, 1996.

[260] S. Osher and R.P. Fedkiw. Level set methods. Technical Report 00-08, UCLA CAM Report, 2000.

[261] S. Osher and L.I. Rudin. Feature-oriented image enhancement using shock filters. *SIAM Journal of Numerical Analysis*, 27(4):919–940, August 1990.

[262] S. Osher and J.A. Sethian. Fronts propagating with curvature-dependent speed: Algorithms based on Hamilton–Jacobi formulations. *Journal of Computational Physics*, 79(1):12–49, 1988.

[263] S. Osher, A. Sole, and L. Vese. Image decomposition and restoration using total variation minimization and the h^{-1} norm. *Multiscale Modeling and Simulation: A SIAM Interdisciplinary Journal*, 1(3):349–370, 2003.

[264] M. Otte and H.H. Nagel. Extraction of line drawings from gray value images by non-local analysis of edge element structures. In G. Sandini, editor, *Proceedings of the 2nd European Conference on Computer Vision*, pages 687–695, Santa Margherita, Italy, May 1992. Springer–Verlag.

[265] M. Otte and H.H. Nagel. Optical flow estimation: Advances and comparisons. In Jan-Olof Eklundh, editor, *Proceedings of the 3rd European Conference on Computer Vision*, volume 800 of *Lecture Notes in Computer Science*, pages 51–70. Springer–Verlag, 1994.

[266] N. Paragios, Y. Chen, and O. Faugeras, editors. *The Handbook of Mathematical Models in Computer Vision*. Springer, 2005.

[267] N. Paragios and R. Deriche. A PDE-based level set approach for detection and tracking of moving objects. In *Proceedings of the 6th International Conference on Computer Vision*, pages 1139–1145, Bombay, India, January 1998. IEEE Computer Society, IEEE Computer Society Press.

[268] N. Paragios and R. Deriche. Geodesic active regions for supervised texture segmentation. In *Proceedings of the 7th International Conference on Computer Vision*, Kerkyra, Greece, September 1999. IEEE Computer Society, IEEE Computer Society Press.

[269] N. Paragios and R. Deriche. Geodesic active contours and level sets for the detection and tracking of moving objects. *IEEE Transactions on Pattern Analysis and Machine Intelligence*, 22:266–280, March 2000.

[270] N. Paragios and R. Deriche. Geodesic active regions: a new paradigm to deal with frame partition problems in computer vision. *Journal of Visual Communication and Image Representation, Special Issue on Partial Differential Equations in Image Processing, Computer Vision and Computer Graphics*, 13(1/2):249–268, march/june 2002.

[271] K.A. Patwardhan and G. Sapiro. Projection based image and video inpainting using wavelets. In *International Conference on Image Processing*, volume 1, pages 857–860, 2003.

[272] T. Pavlidis and Y.T. Liow. Integrating region growing and edge detection. *IEEE Transactions on Pattern Analysis and Machine Intelligence*, 12:225–233, 1990.

[273] D. Peng, B. Merriman, S. Osher, H.-K. Zhao, and M. Kang. A PDE-based fast local level set method. *Journal of Computational Physics*, 155(2):410–438, 1999.

[274] P. Pérez, M. Gangnet, and A. Blake. Poisson image editing. In *Proceedings of the SIGGRAPH*, volume 22, pages 313–318. ACM Press, ACM SIGGRAPH, Addison Wesley Longman, July 2003.

[275] P. Perona and J. Malik. Scale-space and edge detection using anisotropic diffusion. *IEEE Transactions on Pattern Analysis and Machine Intelligence*, 12(7):629–639, July 1990.

[276] C.A. Poynton. Poynton's colour faq (http://www.poynton.com/colorfaq.html). Web page, 1995.

[277] J.M.S. Prewitt. Object enhancement and extraction. In Lipkin, B.S. and Rosenfeld, A., editor, *Picture Processing and Psychopictorics*, pages 75–149. New York: Academic, 1970.

[278] L.G. Roberts. Machine perception of three-dimensional solids. In Tippett, J., Berkowitz, D., Clapp, L., Koester, C., and Vanderburgh, A., editor, *Optical and Electrooptical Information processing*, pages 159–197. MIT Press, 1965.

[279] R.T. Rockafellar. *Convex analysis*. Princeton University Press, Princeton, N. J., 1970.

[280] M. Rosati. Asymptotic behavior of a Geman–Mac Lure discrete model. Technical Report 8, Instituto per le Applicazioni del Calcolo Mauro Picone, 1997.

[281] M. Rosati. Asymptotic behavior of a Geman and McClure discrete model. *Applied Mathematics and Optimization*, 41(1):51–85, January 2000.

[282] A. Rosenfeld and A.C. Kak. *Digital Picture Processing*, volume 1. Academic Press, New York, 1982. Second Edition.

[283] L. Rudin and S. Osher. Total variation based image restoration with free local constraints. In *Proceedings of the International Conference on Image Processing*, volume I, pages 31–35, 1994.

[284] L. Rudin, S. Osher, and E. Fatemi. Nonlinear total variation based noise removal algorithms. *Physica D*, 60:259–268, 1992.

[285] W. Rudin. *Real and Complex Analysis*. McGraw-Hill, 1966.

[286] W. Rudin. *Principles of Mathematical Analysis. 3rd ed.* International Series in Pure and Applied Mathematics. McGraw-Hill, 1976.

[287] W. Rudin. *Functional Analysis. 2nd ed.* International Series in Pure and Applied Mathematics. McGraw-Hill, 1991.

[288] S.J. Ruuth, B. Merriman, and S. Osher. A fixed grid method for capturing the motion of self-intersecting interfaces and related pdes. Technical Report 99-22, UCLA Computational and Applied Mathematics Reports, July 1999.

[289] C. Samson. *Contribution à la classification d'images satellitaires par approche variationnelle et équations aux dérivées partielles*. PhD thesis, Université de Nice Sophia-Antipolis, 2000.

[290] C. Samson, L. Blanc-Féraud, G. Aubert, and J. Zerubia. A level-set model in image classification. In Mads Nielsen, P. Johansen, O.F. Olsen, and J. Weickert, editors, *Scale-Space Theories in Computer Vision*, volume 1682 of *Lecture Notes in Computer Science*, pages 306–317. Springer–Verlag, 1999.

[291] C. Samson, L. Blanc-Féraud, G. Aubert, and J. Zérubia. A variational model for image classification and restoration. *IEEE Transactions on Pattern Analysis and Machine Intelligence*, 22(5):460–472, May 2000.

[292] G. Sapiro. Vector-Valued Active Contours. In *Proceedings of the International Conference on Computer Vision and Pattern Recognition*, pages 680–685, San Francisco, CA, June 1996. IEEE.

[293] G. Sapiro. *Geometric Partial Differential Equations and Image Analysis*. Cambridge University Press, 2001.

[294] G. Sapiro and D.L. Ringach. Anisotropic diffusion of multivalued images with applications to color filtering. *IEEE Transactions on Image Processing*, 5(11):1582–1585, 1996.

[295] C. Schnörr. Determining optical flow for irregular domains by minimizing quadratic functionals of a certain class. *The International Journal of Computer Vision*, 6(1):25–38, 1991.

[296] B.G. Schunck. The motion constraint equation for optical flow. In *Proceedings of the 7th International Conference on Pattern Recognition*, IEEE Publication 84CH2046-1, pages 20–22, Montreal, Canada, July 1984. IEEE.

[297] B.G. Schunck. Image flow continuity equations for motion and density. *Workshop motion: representation and analysis*, pages 89–94, 1986.

[298] J.A. Sethian. Recent numerical algorithms for hypersurfaces moving with curvature-dependent speed: Hamilton–Jacobi equations and conservation laws. *Journal Differential Geometry*, 31:131–136, 1990.

[299] J.A. Sethian. A fast marching sevel set method for monotonically advancing fronts. In *Proceedings of the National Academy of Sciences*, volume 93, pages 1591–1694, 1996.

[300] J.A. Sethian. *Level Set Methods and Fast Marching Methods: Evolving Interfaces in Computational Geometry, Fluid Mechanics, Computer Vision, and Materials Sciences*. Cambridge Monograph on Applied and Computational Mathematics. Cambridge University Press, 1999.

[301] J. Shen and S. Castan. An optimal linear operator for step edge detection. *CVGIP: Graphics Models and Image Processing*, 54(2):112–133, March 1992.

[302] K. Siddiqi, Y. Lauzière, A. Tannenbaum, and S. Zucker. Area and length minimizing flows for shape segmentation. *IEEE Transactions on Image Processing*, 7:433–443, 1998.

[303] S.M. Smith and J.M. Brady. Susan - a new approach to low level image processing. *IJCV*, 23(1):45–78, 1997.

[304] I. Sobel. An isotropic 3 × 3 image gradient operator. *Machine Vision for Three-Dimensional Scenes, H. Freeman editor*, pages 376–379, 1990.

[305] N. Sochen, R. Kimmel, and R. Malladi. From high energy physics to low level vision. Technical Report 39243, LBNL report, UC Berkeley, 1996.

[306] N. Sochen, R. Kimmel, and R. Malladi. A geometrical framework for low level vision. *IEEE Transaction on Image Processing, Special Issue on PDE based Image Processing*, 7(3):310–318, 1998.

[307] P. Soravia. Optimal control with discontinuous running cost: eikonal equation and shape from shading. In *39th IEEE Conference on Decision and Control*, pages 79–84, December 2000.

[308] P. Soravia. Boundary value problems for hamilton-jacobi equations with discontinuous lagrangian. *Indiana Univ. Math. J.*, 51(2):451–477, 2002.

[309] P. Sternberg. The effect of a singular perturbation on nonconvex variational problems. *Archive for Rational Mechanics and Analysis*, 101(3):209–260, 1988.

[310] P. Sternberg. Vector-valued local minimizers of nonconvex variational problems. *Rocky Mt. J. Math.*, 21(2):799–807, 1991.

[311] B. Stroustrup. *The C++ Programming Language*. Addison-Wesley, 1997.

[312] M. Sussman, P. Smereka, and S. Osher. A level set approach for computing solutions to incompressible two-phase flow. *Journal of Computational Physics*, 114(1):146–159, 1994.

[313] D. Suter. Motion estimation and vector splines. In *Proceedings of the International Conference on Computer Vision and Pattern Recognition*, pages 939–942, Seattle, WA, June 1994. IEEE.

[314] B. Tang, G. Sapiro, and V. Caselles. Direction diffusion. *International Conference on Computer Vision*, 1998.

[315] S. Teboul, L. Blanc-Féraud, G. Aubert, and M. Barlaud. Variational approach for edge-preserving regularization using coupled PDE's. *IEEE Transaction on Image Processing, Special Issue on PDE based Image Processing*, 7(3):387–397, 1998.

[316] J.W. Thomas. *Numerical Partial Differential Equations: Finite Difference Methods.* Springer–Verlag, 1995.

[317] A.N. Tikhonov and V.Y. Arsenin. *Solutions of Ill-Posed Problems.* Winston and Sons, Washington, D.C., 1977.

[318] M. Tistarelli. Computation of coherent optical flow by using multiple constraints. In *Proceedings of the 5th International Conference on Computer Vision*, pages 263–268, Boston, MA, June 1995. IEEE Computer Society Press.

[319] C. Tomasi and R. Manduchi. Bilateral filtering for gray and color images. In *Proceedings of the IEEE International Conference on Computer Vision*, pages 839–846, January 1998.

[320] D. Tschumperlé. Fast anisotropic smoothing of multi-valued images using curvature-preserving PDE's. Technical Report 05-01, Les Cahiers du GREYC, February 2005.

[321] D. Tschumperlé. LIC-based regularization of multi-valued images. In *International Conference on Image Processing*, volume III, pages 535–536, 2005.

[322] D. Tschumperlé and R. Deriche. Diffusion PDE's on Vector-Valued images. *IEEE Signal Processing Magazine*, 19(5):16–25, 2002.

[323] D. Tschumperlé and R. Deriche. Vector-valued image regularization with PDE's : A common framework for different applications. In *IEEE Conference on Computer Vision and Pattern Recognition*, Madison, Wisconsin (United States), June 2003.

[324] David Tschumperle and Rachid Deriche. Vector-valued image regularization with pde's : A common framework for different applications. *IEEE Transactions on Pattern Analysis and Machine Intelligence*, 27(4):506–517, April 2005.

[325] J.N. Tsitsiklis. Efficient algorithms for globally optimal trajectories. *IEEE Transactions on Automatic Control*, 40(9):1528–1538, September 1995.

[326] J. Tumblin and G. Turk. Low curvature image simplifiers (lcis): A boundary hierarchy for detail-preserving contrast reduction. In *ACM SIGGRAPH*, pages 83–90, 1999.

[327] P.M.B. van Roosmalen. *Restoration of Archive Film and Video.* PhD thesis, Delft University of Technology, October 1999.

[328] A. Verri, F. Girosi, and V. Torre. Differential techniques for optical flow. *Journal of the Optical Society of America A*, 7:912–922, 1990.

[329] A. Verri and T. Poggio. Against quantitative optical flow. In *Proceedings First International Conference on Computer Vision*, pages 171–180. IEEE Computer Society, 1987.

[330] A. Verri and T. Poggio. Motion field and optical flow: qualitative properties. *IEEE Transactions on Pattern Analysis and Machine Intelligence*, 11(5):490–498, 1989.

[331] L. Vese. *Problèmes variationnels et EDP pour l'analyse d'images et l'évolution de courbes.* PhD thesis, Université de Nice Sophia-Antipolis, November 1996.

[332] L. Vese. A study in the BV space of a denoising–deblurring variational problem. *Applied Mathematics and Optimization,* 44(2):131–161, 2001.

[333] L. Vese and S. Osher. Modeling textures with total variation minimization and oscillating patterns in image processing, journal of scientific computing. *Journal of Scientific Computing,* 19:553–572, 2003.

[334] A.I. Vol'pert. The Spaces BV and Quasilinear Equations. *Math. USSR-Sbornik,* 2(2):225–267, 1967.

[335] J. Weickert. Theoretical foundations of anisotropic diffusion in image processing. *Computing Supplement,* 11:221–236, 1996.

[336] J. Weickert. *Anisotropic Diffusion in Image Processing.* Teubner-Verlag, Stuttgart, 1998.

[337] J. Weickert. Coherence-enhancing diffusion filtering. *The International Journal of Computer Vision,* 31(2/3):111–127, April 1999.

[338] J. Weickert. Linear scale space has first been proposed in Japan. *Journal of Mathematical Imaging and Vision,* 10(3):237–252, May 1999.

[339] J. Weickert and C. Schnörr. A theoretical framework for convex regularizers in PDE-based computation of image motion. Technical Report 13, University of Mannheim, June 2000.

[340] J. Weickert and C. Schnörr. A theoretical framework for convex regularizers in PDE-based computation of image motion. *The International Journal of Computer Vision,* 45(3):245–264, December 2001.

[341] R.P. Wildes, M.J. Amabile, A.M. Lanzillotto, and T.S. Leu. Physically based fluid flow recovery from image sequences. In *Proceedings of the International Conference on Computer Vision and Pattern Recognition,* pages 969–975, San Juan, Puerto Rico, June 1997. IEEE Computer Society, IEEE.

[342] A.P. Witkin. Scale-space filtering. In *International Joint Conference on Artificial Intelligence,* pages 1019–1021, 1983.

[343] C.R. Wren, A. Azarbayejani, T. Darrell, and A. Pentland. Pfinder: real-time tracking of the human body. *IEEE Transactions on Pattern Analysis and Machine Intelligence,* 19(7):780–785, July 1997.

[344] L.P. Yaroslavsky. *Digital Picture Processing. An Introduction.* Springer Verlag, Berlin, Heidelberg, 1985.

[345] A. Yezzi. Modified curvature motion for image smoothing and enhancement. *IEEE Trans. on Image Processing, Special Issue on PDE based Image Processing,* 7(3):345–352, 1998.

[346] K. Yosida. *Functional Analysis.* 6th ed., volume XII of *Grundlehren der mathematischen Wissenschaften.* Springer–Verlag, 1980.

[347] S. Di Zenzo. A note on the gradient of a multi-image. *Computer Vision, Graphics, and Image Processing,* 33:116–125, 1986.

[348] H.-K. Zhao, T. Chan, B. Merriman, and S. Osher. A variational level set approach to multiphase motion. *Journal of Computational Physics,* 127(1):179–195, 1996.

[349] L. Zhou, C. Kambhamettu, and D. Goldgof. Fluid structure and motion analysis from multi-spectrum 2D cloud image sequences. In *Proceedings of the International Conference on Computer Vision and Pattern Recognition*, volume 2, pages 744–751, Hilton Head Island, South Carolina, June 2000. IEEE Computer Society.

[350] S. Zhu and A. Yuille. Region competition: unifying snakes, region growing, and Bayes/MDL for multiband image segmentation. *IEEE Transactions on Pattern Analysis and Machine Intelligence*, 18(9):884–900, September 1996.

Index

Applied Mathematical Sciences

(continued from page ii)

Applied Mathematical Sciences

(continued from previous page)